DIALOGO
di
GALILEO GALILEI LINCEO
AL SER.mo FERD. II. GRAN. DVCA DI
TOSCANA

Stefan. Della Bella. f.

The Nature of Scientific Discovery

A Symposium Commemorating
the 500th Anniversary
of the Birth of Nicolaus Copernicus

Edited by Owen Gingerich

SMITHSONIAN INSTITUTION PRESS
CITY OF WASHINGTON

Library of Congress Cataloging in Publication Data

Main entry under title:
The Nature of scientific discovery.

  (Smithsonian international symposia series, 5)
  "The fifth international symposium of the Smithsonian
Institution organized jointly with the National Academy of
Sciences in cooperation with the Copernicus Society of America."
  1. Copernicus, Nicolaus, 1473-1543—Congresses.
2. Astronomy—Congresses.   3. Science—Congresses.
I. Gingerich, Owen, ed.   II. Smithsonian Institution.
III. National Academy of Sciences, Washington, D.C.
IV. Copernicus Society of America.   V. Series.
QB36.C8N37   1975   520'.92'4   74-18374   ISBN   O-87474-148-3

Smithsonian International Symposia Series

Knowledge Among Men, Paul H. Oehser, editor. Simon and
  Schuster, 1966.
The Fitness of Man's Environment. Smithsonian Institution
  Press, 1968.
Man and Beast: Comparative Social Behavior, J. F. Eisenberg
  and Wilton S. Dillon, editors. Smithsonian Institution
  Press, 1971.
The Cultural Drama, Wilton S. Dillon, editor. Smithsonian
  Institution Press, 1974.

The frontispiece, from the first edition of Galileo's *Dialogue
on the Two Chief World Systems* (1632), symbolizes the
scholarly discourse of the Copernican Symposium in April 1973.
The curious heliocentric instrument held by Copernicus
appeared for the first time in this portrait; a derivative form
was adopted on the United States commemorative stamp first
issued during the Copernican Week (see page 38). (Courtesy
of the Harvard College Library.)

The Fifth International Symposium of the
Smithsonian Institution
organized jointly with
The National Academy of Sciences
in cooperation with
The Copernicus Society of America

Honorary Chairman
Janusz Groszkowski
Past President, Polish Academy of Sciences
and Chairman, All-Polish Committee
of the Front for National Unity

The Smithsonian Institution and the National Academy of Sciences gratefully acknowledge contributions in support of the Symposium from:

Copernicus Society of America
Exxon Corporation
Exxon Engineering and Research Corporation
Hewlett-Packard Company
IBM Corporation
International Telephone and Telegraph Corporation
The Lilly Endowment
Mrs. Paul's Kitchens, Inc.
National Endowment for the Humanities
National Science Foundation
Piasecki Foundation
Alfred P. Sloan Foundation
Texas Instruments Foundation
Time-Life Films
U.S. Office of Education
Xerox Corporation

9

Stephen E. Toulmin
  Philosopher and Historian of Science, Provost of Crown College, University of California, Santa Cruz (now at the University of Chicago)
Fred L. Whipple
  Professor of Astronomy, Harvard University, and Director, Smithsonian Astrophysical Observatory

Smithsonian Office of Seminars

Wilton S. Dillon
  Director of Seminars

Dorothy Richardson
  Assistant Director of Seminars

PREFACE

S. Dillon Ripley

Today, persons of knowledge everywhere owe a debt to Nicolaus Copernicus (Mikołaj Kopernik) and his revolutionary thoughts. Our anniversary celebration has been a stimulus, allowing us to take inventory of the known and the unknown. It has helped to remind us of the short jump in time of the evolution of ascertained facts from fiction. I hope it has reminded us of the need for humility in the quest for knowledge.

The needs of late 20th-century scientists and other intellectuals parallel those of earlier days. We need to solve the problem of the increasing fragmentation of knowledge, just as earlier scientists needed to solve the problem of isolation. We certainly need to develop a world understanding of our total interdependence. At the moment, science, as a human institution, is somewhat threatened, suffering the cyclical fate of doubt or disbelief of other systems of thought or faith. Science, I suppose, has its lean and fat years, its ups and downs—much like religious movements, stock markets, or political parties. Its defenders need great events and revivals to celebrate the proposition that science is necessary and believable. And, as the historians and philosophers of science might say, science is too important to be left to scientists alone.

The 500th anniversary of the birth of Copernicus came like a long-awaited comet, beautifully timed to quicken the sense of fraternity and purpose of scientists in a time of doubt and blame. More than the Copernican legacy, however, is required to cope now with the sense of anticlimax and anxiety felt by citizens in face of the paradox of triumphant men on the moon and the public's stark awareness of the connections between energy, food, transportation, and health on earth.

Margaret Mead has reminded the Smithsonian's National Air and Space Museum that we might recognize ceremonially that Pioneer 10, in its projected leaving of our solar system, becomes in the eyes of other possible creatures in the universe an "unidentified flying object," which we, its makers, can authenticate. While this is food for philosophic reflection, I hope that we can realize the interplay of problem-solving that has gone into this event and its creation of the necessary by-products to the safety and rational standards of life on earth. Can we rouse ourselves to our global problems of finding a solution to the dangerous imbalance of food and population ratios? Knowledge begets knowledge and the by-products of space science are not necessarily exclusive of earth priorities. They should remind us also of the essential proposition that basic and applied research are merely semantic terms, and really part of a seamless web. We need constant revisions in our definitions of what constitutes our universe, in the continuing dialogue of man's existence.

We need simultaneously to reward political leaders who help moralists to fashion and implement an ethic that prompts human competitors and predators to cooperate and share for the survival of all. The idea of self-sufficiency in natural resources within already well-endowed countries deserves no support from those who care about protecting the earth's fragile web of life, sustained by photosynthesis from the publicly owned sources of sunlight, sunlight which figured in the Copernican Revolution. What income tax should we pay for breathing, in contrast to any other peoples? National rights of consumption are not absolute. I share with John Knowles of the Rockefeller Foundation his hope that the American people will provide a model of moral and intellectual suasion for an interdependent world of nation states based on rational austerity and emphasizing the quality, as contrasted with the quantity, of life. Scientists and politicians have much to discover in realms of survival.

The prevention of further social disarray is tied closely with finding leaders and followers who reduce the time lag between early warnings of scientists and engineers and the invention of new techniques of coping and adapting. The strategic politics of science, however, must allow for the serendipity essential to discoveries in science and technology. If nothing else emerges from celebrating the Copernican Revolution it should be the new understanding of the importance of "useless" knowledge.

Inventiveness is a habit of mind, which needs to be cultivated in the domains of science, art, politics, and other human endeavors. Social inventions spring from the same kinds of mental processes as new hypotheses about the workings of nature. During the celebrations of the 100th anniversary of the National Academy of Sciences in 1963, President John F. Kennedy, departing from some speech suggestions prepared by scientists Roger Revelle and Jerome Wiesner, declared:

> Every time you scientists make a major invention, we politicians have to invent a new institution to cope with it, and almost invariably these days, and happily, it must be an international institution.

Such an awareness of science as a joint enterprise with citizens and leaders is all the more important as we face the Copernican millennium with the reminder of John Archibald Wheeler, the Joseph Henry Professor of Physics at Princeton, that

society more insistently today than ever demands answers to its questions. It pushes on the investigator through a highly effective "force structure." The central force-transmitting element is the association of like-minded investigators, the "collegium." This association is as much a source of guidance, pressure to create, and moral support for the individual worker as it ever was; but today it has become more, an instrument of society, both responding to needs and pointing out possibilities.

In the name of increasing and diffusing knowledge, James Smithson's bequest established the Smithsonian and set into motion the institutions and intellectual energies—the spirit and the "infrastructure"—out of which Joseph Henry helped to form one of the world's newest scientific communities, joining those older communities of Europe and Asia. With the birth of the National Academy of Sciences during the Lincoln administration in 1863, Henry gave it a home at the Smithsonian and offered practical proof of the interdependence of science and government, as well as the need for scientists to work close to their organic roots in the arts and humanities.

The Copernican celebrations recorded in this book symbolize the coming together both of institutions and subject matter. The occasion arose from the suggestion of Messrs. Jerzy Neyman and Roger Revelle to me that the Smithsonian and the National Academy of Sciences join together to recognize a common historical interest in astronomy and that our mutual facilities could be drawn upon. The planning and organization of the symposium and festival was a combined effort of the two societies under the aegis of Wilton S. Dillon for the Smithsonian and John Archibald Wheeler for the National Academy. They were of immense help in organizing the collegia and special seminars designed around the commissioned papers, which attested to the universality of the occasion.

President Philip Handler of the National Academy of Sciences and I join with other participants in the celebrations, and now with the larger reading public, in recognizing the synoptic gifts of this book's editor, Professor Owen Gingerich, the Smithsonian astrophysicist and historian of astronomy we share with Harvard

University. This book is but one manifestation of his indefatigable use of the telescope, the computer, and the library—particularly seeking out those hiding places of first editions of Copernicus' *De revolutionibus orbium coelestium;* his skill as a teacher; his fruitful collaboration with Charles Eames, the designer; and a serene use of those scientific, religious, and philosophical insights which prompted him to remark, during the Copernicus celebrations, that we humans are all recycled stellar material.

During the celebration of the Copernicus Year in 1973, the Washington symposium was barely over when Professor Gingerich was flying off to the Vatican Library to discover the original working copy of Tycho Brahe's cosmological notes bound into an edition of *De revolutionibus,* an unexpected record of how Tycho germinated his conception of a non-Copernican model of the solar system. Such fortuitous details of scholarship will be multiplied, no doubt, as the impact of this volume is felt on others beset by that divine impulse to try to find out new things about ourselves and our universe.

Smithsonian Institution
April 1974

Non docet instabiles Copernicus ætheris orbes,
Sed terræ instabiles arguit ille vices.

This imaginative portrait of Copernicus first appeared in Pierre Gassendi's
Tychonis Brahei (The Hague, 1655), and was used as the basis for the
American commemorative postage stamp. The purely symbolic heliocentric
device first appeared in the frontispiece to Galileo's *Dialogues* (1632),
which serves also as the frontispiece to this volume.

A COPERNICAN PERSPECTIVE*

Owen Gingerich

To a modern scientist, science of the 15th century has a strange and unfamiliar quality. When Copernicus was born, men believed that a weightless, crystalline sun whirled daily around the earth. Celestial motions were explained in terms of strivings inherent in matter itself. Professors of physics said that a stone thrown from the hand flew straight until it ran out of "impetus" and then fell directly down to the ground.

In 1543 Copernicus' masterpiece, *De revolutionibus* ("Concerning the Revolutions"), was published, and in the century that followed, the curious, time-honored scientific explanations of antiquity and the Middle Ages gave way to the ideas that form the foundations of today's science. A great change—indeed a revolution—took place in that age, with Copernicus at its forefront. Far more than political struggles, with their attendant fleeting changes, the work of Copernicus and his intellectual successors has shaped our contemporary world.

---

*Adapted from an article written for the *UNESCO Courier*, April 1973.

If indeed Copernicus' contributions triggered the entire sequence of ideas that have become modern science, we may well ask why this is a phenomenon only 500 years old. What prevented the appearance of a Copernicus a century or two before? Was an earlier great unfolding of science prevented by bonds of dogma and ignorance?

Or, in a contrary view, should we consider creative science a delicate and fragile plant that blossoms forth only in particularly germinal conditions? Perhaps it was not mere coincidence that the great astronomer lived as a contemporary of Columbus, Dürer, Leonardo, Erasmus, and Luther. In that case we can make a fascinating inquiry: What factors contribute to the remarkable flowering of science, such as that typified by Copernicus?

The desire to probe these questions, in both a historical and a contemporary setting, motivated this inquiry into "The Nature of Scientific Discovery." Not only in the public symposia, but in the more intimate collegial discussions, men and women of science gathered with scholars of arts and letters to examine the nature of creativity in the sciences and its relation to the cultural matrix in which it is embedded. This volume attempts to preserve some of these deliberations, together with the flavor of the 500th anniversary celebrations of the Copernicus Week in Washington in April 1973.

Because the name of Copernicus is invoked repeatedly throughout this quinquecentennial volume, I take this opportunity to place his work within a historical perspective.

The geocentric astronomy of Copernicus' day had served well for over a thousand years; it meshed perfectly with man's view of himself and with the primitive physics of Aristotle. To be sure, learned prelates recognized that Easter was coming too early in the calendar year, and a few astrologers knew that the planets were sometimes found several degrees from positions predicted by tables based on the venerable Ptolemaic theory. But if there were a crisis in astronomy, Astronomia had just as much of a problem on her hands after Copernicus as before, for the calendar remained the same and planetary predictions were improved only a little.

No one knows precisely how or when the Polish astronomer first seized upon the idea of a sun-centered system; in any event, by 1515 he had circulated a précis of his innovation, and within a few years was hard at work on his manuscript of *De revolutionibus*—a book that encompassed both his new cosmology and a careful reexamination of old and new planetary observations. Nor do scholars know *why* he adopted a heliocentric view of the cosmos, for the observations of his day could neither prove nor disprove his idea. Nevertheless, there is a clue in his own writings. A euphoric sense of beauty pervades the entire composition of the book: "For in this most beautiful temple, who would place this lamp in another or better position than that from which it can light up the whole thing at the same time? . . . We thus discover a marvelous commensurability of the universe and a fixed harmonious linkage as can be found in no other way . . . So vast is this divine handiwork of the most excellent Almighty." Surely an aesthetic vision gripped Copernicus and guided his analysis of the heavens.

Yet it is a curious and nearly forgotten fact that the Copernican Revolution was nearly stillborn. Two decades after he had begun his book, Copernicus was running out of steam. By then in his sixties, he had written the most profound astronomical treatise in a millenium, but the small technical details lacked a final polish and inner consistency. His manuscript was a work of beauty, with exquisitely drafted figures and two-color tables, but these paper folios had apparently been designed as the end product itself.

His post as canon of the Frombork cathedral in Prussian Poland had given him financial security and time for contemplation, but it deprived him of stimulating intellectual companionship. In short, there was no one to talk to about his remarkable book, which had been so long in the making.

Although Copernicus had found printed works essential in his own studies, printing was still a comparatively new invention, one that had begun to flourish only during his own lifetime. There was as yet no printer in Frombork, and the aging astronomer apparently had no intention to publish his work elsewhere. *De revolutionibus* seemed destined to gather dust in the cathedral library, to be forgotten and ignored. Such a fate seems incredible to us today—yet it must have overtaken scores of astronomical achievements of the Middle Ages.

In 1539, during Copernicus' last years, a young mathematician from Germany appeared in Frombork, eager to learn the details of the astronomer's ideas. Georg Joachim Rheticus, already a professor at Wittenberg, though just in his mid-twenties, had heard rumors of the innovative Copernican astronomy. Although the young Rheticus came from the hotbed of Lutheranism, the Catholic Copernicus received him with courage and cordiality. Caught up by the enthusiasm of his young disciple, Copernicus made last-minute revisions to his opus and then entrusted a manuscript copy to Rheticus for publication. Rheticus took the manuscript to a printer in Nuremberg, who finished the production of several hundred copies in 1543. These were sold to scholars and libraries throughout Europe. Thus it came about that the new technology of printing was able to play an absolutely essential role in preserving and disseminating the new astronomy.

The readers of Copernicus' treatise gladly accepted his new planetary observations and his careful attention to the details of planetary orbits, but the heliocentric world view itself found little support. In our own age of space exploration, when men have looked back upon the spinning earth suspended in the skies, the Copernican system seems natural, almost obvious; but to most astronomers of the late 1500s, imbued with the physics of Aristotle, a moving earth had little appeal. Instead, they chose a primitive form of relativity. Keeping the earth as their fixed reference framework, they viewed the Copernican system as a clever mathematical model, somewhat more complicated than the old Ptolemaic system, but not a true description of the physical world. There was certainly nothing obvious about a moving earth and a fixed, central sun; Galileo later remarked that he could not admire enough those who had accepted the Copernican system *in spite of* the evidence of their senses. Nevertheless, Copernicus' *De revolutionibus* acted like a delayed time bomb. Around 1600 two great scientists, each in his own individual way, grasped a truth in the heliocentric system that went beyond geometrical model building. Johannes Kepler saw in the sun-centered spacing of the planets a harmonious, aesthetic relation that could be expressed mathematically. He envisioned a force emanating from the sun and built a "new astronomy or celestial physics based on causes."

In Italy, Galileo Galilei turned the newly invented telescope to the heavens, where he found one astonishing surprise after another. The moon, filled with mountains and plains, traversed the skies as another earthlike planet. Jupiter, with its own retinue of moons, became a miniature Copernican system. To Galileo the universe was comprehensible as a unity only if the earth was a planet revolving about the distant fixed sun.

Both Kepler and Galileo differed strongly from an anonymous introduction that had been added to *De revolutionibus* when it was printed. In that introduction the Lutheran theologian Andreas Osiander stated a widely held philosophical opinion: Astronomical theories were mathematical models intended for predictions of astronomical phenomena, and whether they were ultimately true or false was irrelevant. Such a view is logical and self-consistent, but Kepler and Galileo were convinced that their astronomy gave a true picture of the universe. This claim, plus their opinion that in matters of science the Bible simply spoke in terms of the common man so as to be understood, brought them into conflict with the Catholic Church, and consequently several of their works were placed on the *Index of Prohibited Books*.

To a 20th-century scientist familiar with the concept of relativity, it may seem ultimately irrelevant whether the earth or the sun is the chosen fixed reference point. But the 17th-century collision between religious dogma and an innovative world view has had a profound impact on mankind's views concerning the source and nature of truth about our physical world. And in the 17th century, it *did* make a difference whether the universe was conceived in geocentric or in heliocentric terms, because only a heliocentric solar system leads onward to Newtonian physics. In turn, Newton's laws of motion and his law of universal gravitation describe the orbits of satellites and space probes. Thus, there is a direct line from Copernicus through Kepler, Galileo, and Newton to the marvels of our own space age.

No doubt Copernicus would have been astonished to learn that the world would link his birthday to a celebration of modern science. An unwitting revolutionary, his goal was to return science to a purer state, conceived in terms of the perfect circles of the ancient Greeks. He sought a view "pleasing to the mind," and he gave to the mind's eye a fruitful new way of looking at the cosmos.

Still, our earlier question comes back to haunt us. Why did this new view wait until the early 1500s? The answer lies not so much in the science as in the society and in new-found patterns of communication. The invention of printing and the rise of universities encouraged the flow of information and new ideas. The discovery of America, while Copernicus was a student at Cracow, helped demonstrate the inadequacy of traditional knowledge. A seething intellectual milieu, absent a century or two earlier, characterized the age. Furthermore, Copernicus had a patron—the Frombork cathedral chapter—that enabled him to travel to Italy for graduate study and that freed him from financial worries. More importantly, he had the time and freedom to contemplate and choose innovative views. In that age of change, a new mobility of ideas brought Copernicus fresh knowledge that he required for building his system, and at the end of his life this same combination of travel, freedom, and printing saved his work from oblivion.

The new-found freedom of inquiry, combined with the requisite intellectual resources and the discipline to work out the consequences and to test the data—this is without doubt more important for the rise of modern science than the specific idea of a sun-centered cosmos. We learn more for our own age from Copernicus' persistence, his eagerness to seek learning beyond his own provincial boundaries, and his willingness to share with those outside his own religion and nation than we learn from his formidable astronomical tome or from his paeans to the heliocentric cosmology.

From his "remote corner of the world," Nicolaus Copernicus set into motion not only the earth, but the entire spirit of inquiry that has so richly increased our understanding of the universe. But the ultimate reason for the anniversary celebration—and indeed its challenge—is a rededication to preserve the fragile freedom of inquiry and the resources that make inquiry possible.

A COPERNICAN BIBLIOGRAPHY

Original Editions

Copernicus, Nicolaus. *De revolutionibus orbium coelestium libri sex.* First edition. Nuremberg, 1543.

———. *De revolutionibus orbium coelestium libri sex.* Second edition. Basel, 1566.

———. *Astronomia instaurata, libris sex comprehensa, qui de Revolutionibus orbium coelestium inscribuntur.* Third edition, edited by Nicolaus Mulerius, Amsterdam, 1617.

Modern Facsimile

Copernicus, Nicolaus. *The Manuscript of Nicholas Copernicus'*
'On the Revolutions' Facsimile. In *Nicholas Copernicus Com-*
plete Works, volume 1. London-Warsaw-Cracow, 1972. [Color
facsimile of Copernicus' original manuscript now in the Library of the Jagiellonian University, Cracow.]

Modern Translations

Copernicus, Nicolaus. "On the Revolutions of the Heavenly
Spheres." In *The Great Books of the Western World,* volume
16, translated by C. G. Wallis. Chicago, 1952.
————. *On the Revolutions of the Heavenly Spheres.* In *Nicho-*
las Copernicus Complete Works, volume 2, translated by Edward Rosen. London-Warsaw-Cracow, 1974.
Rosen, Edward. *Three Copernican Treatises.* New York, 1971.
[Contains English translations of Copernicus' *Commentari-*
olus and *Letter against Weiner,* and Rheticus' *Narratio Prima,*
a bibliography of articles about Copernicus from 1939 to 1970,
and, in this edition, a 96-page biography of Copernicus.]
Swerdlow, Noel M. "A Translation of the Commentariolus with
Commentary." *Proceedings of the American Philosophical So-*
ciety, vol. 117, no. 6 (1973), pp. 423-512.

Symposium Volumes

Beer, Arthur, and Kaj Aa. Strand, editors. *Vistas in Astronomy,*
volume 17. London, 1974. [Fifteen papers from a Copernicus
conference held in Washington, D.C., in conjunction with the
American Association for the Advancement of Science meeting, December 1972.]
Bieńkowska, Barbara, editor. *The Scientific World of Coperni-*

cus. Dordrecht, 1973. [Nine articles including a Copernicus biography by J. Dobrzycki.]

Dobrzycki, Jerzy, editor. *The Reception of Copernicus' Heliocentric Theory.* Dordrecht, 1972. [Eleven papers organized by the Copernicus Committee of the International Union of the History and Philosophy of Science and published originally in *Studia Copernicana,* volume 5, Wrocław, 1972; four additional papers appear in *Studia Copernicana,* volume 6, Wrocław, 1973.]

Westman, Robert S., editor. *The Copernican Achievement.* Berkeley-Los Angeles, 1975. [Nine articles plus commentaries from a conference held at the University of California at Los Angeles in November 1973.]

Books

Adamczewski, Jan. *Nicolaus Copernicus and His Epoch.* New York, 1974.

Armitage, Angus. *Copernicus, The Founder of Modern Astronomy.* New York, 1957.

Koyré, Alexander. *The Astronomical Revolution.* Paris-London-Ithaca, 1973.

Kuhn, Thomas. *The Copernican Revolution.* Cambridge, Massachusetts, 1957.

Articles

Gingerich, Owen. "A Fresh Look at Copernicus." Pages 154-178 in R. M. Hutchins, M. J. Adler, and J. Van Doren, editors, *The Great Ideas Today 1973.* Chicago, 1973.

_____. "Copernicus and Tycho." *Scientific American,* vol. 229 (December 1973), pp. 86-90, 95-101.

Neugebauer, O. "On the Planetary Theory of Copernicus." Pages

89-103 in A. Beer, editor, *Vistas in Astronomy*, volume 10. London, 1968.

Ravetz, Jerome R. "The Origins of the Copernican Revolution." *Scientific American*, vol. 215 (October 1966), pp. 88-95, 97-98.

Rosen, Edward. "Copernicus, Nicholas." Pages 401-411 in *Dictionary of Scientific Biography*, volume 3. New York, 1971.

CONTENTS

II THE SYMPOSIUM

III THE COLLEGIA

Collegium I: Science and Society in the Sixteenth Century

Session II: Wednesday, 25 April 1973

Collegium III: Science, Philosophy, and
Religion in Historical Perspective

Session I: Tuesday, 24 April 1973

Session II: Wednesday, 25 April 1973

Session III: Thursday, 26 April 1973

I The Festival

INTRODUCTION TO THE FESTIVAL

Owen Gingerich

Quasars, black holes, nominalism, the Renaissance magi, and the nature of the search for quarks—these were but a few of the topics touched on in the broad and perceptive discourses at the Washington symposium on "The Nature of Scientific Discovery." But the occasion was far more than a gathering of international scientists and scholars to hear learned lectures on the past and future of the scientific tradition. Rather, it was an entire festival honoring the 500th anniversary of the birth of Copernicus, complete with academic pageantry, contributions from the fine arts, and free-wheeling collegial discussions. This volume documents not only the papers—thought-provoking and wide-ranging—but also the matrix of the Copernicus Week in which the entire symposium took place. There were two innovative musical programs, a first-day stamp ceremony, a stately banquet with a Copernican drinking song straight from a hoary Oxford manuscript, and more.

Nicolaus Copernicus Week itself was the invention of the Congress of the United States. It began with a joint resolution introduced by Congressman Thaddeus J. Dulksi on 3 January 1973, passed by the House on 22 March and by the Senate on 27 March, and finally signed into a Presidential Proclamation on the tenth of April, less than two weeks before the annual meeting of the National Academy of Sciences.

The symposium, jointly organized by the Academy and by the Smithsonian Institution, served simultaneously as the principal scientific program at the Academy's spring meeting and as the Smithsonian's Fifth International Symposium; the conclave therefore represented an enriched blend of the patterns and traditions of both institutions.

The festival began, as is the custom for the Academy, with greetings and a Sunday evening musicale. (The texts of the opening remarks by Philip Handler, President of the National Academy of Sciences, and of the responses begin on page 44 of this section.) The musical program was distinguished by the premier performance of "Copernicus—Narrative and Credo" with music by the contemporary American composer Leo Smit and a text by the British cosmologist Sir Fred Hoyle. The work was specially commissioned by the National Academy of Sci-

Public Law 93-16
93rd Congress, H. J. Res. 5
April 9, 1973

Joint Resolution

87 STAT. 12

Requesting the President to issue a proclamation designating the week of April 23, 1973, as "Nicolaus Copernicus Week" marking the quinquecentennial of his birth.

Resolved by the Senate and House of Representatives of the United States of America in Congress assembled, That the President of the United States is hereby authorized and requested to issue a proclamation designating the week of April 23, 1973, as "Nicolaus Copernicus Week" and calling upon the people of the United States to join with the Nation's scientific community as well as that of Poland and other nations in observing such week with appropriate ceremonies and activities.

Approved April 9, 1973.

Nicolaus
Copernicus
Week.
Designation
authorization.

A PROCLAMATION
by the
President of the United States of America

Nineteen hundred seventy-three marks the 500th anniversary of the birth of Nicolaus Copernicus (Mikołaj Kopernik). This brilliant son of Poland distinguished himself as an economist, physician, mathematician, theologian, soldier, and statesman. But above all, it was his inspired work in astronomy and his theories about the place of the earth in the universe that marked him for greatness and precipitated the flowering of modern science.

In a world of darkness, his only weapons were the light of learning and devotion to truth. The daring, imagination, reason, discipline, and versatility of Copernicus led mankind to a brighter age. It is fitting that we pay tribute to him on the anniversary of his birth, and that we remind ourselves how much a single man, dedicated and unafraid, can do to extend knowledge and enrich human consciousness. This anniversary should also serve to remind us that the study of science is one of man's noblest pursuits.

NOW, THEREFORE, I, RICHARD NIXON, President of the United States of America, in consonance with House Joint Resolution 5, do hereby designate the week of April 23, 1973, as Nicolaus Copernicus Week, and I call upon the people of the United States to join with the Nation's scientific community, as well as that of Poland and other nations, in observing that week with appropriate ceremonies and activities.

IN WITNESS WHEREOF, I have hereunto set my hand this Tenth day of April, in the year of our Lord nineteen hundred seventy-three, and of the Independence of the United States of America the one hundred ninety-seventh.

RICHARD NIXON

ences (through the generosity of the Copernicus Society of America) to commemorate the Copernicus quinquecentennial. The musical program opened with three sections of songs from the Italian and Polish Renaissance, performed by the Gregg Smith Singers, a distinguished choral group from California founded and conducted by Gregg Smith. Concerning the Polish Renaissance, the program notes stated:

The years 1543-80 form the period of the full flowering of the music of the Polish Renaissance. During this time poets increasingly broke away from the tradition of writing verse in Latin and began to use the vernacular. This trend coincided with a breaking away from the medieval theology and the ascetic attitude of life, with more interest in the living, concrete world surrounding man, and less interest in the intangible, mystic other world promised by the Church. Thus the music forms a parallel movement, in the arts, suggestive of the attitude personified by Copernicus in the field of science. To the pre-existing Church-forms of music, the Reformation amplified two new tendencies—use of the vernacular and the weaving of folk elements into the web of the music. Music from the Protestant composers prepared the way for the highly mature work of Mikolai Gomułka (born c. 1535) exemplified by his *Melodie na psaltery polski (Melodies for the Polish Psalter)* (1580). Using Polish texts by Kochanowski, the setings are intended for a wider public. The composer's dedication notes that his melodies are "indeed easy in the performance, without let or hindrance to the plain man."

The musical motifs for the new Copernican work by Leo Smit were drawn primarily from the little known but highly developed music of Medieval and Renaissance Poland, much of which was performed in the first part of the concert. "Copernicus—Narrative and Credo" was scored for narrator (spoken by Sir Fred Hoyle himself), mixed chorus, and an ensemble of nine instruments: flute/piccolo, clarinet, bassoon/contrabassoon, horn, trumpet, trombone, viola, contrabass, and bells. The sections of the piece were: Introduction—Birth—The Uncle—Cracow Student Song—Columbus West—Italian Madrigal—Sorrow and Solitude—The Teutonic War—Papal Disputation—The Book—Death

—Laudemus et Credo. The Gregg Smith Singers and Orchestra were conducted by the composer.

The entire musical program was recorded under sponsorship of the National Academy of Sciences (Desto Records, 1860 Broadway, New York, N.Y. 10023, catalogue number DC7178). Two events highlighted the festivities on the Monday of Copernicus Week. The first was the ceremony to inaugurate the Copernicus stamp. The eight-cent commemorative postage stamp, based on an imaginative portrait of Copernicus that first appeared in 1655, was designed by Alvin Eisenman of the Yale School of Art. The first-day ceremony took place in the auditorium of the National Museum of History and Technology. The affair included a prelude and postlude by the U. S. Marine Band brass ensemble, greetings from Poland by Janusz Groszkowski, the Honorary Chairman of the Copernican Quinquecentennial Symposium, and a brief address by Murray Comarow, Senior Assistant Postmaster General of the United States. Mr. Comarow remarked that although Copernicus is generally perceived as the astronomer who moved the Earth from its starring role at center stage, he was more than that: He was a theologian, an economist, a diplomat, a physician, and a humanitarian. Mr. Comarow pointed out that this was an unusual occasion, for the Citizens Stamp Advisory Committee, which agreed on Copernicus, rarely goes beyond our national borders for stamp subjects. "Today we remind ourselves that there are minds which transcend national boundaries. Some call this the Age of Aquarius. It might better be called the Age of Copernicus. . . . The world will honor Copernicus in many ways this year. Today, distinguished scientists, diplomats and artists join with the rest of us to salute him. This beautiful stamp is our tribute at this ceremony." Following his short address, Mr. Comarow presented special albums containing sheets of the Copernicus stamp to some of the distinguished guests present including Professor Groszkowski and the Polish Ambassador, Witold Trampczynski.

Elsewhere in the National Museum of History and Technol-

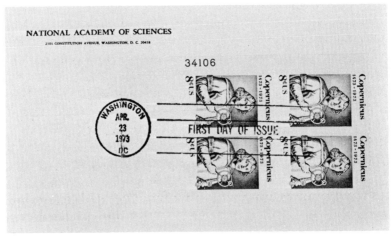

The first-day covers of the American Copernicus stamp. The lower envelope was hand-cancelled with a rubber stamp at the National Academy of Sciences on 23 April; the upper cover was one of 734,000 cancelled by machine.

ogy a special loan exhibition of rare scientific instruments from the Collegium Maius, Copernicus' alma mater in Cracow, was on display. Arranged through the cooperation of Professor Karol Estreicher and through the generosity of the Copernicus Society of America, the loan included the precious gilded "Jagiellonian Globe" from around 1510 (the first globe to show America), Martin Bylica's torquetum dated 1487 (one of a handful preserved from medieval times), as well as Bylica's globe and astrolabe, dated 1480 and 1486, respectively. These instruments were given to the Jagiellonian University in Cracow during Copernicus' lifetime. (Views of the Collegium Maius and of these instruments appear as illustrations 20-22 and 23-25, respectively, on pages 56 and 57.) Also included in the exhibition were two superb 15th-century maces of the University, and several old portraits of Copernicus.

A second highlight of the first day's festivities was the remarkable multiple-screen Copernicus slide show organized by Charles Eames and his associates at the midpoint of the program on Monday afternoon. The presentation consisted of approximately 500 images, shown three at a time on three wide contiguous screens, synchronized with a narration and a musical score composed for the show by Elmer Bernstein. All of the photographs were taken especially for the symposium showing, including many of Copernican documents and of sites in Poland. The original text and selected views from the show are found on pages 51 to 65 of this festival section; a few of the original triplicities are reproduced as reminders of the flow of colorful imagery in the Eames presentation.*

"Symposium" means literally "drinking together," and in that ancient Greek tradition the customary banquet of the National Academy of Sciences became the festive Symposium Dinner on

---

*The Office of Charles and Ray Eames has also made a single-screen film of the show with the original 11-minute sound track; it is available from them at 901 Washington Boulevard, Venice, California, 90291.

Tuesday evening, 24 April. The menu began with borscht, in recognition of Copernicus' Eastern European homeland, followed by a "Salade Kepler" inspired by the following passage from Johannes Kepler's *De stella nova* of 1606:

Yesterday, when weary with writing, I was called to supper, and a salad I had asked for was set before me. "It seems then," I said, "if pewter dishes, leaves of lettuce, grains of salt, drops of water, vinegar, oil and slices of eggs have been flying about in the air from all eternity, it might at last happen by chance that there would come a salad." "Yes," responded my lovely, "but not so nice as this one of mine."

The printed menu also included one stanza of an old Copernican drinking song from England, from a 17th-century manuscript in the hand of the great antiquarian Elias Ashmole and now in the Bodleian Library, Oxford (Ashmole 47, f. 152v.):

Then see that the Glas
Through its circuit do pas
'Til it come where it was
And every nose has byne within it,
'Tyll he end it that first did begin it.
   As Copernicus found
That the Earth did turne round,
Wee will prove so does everything in it.

At the close of the dinner S. Dillon Ripley, Secretary of the Smithsonian Institution, displayed an original portrait of Copernicus by Leonard Baskin, commissioned by the Smithsonian for the quinquecentennial celebration through the generosity of The Lilly Endowment. (His remarks, and reproduction of the portrait, begin on page 67.) The portrait served as the basis of the official symposium poster and, with the aid of the Piasecki Foundation, is available through the Government Printing Office, Washington, D. C. (Stock No. 4706-00011). A Certificate of Merit was later awarded by the Art Directors Society of Metropolitan Washington to Stephen Kraft of the Smithsonian Institution Press for his design of the poster. Mr. Ripley also presented the Smithsonian Institution's Hodgkins Medal ("for contributions to knowledge of the physical environment bearing upon the welfare of man") to Walter Orr Roberts, past president of the Program on Science, Technology, and the Humanities at the Aspen Institute for Humanistic Studies; Dr. Roberts, an astronomer, received the medal for his studies of the sun and solar-terrestrial relationships.

Two other special awards were given at the Symposium dinner, by Edward J. Piszek, President of the Copernicus Society of America, on behalf of that Society. The first medal, prize, and citation went to Professor Edward Rosen of the City College of New York for his long and continuing Copernican studies; he was also an energetic contributor to the first symposium collegium, whose transcript begins on page 309 of this volume. The second award went to Professor Jerzy Neyman of the University of California and Berkeley, who was instrumental in encouraging the National Academy of Sciences to participate in various Copernican commemorations including a special collection of invited papers that he has edited, *The Heritage of Copernicus: Theories "Pleasing to the Mind"* (MIT Press, 1974).

The innovation that characterizes the scientific discovery was paralleled in a group of modern musical works at the end of the symposium festival. Leon Kirchner conducted soloists and a chamber orchestra made up of members of the Boston Symphony Orchestra in a concert of 20th-century music on Wednesday evening, 25 April. The program consisted of Anton Webern's "Four Songs for Singer and Orchestra, Opus 13," Oliver Messiaen's "Oiseaux Exotiques," for piano solo (played by Peter Serkin) and small orchestra, Arnold Schoenberg's "Kammersymphonie für 15 Solo-instrumente, Opus 96," and "Lily," a new work by Kirchner himself. "Lily" was scored for soprano (sung by Diana Hoagland), string trio, woodwind quintet, piano, celeste, percussion, and tape. According to the program notes,

"Lily" is a self-contained segment of a larger theatrical piece entitled "Henderson," but it may be performed as a chamber work. "Henderson" is a theatrical experience, an evening of *Gesangkunstwerk* linked in some strange way to western technological civilization and its possible survival. Mythology, albeit somewhat outdated, plays a central role. In this instance the mythology has been created by Saul Bellow in the novel *Henderson the Rain King*.

Because many of the participants were baffled by the apparent lack of connection between Copernicus and modern music, the following additional notes from an interview with Professor Kirchner may prove helpful:

QUERY: When you talk about music, you very often use visual expressions, such as "landscape of notes." Do you "see" music as something visual?

KIRCHNER: Well, one can look upon it as one can look upon a landscape. In fact it is a very useful way. The structure of music is most interesting in its topography. It is perhaps one of the most compelling designs we have created. Poets and philosophers have talked about this for centuries. Whitehead speaks of the two queens of the intellect: Mathematics and Music. Surely this magical terrain has a lot to do with our conception of the universe. You know when a spider spins its web its really quite a complicated affair. These webs are astonishing in their engineering....

QUERY: Do you feel like a spider when you compose?

KIRCHNER: I don't know whether I feel like a spider, but the most perfect of our species, a Mozart, for instance, in his symphonies, sonatas, quartets, or trios reflects and "spins" as a supreme representative of our species, the vastly intricate relationships that reside in each one of us, and which connect us to the universe. The work of art is not only an abstraction in sight and sound of our own immediate and personal reservoir, but represents a more universal scheme of which we are an ineluctable and infinitesimal part....

On the evening of the Washington concert I imagined a character, "Lily," who remembered a transcendent night and mused (in celebration of Copernicus):

> Venus, moonless planet, shone
> In lunar phases like our own
> As like us full circle round the sun
> She moved and proved Copernican.

On the following pages will be found some of the greetings, responses, and toasts, together with representative pictures from the Eames slide show, that helped make Copernicus Week a particularly festive occasion. Finally, one of the most indefatigable behind-the-scenes organizers of the symposium and its manifold aspects, Wilton Dillon, puts on his social-anthropologist's hat to bring a retrospective perspective to the whole idea of a Copernican Anniversary Celebration.

OPENING CEREMONIES

Sunday Evening, 22 April 1973

INTRODUCTION

Philip Handler, *President*
National Academy of Sciences

It is my high privilege to open our national observance of the quinquecentennial of Mikołaj Kopernik, known to us by the Latin form of his name as Nicolaus Copernicus. In celebrating this occasion, we mark the passage of one-half a millennium in the history of mankind, five centuries which have been extraordinarily distinct from all that went before. While this occasion reminds us of the exponential burgeoning of scientific understanding, and while it marks a specific one of those instances in the saga of human progress in which a cardinal historical figure has been cast up out of humble origins in a relatively obscure locale (in this case the village of Torun in Poland), our principal purpose is to remind ourselves of the continuingly developing power of the human intellect—to understand the natural universe and the phenomenon that is life.

Copernicus was not the first to consider that the sun rather than the earth might be the center of our system, nor indeed did his work in astronomy, at the time, significantly affect either the larger society in which he dwelled or the considerable European intellectual community of the time. But it was his work which, in due course, came to be recognized as the incontrovertible set of arguments with respect to the relations among the planets and the sun. The methods that he used, patient observation and mathematical analysis of the relevant data at his disposal, are those which have been characteristic of science for the succeeding five centuries. From that day forward, *Homo sapiens* has been confronted with the incompatibility of an ever-deepening dual emotion—the incomparable delight of understanding coupled with the lonely pain which such understanding brings as each of us confronts our fleeting, earthly passage, our puny places in the flow of history and in the vast reaches of the cosmos.

It would be erroneous to allege that modern science began with Copernicus. Were there any such "starting date," it should probably mark Bacon's description of the scientific process or Galileo's experiments. But in the subsequent five centuries, the only event comparable to the Copernican Revolution, in respect of its impact on the collective mind of man, was the revelation of biological evolution through the life work of Charles Darwin. Like Copernicus, Darwin also had intellectual predecessors who had reached similar conclusions; and, as in the case of Copernicus, Darwin's predecessors also failed to make their case with sufficient clarity and force of evidence to compel acceptance of their conclusions. Since then, however, the collective mind of man can never be the same. Although the ideas of both were resisted by those who feared that mankind was humbled thereby, yet, with the passage of years, we have learned to find glory as well as humility in the wonder of our history and of the cosmos.

The pace of scientific understanding has so accelerated that the overwhelming bulk of what we now think we know about where we are, and what we are, has been learned within the working lifetimes of those assembled in this hall. And there are those here whose contributions to that march of scientific understanding are among the most brilliant of all.

In this long history, from time to time, the torch of learning has burned more brightly in some corners of the world than in others, occasionally coming to peak intensity by happy concatenation of circumstances, such as the simultaneous appearance of a few remarkably creative individuals in one time and place. One cannot escape the sense that some omen is implicit in the fact that Isaac Newton was born in the year that Galileo died. Whether "the man makes the times," or "the times make the man," each time that torch has glowed most brilliantly, in large measure it has been a reflection of the cultural climate, which made it possible in that time and place—Poland in the Middle Ages, 17th-century England, post-World War II America.

The history of science is the history of mankind fulfilling its purpose. Each of the peoples of the world can number its heroes in that succession; each has contributed to the development of the collective human mind. What is learned by each contributes to all and can never be unlearned. There may be truths that make us uncomfortable, there may be technological capabilities difficult to contain lest they damage more than they benefit. That must be both the glory and the burden of mankind. And we have known it since Copernicus.

RESPONSES

S. Dillon Ripley, *Secretary,*
Smithsonian Institution

The rites of spring have even richer dimensions this Easter night as we celebrate, with music, the ideas of a man who continues to provoke a sense of renewal, a sense of reconstituting the whole in a fragmented world. Thanks to the Copernican anniversary, the quickening of collaboration between scientists, artists, poets, persons of God, and persons of letters provides us—in this temple of science—with a glimpse of what a new renaissance could be. Such are the practical uses of anniversaries and heroes. The Smithsonian and the National Academy of Sciences, bound by a common ancestor, Joseph Henry, have been brought to a new bond with each other by the Father of Modern Science. We feel a new bond with our fellow truth-speakers in Poland, and with the whole community of learning. As we help to launch Copernicus Week in the United States, we give new attention to the magic of how a single human can bring about a change in the consciousness of mankind.

E. J. Piszek, *President,*
Copernicus Society of America

The Copernicus Society of America is most grateful to you, President Handler, to Secretary Ripley of the Smithsonian Institution, and to your devoted staff for permitting us to participate in your historic program celebrating the 500th anniversary of the birth of Copernicus. The Copernicus Society accepts, as an article of faith, that even the vast physical universe is no more of a miracle than the mind and the spirit of man.

The Copernicus Society believes that man's tremendous discoveries in the past three thousand years indicate that man's imagination and intuition are but the forerunners of his actual accomplishments.

It seems to us self-evident that since science and the arts have discovered the noncollision courses of the planets themselves, an anxious world may well hope that science and the arts will make further discoveries by which the nations of the earth may move forever on the noncollision course of permanent world peace.

The Copernicus Society conceives of this symposium, which you initiate tonight, as among the first great steps toward the realization of that hope, which is first in the hearts of all men—permanent world peace.

To the illustrious scholars and distinguished scientists of the world, who have assembled here, and to you President Handler and Secretary Ripley, for including the Copernicus Society of America in the proceedings, permit us to express our deepest appreciation. Thank you.

Janusz Groszkowski,
Honorary Chairman of the Symposium

We have gathered tonight in such a distinguished company to recall the merits that Mikołaj Kopernik—that great and universal scientist, and creator of the new heavens—had for all of humanity. We return our thoughts to the work of his life to receive inspiration for our contemporary scientific activity.

The ideas and examples of great creators will always lie behind the efforts of the scientists of the next generations. Therefore, it is worthwhile to spare no effort to recognize their scholarly achievements, their explorations, successes, and mistakes, in order to enter ever more boldly into the laws of nature and to subject them to the human will and intellect.

Mikołaj Kopernik is and will forever remain a symbol of the free and independent scientific mind, and of the dedication to truth. He searched for it, and he served it. This should be an eminent characteristic of every researcher. Without it there is no progress in science.

The organizers of the Copernican celebrations in Washington, D. C., have found a fine idea to begin the festivities, the composition by Mr. Leo Smit, the renowned American composer, with the narration by Sir Fred Hoyle, an outstanding British astronomer. The "Copernicus" was composed in Italy. It was based upon the motifs of the music of the Polish and Italian Renaissance. It reflects in the best way the international character of our symposium, as well as the universal significance of the work of the great Polish astronomer.

Please accept from me and my Polish colleagues our most cordial thanks and greetings for all of those who contributed to organizing this wonderful evening. Let me also congratulate the creators and performers of the "Copernicus." On behalf of our Copernican Committee, may I wish the symposium and all of the celebrations connected with it every success.

THE EAMES SHOW:
THE AGE OF COPERNICUS

Scenes and text from the show presented by the Office of Charles and Ray Eames in the Academy Auditorium at the symposium session of Monday afternoon, 23 April. The production was made possible by a grant from the IBM Corporation.

Photographs: Charles Eames

Consultant: Owen Gingerich

Production: Jehane Burns and Jeannine Oppewall

Musical Score: Elmer Bernstein

Layout: Harold Lind

1

It was close to the northernmost coast of Europe, in the city of Torun, the birthplace of Nicholas Copernicus

that the King of Poland and the Teutonic Knights signed and sealed the Peace of 1466, which made West Prussia part of the Polish territory.

3

6

7

Copernicus grew up far from the
centers of Renaissance innovation; and he was schooled in a
world that still fitted the classi-
fications of Aristotle.

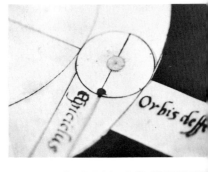

The sciences, and astronomy not least, were beginning to respond to the opportunities offered by the printing press. Copernicus and his contemporaries found themselves with new ways of access to the past. In their lifetimes they put together a completely renewed picture—to us a much more familiar one—of what nature is and what learning and observation should be.

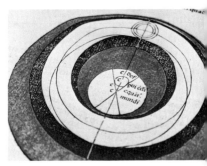

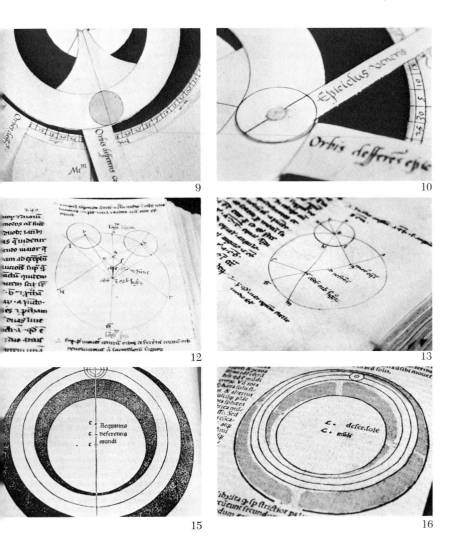

9

10

12

13

15

16

17

20

23

19

22

25

The training of the new generation had been throughout Europe much the same as Copernicus received in the university town of Cracow.

After leaving the Collegium Maius, and before going on to his lifetime post as a canon of the Cathedral of Frombork, Copernicus made a journey to study medicine and canon law in the university cities of northern Italy.

"I began to be annoyed that the philosophers had discovered no sure scheme for the movements of the machinery of the world. Therefore I also began to meditate on the mobility of the Earth."

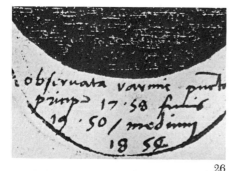

26

27

28

At Frombork Copernicus had time, security, and a tower room to work in; and it was here that the bulk of his astronomical research was completed.

29

30

33

36

32

35

Copernicus had administrative duties on the Frombork cathedral estates; and he put his knowledge of medicine and law at the service of the local community.

38

Dixit Deus:
Fiant luminaria in firma
mento cœli, ut diuidat diem ac
noctem, & sint IN SIGNA &
tempora, & dies & annos,
ut luceant in firmamêto
cœli, & illuminent
terram.
Geñ.I.

NORIMBERGAE APVD IO. PETREIVM. ANNO M. D. XXXIIII.

39

Copernicus seems to have intended
not to put his ideas into print—until,
in 1539 a young professor of mathematics, named Joachim Rheticus,
made the journey from Wittenberg
to interview him—bringing as gifts
some of the latest and finest printed
books in the field of astronomy.

40

41

42

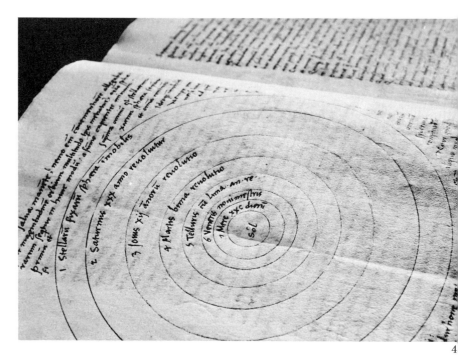

4

Rheticus studied the manuscript of
the *De revolutionibus* as Copernicus'
guest; and it was Rheticus who persuaded Copernicus to let him carry
away the completed manuscript—
after some last retouches—to the
printer.

45

1. Nicolaus Copernicus, late 16th century, Municipal Museum, Torun.
2. Copernicus' birthplace, Torun.
3. Bovillus, *Liber de intellectu* (Paris, 1511). From Copernicus' copy in the Uppsala University Library.
4. Torun, from Hartknock, *Alt und neues Preussen* (Thorn, 1684).
5. Facsimile of the Peace of 1466, Copernicus Museum, Torun.
6. Peckham, *Perspectiva communis* (Venice, 1504). From Copernicus' copy in the Uppsala University Library.
7. Reisch, *Margarita philosophica*. First edition in Freiburg, 1503. Collection of R. S. and M. B. Honeyman.
8-10. Illustrations for Peurbach's *Novae theoricae planetarum*, 15th-century manuscript. Honeyman Collection.
11-13. Ptolemy, *Almagestum*, vellum manuscript in Latin, France [?], 13th century [?]. Honeyman Collection.
14-16. Peurbach, *Novae theoricae planetarum*, printed in Venice, c. 1490. Honeyman Collection.
17-19. St. Mary's Cathedral, Cracow.
20-22. Collegium Maius, Copernicus' *alma mater* at the Jagiellonian University, Cracow.
23-25. Old instruments in the Collegium Maius: the Jagiellonian globe, c. 1510; torquetum of Martin Bylica, 1487; astrolabe of Martin Bylica, 1486.
26. Eclipse observation recorded by Copernicus in his copy of Stoeffler's *Calendarium Romanum magnum* (Oppenheim, 1518). Uppsala Observatory Library.
27. Ptolemy, *Almagestum*. Honeyman Collection.
28. Detail of an astronomical table written in the back of Copernicus' copy of his Alfonsine *Tabulae astronomicae* (Venice, 1492). Uppsala University Library.
29. Nave of the Frombork Cathedral.
30-32. Scenes from northern Poland, November 1972.
33-35. Frombork Cathedral, gate house, and wall.
36-38. Scenes from nothern Poland, November 1972.
39. Detail from the title page with Rheticus' presentation inscription, Apianus, *Instrumentum primi mobilis* (Nuremberg, 1534). Uppsala University Library.
40. Books presented by Rheticus to Copernicus, Uppsala University Library.
41. Books used by Copernicus, including Ptolemy's *Almagestum* (Venice, 1515), with his manuscript diagrams on the outer margins. Uppsala University Library.
42. Book printing from Sachs, *Eygentliche Beschreibung aller Stände* (Frankfurt, 1568). Harvard College Library.
43. Autograph manuscript of Copernicus' *De revolutionibus*. Jagiellonian University, Cracow.
44. Copernicus' autograph from his Pontanus, *Opera* (Venice, 1501). Uppsala University Library; the "p" with a stroke through its tail is the standard Latin abbreviation for "per."
45. Copernicus, *De revolutionibus*, first edition (Nuremberg, 1543).

CEREMONIES AT THE SYMPOSIUM DINNER

BASKIN'S PORTRAIT OF COPERNICUS

S. Dillon Ripley

Celebrating the 500th anniversary of the birth of Copernicus is indeed a novel and new experience. It has never happened before, and never shall again. Yet future archivists and historians of Copernicus Week in 1973 will find some residues, some continuities, between this and other celebrations, at least insofar as the Smithsonian is concerned. They might note, for example, that our previous international symposia have included a dinner, and that sharing food and toasting scholarly achievements are ways of linking the universal community of learning, all of which I mentioned during the opening ceremony at the Academy on Easter night.

In 1969, the Vice Chancellor of the University of Ghana, Dr. Alex Kwapong, a classicist, served as chairman of our symposium, "Man and Beast: Comparative Social Behavior." He reminded us of Plato's all-Athenian symposium, which included Socrates as a guest, and of the rediscovery of the importance of questions, the kind of searching for the unknown that characterizes our examination this week of "the nature of scientific

Nicolaus Copernicus by Leonard Baskin.

discovery." Plato put words into the mouth of the comic poet, Aristophanes, spinning a beautiful and humorous fantasy, which he did not expect to be taken too seriously or literally. I am not identifying who among our guests is Socrates tonight, nor making any mention of terminal libations; but I am taking a cue from our distinguished African chairman of 1969 to recall the lightness of spirit that can find expression in both art and science, the sources of creativity that Dr. Bronowski recognized in his paper on Copernicus as a humanist, in which he said: "I want to bring home the sense of the age of Copernicus that the starry heavens are a work of art, and a divine inspiration for the poet and scientist alike."

The Smithsonian is pleased to have commissioned a powerful new portrait of Copernicus as a young man which you have seen on the cover of your program and in the official poster of our celebration. Leonard Baskin, who has done stamps of Thoreau and posters of General Custer, now has turned his ink and wash to Copernicus with a stunning result—some haunting eyes, observing unseen corners of the universe, a nose which provides the optical illusion of a profile even when seen head on, a side of the face dark with night, and the other light with the sun. We need a member of Plato's symposium banquet to phrase the question, which I will put crudely: Did Baskin suggest with the stroke of the brush the transition from the Medieval to the Renaissance periods? One might go to London and ask him. But it is more fun to brood on the ambiguity of symbols, and the artist's own stated preference that he wished Copernicus might not have had the audacity to throw us little Earth people into outer space, leaving us instead snug and comfortable with our egocentrism. What is Baskin's vision? "Man must rediscover man, harried and brutalized, distended and eviscerated but noble withal—and enduring in the posture of love." Baskin seems to be trying to tell us something else. Is it that modern technology, through space exploration, is giving us a whole new basis for self-perception?

EDWARD ROSEN

☆ International dean of Copernican scholars.

☆ Fastitious student of Renaissance astronomy, and arbiter of the highest standards of historical accuracy.

☆ Translator of the *Narratio prima, Commentariolus,* and now, in the new edition of the Polish Academy of Sciences, of *De revolutionibus* itself.

April 24, 1973
Edward J. Piszek, President   Copernicus Society of America

JERZY NEYMAN

☆ Celestial statistician and penetrating investigator of the cosmic distribution of galaxies.

☆ Son of Polonia in Bessarabia, intellectually nurtured in the Copernican homeland, with Berkeley as his adopted home.

☆ Inspiring teacher, Academician and organizer of the Copernican commemoration volume of the National Academy of Sciences.

April 24, 1973
Edward J. Piszek, President   Copernicus Society of America

Citations awarded Edward Rosen and Jerzy Neyman by the Copernicus Society of America at the Symposium Dinner.

Let your fancy run to the bottom of the portrait, and see if you have the same reaction as did Philleo Nash, the anthropologist. He saw the long line, dissolving into infinity at the bottom of the face, as the trailing edge of night and day that cameras, telescopes, and men on the moon can see when they gaze back on earth. Professor Edward Rosen, who is with us tonight, has proven himself as one of the Socratic question-askers among us. He sums up the curiosity of Copernicus in this question: "What is night, and how is it produced?" These are still thrilling questions, and the artist, Baskin, has demonstrated that he can use artistry to help scientists and other citizens in their seeing and discovering things.

A COPERNICAN APPRECIATION AND TOAST

Janusz Groszkowski

On behalf of our Polish Anniversary Committee and the Polish world of learning, I can state with satisfaction that the celebrations devoted to our compatriot in the United States of America have embraced all its states and many different walks of life, and that leading figures in its academic and political life have taken part in them. We particularly appreciate that the highlight of all of these varied undertakings is this symposium on "The Nature of Scientific Discovery." For Copernicus was, first and foremost, a scholar, a thinker, whose love of truth drove him to throw down a challenge to diehard beliefs about the structure of the universe entrenched since antiquity.

The five centuries that have passed since this scholar of genius was born in the ancient city of Torun are a very long span of time if measured by subsequent advances in science and human knowledge. Why is it then that all over the world we are turning back the pages of history and recalling the character and work of this great man? I would say that it has to do with a certain phenomenon of a more general nature. The fact is that Copernicus symbolizes one of those by no means frequent watersheds that mark the twilight of one era and the dawn of a new one, and that have become milestones on the path of progress.

We, too, are living at a similar turning-point in history, which is what all of us perceive the scientific and technological revolution to be. We are living at a time when scientific discoveries are impinging on the world around us with greater force than ever, are swiftly translated into practice, and are presenting all countries and nations, the whole of humanity, with a tremendous challenge of measuring up to the formidable problems of today and tomorrow.

Our knowledge of the secrets of nature, of the earth and outer space, of the phenomena occurring in the universe at large and inside the atom, are being continually enriched by new discoveries. At the same time our accomplishment to date brings home to us how little we now know and how much there remains to be done for present and future generations of scientists. This must account for the spate of interest in the Copernicus anniversary, in the history of science, and the evolution of human thought. To no small extent the edifice of our research today rests on the lasting foundations laid by the Canon of Frombork and his great successors.

Man has now taken wing from his native planet and is using the equipment and space laboratories devised by him to explore our solar system. His telescopes and radio-telescopes are revealing phenomena and processes of which we once had only the barest inkling or which were merely the subject of theoretical conjecture.

Many of the scientists at our symposium, especially the astronomers, are engaged in research that is a direct or indirect extension of the achievements of Copernicus. I hope I shall be forgiven if I cast a certain personal sidelight on this matter. In my particular field, which seems on the face of it far removed from astronomy and which is classed among the technological sciences, I, too, owe a great deal to my compatriot. It would, after all, be impossible in radio communication research to ignore the phenomena to be observed in the upper layers of the atmosphere or the stream of radio waves pouring down from outer space.

As a student of high-vacuum engineering, I have continually found myself wandering into the borderland of electronics and space research. The last war and my part in the resistance movement also brought me face to face, in somewhat dramatic circumstances, with a problem that is connected with the technology and the subsequent space flights. This was in 1943 and 1944 when the V2 rockets were being tested in occupied Poland. It so happened that some of them strayed off course and were detected and captured by the Polish underground. I was assigned the task of making an evaluation of this missile's electronic and steering system in case the components in our hands could not be smuggled out to Britain.

Our country has proclaimed 1973 as Polish Science Year, and one of its climaxes is the 500th anniversary of the birth of Kopernik. We are commemorating the honorable and progressive currents in the history of our learning, culture, and education, and we are fusing the tradition with the tasks of the present and the prospects of the future.

The symposium over which I have the honor to preside is of an interdisciplinary nature. This is as it should be, because Copernicus was not only an astronomer, though his political, administrative, literary and economic interests are less familiar to the layman. A banquet is not the place to go into these aspects of his career and deliver yet another learned paper. Let me only draw attention to his studies in economics. Copernicus was alive to all the hazards of "the debasement of money," and stated and amplified theoretically the law that "bad money drives out good." Many of his ideas in this field still retain their validity.

A closer look at the biography of Copernicus conjures up a picture of a man who was a scholar to his fingertips, an investigator who relied on modern methods of reasoning and deducing, who was a patriot dedicated to his country, and who shirked no civic duty, but remained modest and unassuming. It would be wrong, however, to conclude that he was some kind of medieval ascetic. Quite to the contrary: he was cast utterly in the rich mold of the Renaissance. Knowing his life, one could say a little facetiously that if he were alive today, he would not only be exploring space, but bringing the same passionate concern to protection of the environment, to the preservation of the natural resources of our planet, and to their rational management. We would do well to know him better in all his greatness, in all the facets of the background of the age in which he lived and worked, an age that was so fascinating and of such consequence in the history of the human race.

We scientists of the latter half of the 20th century feel a responsibility for this race. For its present and future, its progress, well-being and happiness, for peace and brotherhood among nations whatever the social and political divisions between them, for the earth, which we jointly inhabit.

The Polish astronomer, whose memory is being honored this year by countries and nations on all continents, belongs to the whole of humanity to the same extent as the universality of his theory.

Ladies and gentlemen, and distinguished members of the Academy!

It is with a great honor and pleasure to me to participate in this 110th annual meeting of the National Academy of Sciences, devoted to the celebration of my great compatriot. I am all the more happy because ten years ago, I had the opportunity on behalf of Polish science and our Academy to address the ceremonial meeting of the 100th anniversary of the foundation of the National Academy of Sciences.

Allow me now to discharge the agreeable duty of expressing my warmest thanks to the National Academy of Sciences and to the Smithsonian Institution, and to the U.S. National Commission for UNESCO and the Copernicus Society, which have worked with them, for organizing this symposium and the numerous companion events and functions that have been arranged to pay tribute to Copernicus.

Allow me at this dinner also to express my sincere thanks to those Polish-American citizens who have put so much heart, enthusiasm, and hard work into the commemoration of the Mikołaj Kopernik anniversary in their American homeland. Among them there is a large group of scholars and scientists, frequently of great distinction. Many of them are attending this symposium; others have helped to organize Copernican celebrations in various parts of the United States.

A few days ago the President of the United States proclaimed that the week of 23 April be celebrated throughout the United States in honor of the 500th anniversary of the birth of the great Polish astronomer. The joint communique issued in the course of President Nixon's visit to Poland last year included a reference to the commemoration of this Copernican anniversary and to the necessity of expanding Polish-American cooperation in science and technology. This has already achieved a notable success in the Copernican Center in Warsaw, funds for which were contributed also by the American People.

Allow me now to propose a toast to the health of the organizers of our symposium, to the progress of science and technology, to the personal happiness of all the participants of our present meeting.

LATENT AND MANIFEST MESSAGES OF CEREMONY

Wilton S. Dillon

Scientific communities, no less than families, clans, tribes, and religious associations, need their feast days, saints days, or other special ceremonials to honor the dead. Such honoring becomes something special if one pays tribute to a paramount hero famed for some great deed or virtue that later generations ought to emulate. In the pantheon of heroes of science, Copernicus still looms as a primeval figure, whose importance deserves fanfare even if the Western calendar had not reminded us that 1973 marked the 500th anniversary of his birth. Memorable dates, even if arbitrary, serve as convenient excuses for taking stock of past, present, and future, and recognizing that preliterate or traditional people are not the only ones with origin myths that need transmission. Science is a human institution.

As an anthropologist interested in the sociology of knowledge, I welcomed the task of helping to design the rituals and substance of the major Copernicus quinquecentennial celebration in the United States. Sir Edmund Leach inspired me to realize that the Smithsonian Institution and the National Academy of

Sciences, pillars of the scientific community in our republic, and direct heirs and descendants of European scientific thought, had a noble obligation to proclaim that patrimony. Rites, rituals, ceremonies, anniversaries serve to remind "the congregation," Sir Edmund wrote, just where each member stands in relation to every other member, and in relation to a larger system.[1] Ritual serves as social communication among nomads, agriculturalists, and scientists alike. Logic, empirical investigation, and an iconoclastic or critical view of myth need not be abandoned to recognize that scientific congregations have their "nonrational" elements. The language of gesture and costume, sight, sound, taste, and smell, help to reinforce the ties that bind participants in the scientific endeavor to each other, and to their patrons and consumers.

With a distinguished Polish scientist, Janusz Groszkowski as our Honorary Chairman, and men and women of knowledge from all quarters of the earth, we gathered to break intellectual bread in Washington's museums, churches, hotels, academies, libraries, and concert halls along the axis of a capital designed by a Frenchman. The ceremonials and the ideas seemed genuine manifestations of Michael Polanyi's description of "the republic of science," as a movable congregation independent of geography, "an international community of scientists with a culture and structure of its own, a communications network, a habit of reliance on the work of others without regard of nationality, extensive international meetings and projects, and common standards of achievement and recognition, all lending themselves to an ease of understanding and establishment of common interests and goals."[2] But it was not intended as a gathering of insiders, but a convocation of diverse constituents interested in the public implications of the life of the mind.

If politics is theater, as described by Conor Cruise O'Brien in an essay, "Actors, Roles, and Stages,"[3] this springtime tribute to Copernicus demonstrated that science, too, has its moments of stagecraft; that any secular activity can be stylized into dramatic performances; and that my fellow anthropologist, Victor Turner, should have been on hand with field workers to study new instances of *"communitas"* as a part of the ritual process.

Academic gowns worn "opening night" by the heads of the Academy and the Smithsonian, special guests, patrons, and participants in the symposium did not suggest distinctions between the priests and the laity, but served as a visual reminder of the continuity in the traditions of science, symbolic links to both the universities and the churches of Medieval and Renaissance Europe, and even earlier Greek, Chinese, or Indian sages who thought about or studied the cosmos. The original music and lyrics by Leo Smit and Sir Fred Hoyle, performed after the rite of greetings and salutations, represented the unity of science and the arts, and suggested the ancient Greek symbiosis between mathematics and music.

Symbolic behavior communicates various messages. Some are direct, and some have to be decoded. The learned papers read by the contributors, savants of various academic disciplines and hybrids, were less symbolic and more straightforward examples of discourse, which readers will find in this book. But it is important to recognize the historical and cultural context out of which our celebration grew, and the implicit intentions of its patrons and organizers. I will speak, at least, of my own impressions of *Zeitgeist*, and the messages I hoped the ceremonies would help to communicate.

I came to regard the quinquecentennial as a festival of renewal, not only because of the coincidence of Copernicus Week starting on Easter Sunday, but because I felt that future historians of science would have no trouble describing the early 1970s as a period of disenchantment with science and technology, a "time of troubles" in which public skeptics and internal critics of scence were contributing to bites on the hands that fed us modern civilization. Roger Revelle, in suggesting the celebration with S. Dillon Ripley and Jerzy Neyman, lamented a new self-hate he saw developing, which would erode belief in reason, rob science of new recruits, financial support, and public understanding of how science and technology might help solve the problems of environmental quality—for which some were blaming scientists and engineers for having "caused." Would not a close look at the processes of scientific discovery reveal those redeeming virtues of human curiosity used for human melioration rather than war and pollution? [4] Would not an analysis of creativity show that careers in physics or literature require similar uses of intuition, inspiration, and wonder?

Apparently, one of these implicit messages reached Kendrick Frazier, editor of *Science News,* who saw in the "mind-expanding week" of the Copernican festival a basis for writing: "One observation to emerge is that science is much more closely related to the humanists than scientists readily admit or the public perceives. This misconception is no doubt due in great part to the misguided wish of many scientists to see and portray themselves as dealing purely in the cold, impartial world of objective fact rather than in the more subjective processes of their colleagues in art, literature, poetry and music." Quoting Sir Peter Medawar about "artistic science," Frazier added that creativity is central to both science and the humanities, a link that ties science with the rest of culture "in a way that is at once ennobling and humbling to science." [5]

The symposium and its flourishes—Renaissance dancers performing in front of priceless scientific instruments from the Copernican student period in Cracow, the reading of a presidential proclamation, the Copernicus stamp ceremony, the unveiling of Leonard Baskin's new portrait of Copernicus, the eating of Torun honeycake at a party for American student pilgrims returning from places important in Copernicus' career, the installation of the orbiting satellite telescope, "Copernicus," the launching of the BBC-Time and Life-Jacob Bronowski film, "Ascent of Man," the reading of Robert Harding's poem, "Ancient Messengers," the almost daily rhetoric on the Copernican Revolution in the *Congressional Record,* to cite a few elements of the festival—as an ensemble were designed to communicate two other significant messages: (1) The persistence of the Copernican spirit may be analogous to the survival of the Polish nation and the overseas carriers of Polish culture who make up Polonia; and (2) though identified with Poland, Copernicus transcends ethnic and national origins, and is increasingly useful as a symbol of wholeness and versatility in an age of specialization. Ethnicity is a prescientific phenomenon. Yet ethnic pride, the celebration clearly demonstrated, is a factor quite inseparable from the celebration of the birth of the Father of Modern Science. Torun, not a space platform, was the scene of his birth, and Cracow the scene of his early studies. Caught historically between Swedish, German, and Russian giants, Poles of Poland and Polonia joined the scientific community to revitalize the Copernican legend, and to remind themselves, and the world, in the process, that distant heroes can bring modern hope on a larger stage.

Splendid as they were, our rites of passage for Copernicus were hardly adequate as homage to the man who asked the deceptively simple question, "What is night, and how is it produced?" and, in seeking the answer, both honored and humbled the small planet that produced him. We academic ritualists have much to learn before we can invent new ceremonies freighted with latent and manifest messages, commensurate with the importance of the new dimensions we attach to heroes we want to bring back from the dead. Perhaps the new encounters with Chinese civilization and its science—too long cut off from the world community—will give Western science inspiration for how to give immortality to departed ancestors.

Notes

[1] Edmund Leach, "Ritual," *International Encyclopedia of the Social Sciences* (New York, 1968), vol. 13, pp. 520-526.

[2] Michael Polanyi, "The Republic of Science: Its Political and Economic Theory," *Minerva*, vol. 1, no. 1 (Autumn, 1962), pp. 54-73.

³ Wilton S. Dillon, editor, *The Cultural Drama: Modern Identities and Social Ferment* (Washington, D.C., 1974). See also Victor Turner, *Dramas, Fields and Metaphors: Symbolic Action in Human Society* (Ithaca and London, 1974).

⁴ Detlev W. Bronk has analyzed the rising clamor against science and technology as causes of ills of modern life in an essay, "The Nature of Science and its Humane Values," in Taylor Littleton, editor, *The Shape of Likelihood: Relevance and the University* (University, Alabama, 1971), pp. 21-40, originally one of the Franklin lectures on the sciences and humanities at Auburn University. "During thirty years on the frontiers between science, technology and public affairs . . . I have found three [causes of disdain of science]: Lack of understanding of the nature of science, loss of individual identify in the creative use of science, the misuse of science and technology."

⁵ His editorial, "The Nature of Scientific Discovery," appears in the 5 May 1973 issue of *Science News*.

II The Symposium

CONTRIBUTORS TO THE SYMPOSIUM

Janusz Groszkowski, *Polish Academy of Sciences*
Honorary Chairman

Jacob Bronowski
    Director of the Council for Biology in Human Affairs, The
    Salk Institute
Charles Eames
    American architect and designer, creator of furniture and
    photographic documentaries
Owen Gingerich
    Professor of Astronomy and of the History of Science, Har-
    vard University, and Astrophysicist, Smithsonian Astrophysi-
    cal Observatory
A. Rupert Hall
    Professor of the History of Science and Technology, Imperial
    College, London
Werner Heisenberg
    Director Emeritus, Max Planck Institute for Physics and As-
    trophysics, Munich
Gerald Holton
    Professor of Physics and Associate of the History of Science
    Department, Harvard University

Heiko A. Oberman
Director, Institute of Reformation Studies, University of Tübingen
Maarten Schmidt
Professor of Astronomy, California Institute of Technology
Owsei Temkin
William H. Welch Professor Emeritus, Institute of History of Medicine, Johns Hopkins University
Stephen Toulmin
Philosopher and Historian of Science, Provost of Crown College, University of California, Santa Cruz (now at the University of Chicago)
John Archibald Wheeler
Joseph Henry Professor of Physics, Princeton University

OPENING REMARKS

Janusz Groszkowski

In assuming my function as honorary chairman of this symposium on the nature of scientific discovery, I should like to express my gratitude and thanks for the honor that falls to me of presiding over such an excellent gathering.

When Christopher Columbus' ships set sail on 14 August 1492, on their conquest of the ocean that ended several months later with a landfall in the Caribbean, far away in Cracow the young Copernicus was leafing through a textbook of medieval astronomy with no less consuming a passion for knowledge, and he was harboring his first doubts about the Ptolemaic model of the universe. Though the motives of these two towering figures in history were different, there is an invisible thread between them. The first extended man's geographical horizons; the second reached out for even greater understanding, projecting the power of his intellect beyond the boundaries of our planet and paving the way to the encompassment of the cosmos itself. Both were men of destiny and both were sons of the Renaissance, which molded their temper and furnished the intellectual and psychological spurs to bold, enduring enterprise.

News of Columbus' discovery reached Cracow in next to no time, even though there was then no telephone, telegraph, radio, or television. In those days Poland, and particularly Cracow and his native Torun, were agog for developments in the world, for new discoveries, for new scientific theories. This intellectual curiosity, anchored to the thorough mathematical and astronomical foundations provided by the Cracow Academy, which was in the forefront of these sciences in Europe, led Copernicus to Italy, the birthplace of the Renaissance. It was there, in Bologna, Padua, Ferrara, in the Rome of the Borgias, that he encountered the exciting life of *Rinascita Italia* and deepened his knowledge. Perhaps in Italy, or perhaps in his homeland, he confirmed his revolutionary theory. For a long time he vacillated before publishing his epoch-making work. Printed in Nuremberg, it was delivered to him on his deathbed. He was aware of the storm it would raise and said as much in the preface to *De revolutionibus*. But like all truth, the heliocentric theory, too, carried the day.

That this symposium observes the 500th anniversary of Mikołaj Kopernik gives a particular satisfaction to my country, because that great astronomer was born, lived, and flourished on Polish soil. The celebration, coinciding with the annual meeting of the National Academy of Sciences, embraces various academic disciplines, in recognition of the versatility of Copernicus. The main objective, however, is to examine the consequences of the revolutonary concept of Copernicus on the ensuing events that occurred during the last quincentenary: on the progress of sciences and technologies, arts and religion, and, in general, on the development of mankind and humanity—all this in order to better understand our present world and to cope more effectively with its problems.

Professor Olaf Pedersen of the International Union of History and Philosophy of Science, in his papers at the UNESCO celebration last March in Paris, said: "Our newly acquired insight is merely another proof that science knows no boundaries. A new link has been established between East and West, and Nicolaus Copernicus appears in a fuller light as a truly universal representative of that ever probing and questioning spirit of mankind which is the common property of all civilizations." I fully agree with this statement. Nicolaus Copernicus, Polish scientist, belongs as well to all of humanity, as his theory is universal.

The resolution of the General Conference of the UNESCO gave it public utterance; it invited member states to expand international scientific cooperation for the benefit of humanity as a whole, in a spirit of peace and progress. This imposes upon us a responsibility for the destiny of the earth we inhabit, for the peace that reigns on it, for the maintenance and reasonable exploitation of natural resources, for our attitude towards the environment, and for the progress of science.

The future of mankind depends only on us, on the wisdom and the sense of responsibility of the peoples, of the states, of the governments.

¶THEORICAE NOVAE PLANETARVM GEORGII
PVRBACHII ASTRONOMI CELEBRATISSIMI
¶DE SOLE

Ol habet tref orbes a fe inuicem omniquaꝗ diuifos
atꝗ fibi contiguos. quoꝝ fupmus fecundu fupficie
conuexam eft mundo concentricus: fecundu coca/
uam aut eccentricus. Infim⁹ uero fecundu cocauā
cocentric⁹: fed fecundu conuexā eccentricus. Ter/
cius aut i hoꝝ medio locatus tam fecundu fupficie
fua conuexā q̃ concauā eſt mudo eccentricus. Dicit
aut mundo concentricus orbis cui⁹ centrum eſt ce
trum mudi. Eccentricus uero cui⁹ centru eſt aliud
a centro mundi. Duo itaqꝗ primi funt eccentrici fecundu qd: & uocant orbes
augem folis deferentes. Ad motum enim eorum aux folis uariatur. Terci⁹
uero eſt eccentricus fimpliciter: & uocatur orbis folem deferens. ad motum
enim eius corpus folare infixu fibi mouetur. Hi tres orbes duo cetra tenet.

¶THEORICA SOLIS.

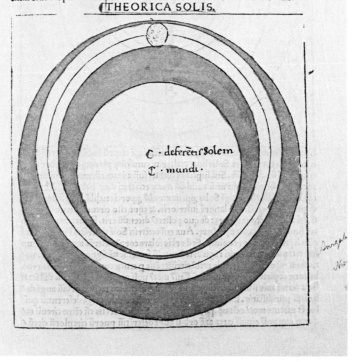

The first printed astronomy text, Georg Peurbach's *Theoricae novae planetarum,* printed by Regiomontanus at Nuremberg in 1473. (Courtesy of
Yale Historical Medical Library.)

INTRODUCTION:
THE NATURE OF SCIENTIFIC DISCOVERY
IN THE SIXTEENTH CENTURY

A. Rupert Hall

Our concern this afternoon is with the age of Copernicus, the nature of the society in which he lived and the chief stresses to which it was subjected. It is my pleasant role to introduce this discussion, and since it is Nicholas Copernicus, the paradigm of scientific innovation, whom we honor at these meetings, I thought I would begin by attempting to give a general picture of the character of scientific discovery in his age.

First, a few stage-setting remarks. Copernicus (1473-1543) was a little younger than Leonardo da Vinci; and a little older than Martin Luther. He was an exact contemporary of Machiavelli. The earlier pioneers of the Italian Renaissance in literature and the arts—such men as Poggio Bracciolini, Alberti, Brunelleschi—were all gone before Copernicus was born. Their names remind us that science was in a state of relative eclipse at the end of the 15th century. By the year 1550 one could recite the names of Copernicus and Vesalius, among other lesser ones, as those of outstanding contributors to a "renaissance of science," but any reference to a scientific renaissance when Copernicus was young would be, to my mind, premature. For example, before the birth of Copernicus the first printers or books had made available no single technical work of scientific learning; but in 1473 itself appeared Avicenna's *Canon of Medicine,* Gerard's *Theorica planetarum,* and above all the *Theoricae novae planetarum* of Georg Peurbach. For astrologically linked medical reasons, astronomy was by far the most popular of the exact sciences, and this last book, the first work of the scientific renaissance, was studied by Copernicus. Most of the publication of the scientific classics—the writings of Aristotle and Galen, of Celsus, Ptolemy and Lucretius, of the 13th-century scholastics and the 14th-century nominalists, of Euclid and even (barely) Archimedes—occurred during the lifetime of Copernicus, embracing as it did nearly all the first great century of printing. At last came the issue of Ptolemy's *Syntaxis* at Venice by Petrus Liechtenstein in 1515.

Although both the immediate predecessors of Copernicus in astronomy were, like himself, northern Europeans, during his youth Italy was still the pivot of learning, art, and civilized living. Italy bore the visible imprint of Rome, it had been the recent refuge of Greek scholars, it possessed incomparable libraries. Italy drew Copernicus as it drew Vesalius, as it was to continue to attract William Harvey and John Milton. But here we come to a curious paradox. If you or I could revisit Italy about the year 1500 (if only in television fantasy) we should rush to Medici's Florence, the Sforza's Milan, the Borgia's Rome; eagerly we would pursue Leonardo, Michelangelo, Raphael, Benvenuto Cellini; we would feast our eyes on marble still partly uncut, on paint not yet dry. But the majority of northern Europeans who went to Italy in Copernicus' time did not do so for our reasons, not for its antiquity and art, but for reasons that had long held good. They went in search of degrees. They went (as Copernicus himself did) to study law, medicine, or philosophy. In the two former disciplines Italy enjoyed a traditional excellence; in philosophy, where Oxford and Paris had once ruled,

Padua was now in the ascendant and foreign students were to roll the long names of Pomponazzi, Zabarella, and Piccolomini through their theses. That Copernicus was ever touched by the artistic renaissance in Italy is, I suppose, dubious; that he was touched by its humanism and hermeticism we may be fairly sure. Newton and many other early post-Copernicans referred to Copernicus as the "reviver of the Pythagorean hypothesis." When we find Copernicus himself in Book I of *De revolutionibus* quoting Cicero and Plutarch on the beliefs of the Pythagorean philosophers or recollecting that Hermes Trismegistos had called the sun a "Visible God," we observe his reflection of those aspects of contemporary Italian philosophy that had, under the guidance of Pico della Mirandola and others, moved far from the straight path of Aristotelianism. But we should be reluctant to hang much weight on such allusions, or to picture Copernicus as a northerner captivated by the more brilliant and esoteric treasures of the warm south.

I do not wish to raise now the question of whether Copernicus was medieval or modern in his astronomy, or indeed discuss the extent of his technical originality. It is enough for me to remind you that whatever his world-class, by the standards of 16th-century Europe Copernicus was an extremely rugged professional—the best technically equipped astronomer who had lived in Europe for many centuries. The man who wrote that mathematical science is for mathematicians and called Lactantius a geometrical illiterate was not, I think, profoundly influenced by the philosophical currents fashionable in court circles. Whether, as Dr. Ravetz suggests, Copernicus was originally and chiefly moved by the technical incompetence of Ptolemaic astronomy as then interpreted, or whether rather it was the incongruity of its consequences that most struck him, either way it was something in the nature of the existing mathematical system. Copernicus was no metaphysical cosmologist. He had no affinity with Nicholas of Cusa before or with Giordano Bruno after. He was, in a way, a plain man, not one of teeming imagination. Whatever its contribution to physical hypotheses of the Universe—think of William Herschel—I doubt whether imagination has ever played much part in mathematical astronomy.

If he was not highly imaginative, then surely Copernicus resembled the majority of his learned contemporaries. The imagination of the humanists did not extend beyond discovering and printing Greek and Roman authors; they wished young men to enrich their minds by reading in both these tongues but it was far from the scholar's aim to transform thereby the morals, religion, or social behavior of their charges. The imagination of mathematicians hardly stretched beyond trigonometry and simple algebra; Apollonios and Diophantos still exceeded its limits. When Copernicus was young, the imagination of navigators and geographers hardly yet extended so far as to distinguish Cathay from the Caribbean; only the imagination of gold in the eyes of greedy men was boundless. As for the physicians, their imagination could discern no medicine outside the works of Galen; and for the philosophers nature was still writ not in numbers but in the eight books of Aristotle's *Physics*. True, they knew of Calculator as well, and Heytesbury and John of Dumbleton—or at any rate they had found these sonorous names in the books of others like themselves—but certainly nothing really new had happened in Aristotelian commentary between the death of Nicole Oresme in 1382 and the end of the 15th century.

Let us pause here for a moment to consider what kinds of progressive thrust may have come through the writings of the philosophers in Copernicus' lifetime. For, to speak in very broad terms, I think we may expect to find the advance of science at this time occurring on three fronts: the philosophical, the astronomical (almost exclusively represented by Copernicus, himself), and the anatomical. I will consider natural philosophy first, move to anatomy next, then return to, and conclude with, astronomy.

Those historians who have seen the Aristotelian commentators of the 16th century as constructing a royal road towards "modern science" have divided into two schools: one school emphasizes *method*, believing, in the words of J. H. Randall, Jr., that

the conception of the nature of science, of its relation to the observations of fact and of the method by which it might be achieved and formulated, that was handed on to his successors by Galileo, was . . . the culmination of the cooperative efforts of the generations of scientists [*read* philosophers] inquiring into methodological problems in the universities of northern Italy.[1]

The other school, after the example of Pierre Duhem, emphasizes continuity in the history of the science of mechanics; or as Marshall Clagett puts it, sees Galileo as "clearly the heir of the medieval kinematicists." [2] The former might be characterized as seeing *induction* as the essence of modern science, the latter as seeing *deduction* in the leading role. They are at one in seeing Galileo as the target of their historical arrows, as the gate-keeper to modern science, even though they differ in their judgment of the nature of his achievement. Now in view of what I have said, someone might ask: Might it not be a plausible hypothesis to suppose that Galileo was *both* an inductivist and a deductivist, and that it was precisely in combining these two approaches to science that his strength lay? For no less an authority than Isaac Newton assured his readers that the true method of science does employ both induction and deduction, resolution and composition—using indeed the very medieval terms. Professor Randall also supports the alliance, writing that when "mathematical emphasis" had been "added to the logical methodology of Zabarella, there stands completed the 'new method' for which men had been so eagerly seeking." [3] If he will allow Galileo a few theorems, I think Professor Clagett might in return permit Galileo to make a few experiments.

Do we, then, really need to inquire further? I'm afraid we do. Many of us are not really happy about the suggestion—to put it crudely—that the philosophers placed Galileo's hand on the handle of a logical machine; so that he only had to feed in a steady flow of facts, turn the handle, and read off a new science from the printout. Historians of science feel uncomfortable, too, that Copernicus, Vesalius, and Paracelsus (for example) should seem totally irrelevant as predecessors of Galileo, or Kepler, Harvey, Descartes as contemporaries. Moreover, we find that philosophers like Zabarella and Pomponazzi, estimable men as they were, were not concerned with real scientific problems. As Randall says, Zabarella "drew his illustrations largely from Aristotle's biological subject-matter." A famous syllogism proves that Socrates is mortal; did we ever suppose he was not? I rather doubt whether (to borrow Popper's title) you can construct a valid logic of scientific discovery unless you have first observed a scientist actually making discoveries; even so, the philosophic discussion has not stopped yet. I do not think that the 16th-century Italians were in the happy empirical position of having a living, fast-changing science around them. They could discuss the *demonstration* of propositions profitably, but that is a very different thing. Consider a rather simple biological proposition offered to us by Robert Grosseteste in the 13th century: "All animals possessing horns lack teeth in the upper jaw." [4]

We note first Grosseteste's sensible opinion that if you would examine this generalization, it is a good move to open the mouths of as many horned animals as you can find, and look inside. Knowledge is more than descriptive correlations; we crave (rightly or not) something called an explanation. What does Grosseteste have to offer? It is, in part, this: "The cause of having horns is not having teeth in both jaws, and not having teeth in both jaws is the cause of having several stomachs."

True, we are beginning to build up a picture of a cow in our minds, but the line of thought here (e.g., ruminants don't need to chew much, therefore no teeth, therefore excess bony matter collects on top of head) strikes us as a bit precarious. Wouldn't it be as good reversed, for example, so as to read: horns for defense, no bony matter left for teeth, hence ruminant stomachs?

I do not mean to belittle the importance of abstract methodological discussions in intellectual history or even in the history of science, though the problems inserted to illustrate the method have often been trivial. Certainly science does not lack universal generalizations, which we often call laws. I believe that there is a constant and fertile interraction between those who are trying to solve problems and those who try to rationalize the methods by which problem-solvers work. I do not believe that Galileo or any other problem-solver proceeded by mechanically applying an abstract method; yet I do believe that the ways used by successful problem-solvers to justify their solutions and their description of the methods employed to attain them is influenced by their knowledge of the *rationale* of problem-solving. Therefore, one generation's rationalizing feeds back into another generation's doing in a roundabout way; *but*, the important thing in science is that the problems are real. I shall return to this, but I come next, with no less hardihood, to mechancis.

In the decades before Copernicus' birth, and during his lifetime, there was a great advance in practical mechanics, set pictorially before us by Taccola, Francesco di Giorgio Martini, Leonardo da Vinci, Georgius Agricola, and many more. We cannot date or document its stages precisely, but it seems likely that the advance was swift. Before Copernicus' death it was within a few steps of steam power and the industrial revolution, but these steps took time to accomplish partly because of adverse religious, political, and demographic factors. Now, whereas comparable developments in civil engineering are clearly reflected in the writings of Simon Stevin, so that we would know from Stevin's books alone how successful the Dutch had been in land drainage, canal navigation, and so on, very little of the mechanical engineering achievement is reflected in the writings of the Italian natural philosophers, not excepting Galileo. The most obvious technical matters you would learn from Galileo are that men had eyeglasses, that the Venetians built ships (ships that were, in fact, completely out of date) and that cannon were important in war. Perhaps this is partly because Galileo was

Italian, not a central European.

The antecedents of what we call the science of mechanics lie buried in a very few pages out of exceedingly large volumes devoted to Aristotelian physics. Here and there in the writings of such 16th-century philosophers, in those of Cardan, for example, you will find some hints that technically the world was not quite as it had been in Aristotle's day. Now, just as the more popular, short expressions of astronomy after 1543 made no reference to Copernicus or dismissed him in a word or two, so the popular short accounts of natural philosphy simply gave a banal précis of Aristotle. The professional scholars considered and often embraced a newer treatment of motion whose medieval development from different origins has been magisterially treated by Pierre Duhem, Marshall Clagett, and many more. Towards the end of Copernicus' lifetime, but not (as I have indicated) earlier, this by now respectable and established tradition of critical Aristotelianism was enriched by further new ideas though (I must insist) these were essentially qualitative and conceptual, while the kinematical, mathematical elements in the tradition remained unmodified. I will very briefly indicate the three chief instances. In 1537 the Italian mathematician Niccolò Tartaglia extended the existing verbal account of projectile motion to firearm missiles, pointing out that the trajectory was a continuous curve because the missile *fell* from the moment it left the barrel. Secondly, in 1555 the Spanish Aristotelian Domingo de Soto, in a book often reprinted, identified the free fall of bodies as an example of *uniform acceleration*. Thirdly, in 1553 the Italian philosopher Giovanbaptista Benedetti first propounded the hypothesis that a body's speed of fall is proportionate to the difference between its density and that of the medium through which it falls, irrespective altogether of the weight of the body. This hypothesis was taken up by a number of writers including Galileo.

Clearly, minds were stirring. In discussions of methodology, in discussions of motion, we discover in the post-Copernican

Title page with projectile motion from Niccolò Tartaglia's *La Nova Scientia,* first published in 1537. (Permission of Harvard College Library.)

period the first major innovations for some two centuries. As for Galileo, the Wagnerian figure of this period, of course no one can deny that he was shaped by his education, his philosophic studies, his discourse with colleagues. Recall his claim to be philosopher first, mathematician a long way behind. Galileo no more talked kinematics from birth than muted orphans talk Hebrew. But whereas Galileo finally became the solver of a precise problem ("If a body moves 4 miles in 4 hours, how long will it take to move 9 miles at the same uniformly increasing speed?"—to which the answer is 6 hours), the earlier philosophers failed to shake off the foggy generality inherent in their tradition. Tartaglia had no idea in the world how to *define* his curve, or perhaps that such a curve could be defined. De Soto made nothing about his statement about free fall—believe it or not he uses it simply as an example of a class, as one might say an orange is a citrus—and no one seems to have thought his example significant. As for the Benedetti buoyancy theory, Galileo in the end saw that it was an incidental rather than a basic truth, and certainly was of no help to anyone trying to formulate the fundamental concepts of motion.

Fortunately, I don't have to make a case for thinking that Galileo played a decisive role in the history of science either because he combined deduction with induction, or because he was a genius, or because he was a Platonist, or indeed offer any hypothesis at all. I will point out, however, that within the tradition of mechanical science (i.e., commentaries on Aristotle's *Physics*), there were at least 20 better experts than Galileo in, or a little before, his own time. Some were his own colleagues at Padua; some were men of considerable ingenuity, who said things that Galileo also said. But the medieval kinematic tradition did not produce through them what Galileo produced, a new science of motion. More, therefore, than this tradition alone was required. If it was necessary for a scientific revolution, it was certainly not sufficient.

Now for a very few words on the anatomical front. One who

digs beneath the surface may yet find that the problems of theory-making in physiology do bear some analogy to the similar problems in mechanics, and that discussions about the physician's method run parallel to those in natural philosophy.[5] In the medical science of the 16th century, far more clearly than in its natural philosophy, we see the supreme importance of the inescapable concrete problem. In the science of motion one could imagine all sorts of silly things—if a stone falls one mile in one hour, how far will it fall in the first minute?—but no one really cared whether it takes an hour or ten seconds for a stone to fall a mile. In medicine and anatomy it was different. Patients have pain in the heart or deep in the lungs, their breath hurts, they breathe gaspingly, they die. What are breathing, the heart, the lungs, the precious red blood, what are they for? You really can't escape, from ancient Egypt onwards, giving *some* sort of answer to these questions, and the answer can't help having some sort of empirical content.

Physicians like philosophers could, of course, endlessly dispute out of their books about the theory of humors, bloodletting, and the action of drugs, for none of these things could be made demonstrable; but if it was a question of whether the intraventricular septum is normally perforated, or whether the vessels below the human brain form a *rete mirabile*, or whether the right kidney is higher than the left, everyone knew and had known that the way to be sure was by ocular, anatomical inspection. Sages might say it was unnecessary to look; those who did look might see only what their predecessors had already described; yet the inductive method in anatomy is really inescapable. The steps from anatomy to physiology is, it is obvious, a far from steady one. It did not require a tremendous amount of new thinking to quibble with Galen—once the medical humanists had set Galen into print—about the spleen, or the reality of black bile, or even the pulmonary passage of the blood, though in fact even this did not occur in Copernicus' lifetime. Quibbles did not weaken the highly systematic, theoretical character of Galenic physiology. Against these systems induction was too weak a logical tool. Dissection furnished important facts about the vascular system unknown to Galen; induction suggested that Galen's account of the functioning of this system was at least incomplete, but a fundamentally new concept of the blood was needed as the foundation for a new account, from which all the phenomena old and new could be neatly deduced.

Of course this has no direct relevance to Copernicus, who was long dead when Harvey went to Padua, even if Harvey was to write of the heart in language reminiscent of that which Copernicus had used of the sun; and even if both adopted a circulation. Indeed, it would seem obvious to remark that in 16th-century astronomy theory preceded facts, while in biological science the case was reversed. Again, one notices that the alliance of astronomy with that great creative trunk of 17th-century science, post-Galilean mechanics, led to the *Principia* of Newton, whereas the alliance of medicine and mechanics yielded only Borelli's hypothetical studies of vascular flow and Stephen Hales' *Haemostaticks*. Nevertheless, for all their obvious and deep differences, both historical and philosophical, these two contemporaneous lines of investigation have in common not only a common tradition of logical precept about investigation into nature, but, what I take to be more important, a concern for inquiry into real problems; that is, both affected the transition from philosophy to natural philosophy, or science.

I propose to conclude with a few words on the epoch-making role of Copernicus in promoting this transition, and extending it widely beyond astronomy. He could so easily have been a narrow reformer. Suppose he had really written the Osiander preface; if his views of astronomy had been Osiander's, that is, Ptolemy's and Simplicius' and Duhem's, Copernicus would have been no more than the author of a useful algorithm.[6] In fact, it is very likely that Copernicus' name would have been forgotten (as the names of the authors of so many ingenious bits of geometry have been forgotten) had he simply been the inverse of

Ptolemy. Copernicus is known to mankind; how many recognize the names of Nasir ad-Din at-Tusi or Ali ibn Ibrahim ibn ash-Shatir, great astronomers both?

Professor Grant has already stressed the importance of Copernicus' role in making the claim that science is concerned with reality.[7] This is indeed a double claim, for there is first the assertion that mathematical science is not a convenient assembly of fictions to save the phenomena, but rather provides as authentic an understanding of the phenomena as any other method of philosophy; secondly, there is the assertion that philosophy provides not necessarily a *true*, but the best understanding of, phenomena that we can at any time obtain. As Copernicus himself put it, idle babblers, ignorant of mathematics, enjoyed no right to pronounce against his work on the basis of a Scriptural passage "basely twisted to their purpose."

What Grant has said about Copernicus is very true. However, I would add this: Copernicus is not merely taking or defending a philosophical position, which we might call mathematical realism as against mathematical conventionalism, though this would be important enough. He is linking this philosophical position indissolubly with a major issue concerning knowledge, concerning the description of the universe. Copernicus' claim is that this new philosophical position is one that really has what philosophers sometimes curiously call "cash value": It isn't a question of playing a logical game one way or another, but of being able to satisfy ourselves—as well as men can ever be intellectually satisfied of anything—as to the nature of the universe. Copernicus' claim is that if we concur with him we cannot only form a far more tidy, consistent, interlocking picture of the universe, but we are also entitled to believe that this picture is (under heaven) true, not a useful fiction or a likely story. Perhaps you may think this a somewhat naive view of scientific ontology. No matter, it is what it permits that is important; and what it particularly permits, nay, forces on everyone, is the conviction that the question, "Copernicus, right or wrong?" is one to which every man must know the answer. It isn't mere academic drivel.

This conviction extends to mechanics also. As Copernicus himself well knew, as first Tycho Brahe on one side and then Galileo on the other set it out in every detail, the motion of the earth was a problem in physics no less than in astronomy. All those questions about falling stones also ceased to be academic drivel; they could throw a decisive light on the nature of the universe, but only if mechanics became mathematically real rather than conventionalist. Just as (in the 16th century at any rate) you could not just *look* to see if the earth moves or not, so you could not find out about falling stones by looking; in both cases it was a question of the consistent interweaving of observation and theory. Copernicus made it essential for the philosophers of the late 16th century to treat the science of motion not as a happy intellectual exercise, but as a peephole on reality. No philosopher apprised of the Copernican issue could have spoken of the descent of heavy bodies as lightly as de Soto did; and, of course, it was Galileo who responded most strongly and effectively to the challenge that Copernicus had posed. If Galileo converted mechanics into a branch of natural philosophy, it was Copernicus who demanded that this be done, so that men could distinguish that motion that belongs to the essential order of things from that which is merely adventitious; or in Newton's words:

"*Motus autem veros ex eorum causis, effectibus, & apparentibus differentiis colligere, & contra ex motibus seu veris seu apparentibus eorum causas & effectus . . . .*" ("To establish the true motions from their causes, effects, and apparent differences, and conversely [to determine] their causes and effects from the true or apparent motions.") [8]

This, surely, was the greatest of Copernicus' achievements, for by turning science strongly to real problems, he made the development of modern science possible.

Notes

[1] J. H. Randall Jr., *The School of Padua and the Emergence of Modern Science* (Padua, 1961), p. 16.

[2] Marshall Clagett, *The Science of Mechanics in the Middle Ages* (Madison, 1959), p. 666.

[3] Randall, *op. cit.*, p. 66.

[4] A. C. Crombie, *Robert Grosseteste and the Origins of Experimental Science, 1100-1700* (Oxford, 1953), p. 67 ff.

[5] My student, Mr. Andrew Wear, has been examining such questions as these and I have learned much from him.

[6] Pierre Duhem, *To Save the Phenomena: An Essay on the Idea of Physical Theory from Plato to Galileo* (Chicago, 1969).

[7] Edward Grant, "Late Medieval Thought, Copernicus, and the Scientific Revolution," *Journal of the History of Ideas*, vol. 23 (1962), pp. 197-220.

[8] *Principia*, Scholium following Definition VIII, *ad fin.*

SCIENCE AND SOCIETY
IN THE AGE OF COPERNICUS

Owsei Temkin

Those of us who can look back over the past seventy years may well claim to have witnessed a period unique in the history of mankind. If we look for another seventy years equally fraught with momentous events, none perhaps can compete with the age of Copernicus, the seventy years from his birth in 1473 to his death in 1543.

The discovery of a new continent,[1] the first journey around the earth, the rise of Spain to a world monarchy on which the sun never set,[2] the wars that put the Ottoman Turks before the gates of Vienna, that carried foreign armies into Italy, ruined the country, and sacked Rome herself in 1527, the unrest of the peasants and of various factions in the cities, the economic preeminence of Italian and then of German cities (especially Nuremberg and Augsburg), the influx of silver from the New World which was concomitant with a price-revolution, the spread of books through the printing presses, books in the vernacular, in Latin

---

A discussion of this paper begins on page 385.

(the language of clerks, philosophers, doctors, and lawyers), and in Greek (the proud possession of the scholar), and finally the Reformation—What other age indeed could compete with ours? Then, as now, people reacted to events, to those created by themselves, as well as those imposed upon them.[3] The increasing importance of siege artillery gave rise to fortifications with bastions from which the defender's cannons could sweep the attacker.[4] Following the foreign invasions, Italian engineers became experts in the planning of fortifications. The firearms themselves also received due attention. In a book of 1537, Niccolò Tartaglia tells of a bombardier's question of how to aim a cannon "for its farthest shot." Tartaglia, "by physical and geometrical reasoning," proved that an elevation of 45 degrees above the horizon would be optimal.[5] The problem led Tartaglia to theoretical considerations about natural and violent movements, the former represented by freely falling bodies, the latter by projectiles,[6] to the invention of various explosives and other things. He "was going to give rules for the art of the bombardier and bring this to the greatest possible detail . . .," when the thought struck him that he was engaging in a matter "cruel and deserving of no small punishment by God." He stopped, destroyed what he had done, and was full of regrets. But when the Turks threatened Italy, Tartaglia reconsidered: "Seeing that the wolf is anxious to ravage our flock . . . it no longer appears permissible to me at present to keep these things hidden."[7]

The voyages of exploration naturally led to charting and map making, and these in turn were connected with astronomical observations and astronomical instruments. The close connection between geography, which had to define the latitude and longitude of localities, and astronomy is seen in the term cosmography, which could refer to the terrestrial world as well as the cosmic.

Another practical aspect of astronomy was calendar making, and the German calendar by Regiomontanus may be taken as an example of what interested its users. It begins with 1475, the year before Regiomontanus' death, and takes Nuremberg for its geographic base. It allows calculation for many years to come and for many towns and countries. Latin dates corresponding to the days of the month, the dates of the holidays, the times of the new moon and the full moon, the positions of sun and moon, all can be calculated. The large space given to eclipses of sun and moon is surprising. The location of the moon within the signs of the zodiac was important, because these signs were credited with qualities, and they were correlated to the various parts of the body. The reader was informed whether the lunar position was, or was not, favorable for bleeding. For instance: "As the doctors tell us, the Ram (Aries) is hot, dry, and fiery. Man's head is associated with it, and it [i.e., the time, not the head] is convenient for bleeding."[8] He also informs the reader that young people are usually bled at the time of the waxing moon, old people at the time of the waning moon.

Now Regiomontanus, besides being one of the greatest astronomers and mathematicians of his day, also was a very learned man. His calendar appeared in Latin and German versions.[9] The calendar proved very popular, though the demands it made on the interests and the intelligence of the reader must not be underrated; they are quite comparable with the trust our government puts in the citizen's ability to fill out his tax forms correctly.

The importance of mining was reflected in the work of the physician and classical scholar Georg Agricola.[10] His first book on mining was a dialogue between an employee at the mines of Joachimsthal, now Jachymov, in Bohemia, and two doctors of medicine. The employee, Bermannus, whose name gave the book its title, is the expert who demonstrates various minerals to the others. Together they discuss their properties and their relationship to the minerals described by Dioscorides, Vitruvius, Pliny, and Galen. Besides being interested in mathematics, geometry, poetry, and astronomy, Bermannus reveals himself as a Latin

p 3 2. Reſtat

Methods of descending into mines, from Georg Agricola, *De re metallica* (Basel, 1556). (Permission of Harvard College Library.)

scholar of no mean rank—but he does not know Greek and is not at home in the writings of the Arabs.[11] The book is a preliminary study to Agricola's more famous treatise; it aims at arousing interest in minerals and their medicinal properties. The ancients knew many of them—but now ignorance prevails; neither apothecaries nor physicians have any idea of what the substances they read about are like. Moreover, German mines contain things of which the Greeks were ignorant, and which should be made known.[12]

Bermannus is a realistic book. It names persons who owed their wealth to the mines; it mentions the yield in figures, it mentions new machines; it touches on the history of the mining towns: twelve years before, there was only one house in Joachimsthal next to an old dilapidated pit—now the town was buzzing with miners and their activities.[13] The same realistic attitude prevails toward the minerals: their identification is the main point. The books of the ancients and of the Arabs are the existing reference works giving basic information. The Arabs knew substances unknown to the Greeks, and now there are others to be added. Agricola complains about the repetitiousness in quoting the ancients; a good compendium from their best authors would suffice; only new things are worth saying about medicine.[14] Agricola's wish to see medicine brought back to the high standard of antiquity means a return to a progressive study of things as they are, without indulgence in subtleties.

When we think of the period between 1473 and 1543, our first association probably will be with beautiful pictures, statues, and buildings, rather than with cannons, calendars, mining, and other technical matters.[15] Indeed, in the artist, so we are told, we meet the predecessor of the modern scientist.[16] The artist wished to show things as they "really" were, and he was thus led to ask what things were really like and how they could be represented so as to be seen as they are.[17] In particular, the artist studied the human body, and he studied proportions and perspective, both of which rest on mathematics. And since the artist of that age also was architect and engineer, he appears as the early embodiment of the triad: knowledge of nature, mathematics, and technology. We have only to think of Leonardo da Vinci, who left anatomical drawings unequaled in their truthfulness by anything anatomists of his age had to offer, who recommended himself as an army engineer to the Italian princes, and whose imaginative power ranged far into the fantastic. We remember that beyond the Alps, and somewhat later, Albrecht Dürer, born two years before Copernicus, not only worked on fortifications, but also wrote on human proportions and on mathematics. Recent studies on the seventh book of the *Trattato di architectura* by Francesco di Giorgio Martini have revealed drawings of various mechanical devices for civilian use, especially mills, the main inanimate power machine, drawings from which later authors borrowed freely.[18] Most artists of the time lacked formal classical education.[19] They learned their craft from their masters but, conscious of their knowledge, many of them "fought the great battle for the recognition of painting as one of the 'liberal' arts," to quote the late Erwin Panofsky.[20]

All this agrees well with the discovery by Andreas Vesalius that Galen's anatomy was based on animal dissections, and that human anatomy had to be based on a careful study of human cadavers. By a curious coincidence, Copernicus' *De revolutionibus* offering the new view of the macrocosm and Vesalius' *Fabrica* containing Galen's refutation and many new anatomical discoveries about the microcosm, man's body, appeared in the same year. Vesalius demonstrated that anatomical truth about man had to be derived from autopsy, i.e., inspection of human bodies, as Agricola was teaching with regard to mineralogy, and Fuchs and others with regard to botany.[21] Text and illustrations of the *Fabrica* view the body realistically in contrast to its former schematic representations. The printing press assured that all readers could see the same illustrations, and Vesalius himself emphasized the social ties of anatomy. Contempt for manual work in medicine, i.e., surgery and anatomy, he thought, was

PICTORES OPERIS,
Heinricus Füllmaurer.   Albertus Meyer.

Artists preparing pictures of plants, from Leonhart Fuchs, *De historia stirpium commentarii* (Basel, 1542). (Permission of Harvard College Library.)

responsible for their decline.[22] He too hoped for their restoration to the glory of the ancients, and even more than Agricola he returned to antiquity via a devastating critique of ancient authority, in this case, Galen.

What we have done so far is to choose a few works we could easily call "scientific," because they were mainly occupied with natural facts and mathematics, and relate them to the needs of the time. This relationship was technological in a very broad sense; directly or indirectly these books intended to serve political, economic, medical, and esthetic needs of man. To us, the technological aspect seems characteristic of science. It pleases us when we think of the good science can do, and it frightens us when we think of the evils.

In the age of Copernicus technology was also associated with work and ideas that do not fit into our picture of science and which, for brevity's sake, we shall call magic. Magic was woven of many threads: the old belief in the occult forces of nature and the equally old belief in the influence of the stars; the old fear of witches and devils, powerfully documented in the *Malleus maleficarum*, of 1485, the *Witches' Hammer*, that made discovery of witches a fine art; the study of the Hermetic Books of late antiquity, which Marsilio Ficino introduced into Neo-Platonic philosophy; the study of the Cabala; and a whole body of popular beliefs taken seriously by intellectual leaders of the time. When about 23 years old, the arch magician Agrippa of Nettesheim wrote a book *De occulta medicina (On Occult Medicine)*, which he published later, in 1533. Agrippa acknowledged himself to be a magus (a magician), i.e., one of those who thought it possible to ascend from the world of elements to that of the stars, and hence to the intellectual world and to God Himself. The magician learned virtues of things, which empowered him to prognosticate and to perform astonishing deeds. Agrippa denied being a sorcerer; he freely admitted that the study of magic taught forbidden things. "But those things which are for the profit of men—for the turning away of evil events, for the destroying of sorceries, for the curing of diseases, for the exterminating of phantasms, for the preserving of life, honor, or fortune—may be done without offense to God or injury to religion. Who will not deem them just as profitable as necessary?"[23] The magician's confidence in the ability to operate with the magic forces that build the universe has led a modern philosopher to say that "magic . . . was a set of rules for gaining power over the world, and that was also Bacon's program and has remained the program of the applied scientists, engineers, physicians, and advertising men."[24]

Nevertheless, the temptation may be great to dismiss all magic, both black and white, and all divination including astrology, and all belief in demons and witches by lumping them together as pseudoscience or superstitions. We can do this for our own times, but we cannot do it for the age of Copernicus.[25]

Where would we draw the line? Bermannus speaks of the demons in the mines, some harmless and some evil.[26] Regiomontanus relates the astrological significance of the signs of the zodiac. If we eliminate these matters as "superstitions," the remainder may be a pure distillate of science as we understand it. But our treatment of science and society would become unreal. We cannot disregard Paracelsus, the most colorful medical figure of the age of Copernicus, simply because he too was a magus; nor can we divide his work and say these books and chapters we accept, those we disregard. The original Doctor Faustus lived in the age of Copernicus, and his spirit refuses to be exorcised.

What then was meant by science? Linguistically, the answer is simple. *Scientia* meant knowledge, more particularly, knowledge methodically acquired and taught, and there were many *scientiae*, many kinds of knowledge. But what we call science did not exist as a separate discipline; its use for the time is essentially anachronistic, as has been pointed out by others before me.[27]

There existed the mathematical disciplines, including astronomy; there existed natural philosophy that ranged from a knowledge of the elements to that of the soul as a physiological and psychological entity; and there existed a metaphysical realm from which physics was not clearly separated. There were physicians who, as the name indicates, were students of nature (*phusis*), there were mathematicians and astronomers or astrologers, there were teachers of philosophy; but "scientists" as a group or profession did not yet exist. In recent years much has been written about the preoccupation with methodology, especially in Padua, where arts and medicine were closely associated, and where the so-called scientific method, later applied by Galileo, is said to have been advanced theoretically. Professor Hall has already analyzed these discussions. Whatever their later influence may have been, in the age of Copernicus these debates hardly led to scientific discoveries, though they found an echo in Copernicus himself.[28]

Discovery is inbuilt within the very notion of modern science where knowledge is not allowed to remain static. Experiments are devised so as to give new answers to what is hypothetical or to establish facts that will verify or falsify theories. A theory, be it ever so beautiful, is thought unsatisfactory if it is without experimental or observational consequences. But in the age of Copernicus, knowledge, including knowledge of nature, was still taught as a traditional body, largely on the basis of authoritative texts. In the *Margarita philosophica* of Reisch, an encyclopedia in Latin, (i.e., for the educated), the subdivisions of natural philosophy consist mainly in the titles of Aristotelian books.[29]

For those who could not read, knowledge of nature was the practical knowledge required for tilling the land, raising cattle, and practicing one's craft, spiced by homely wisdom, the preacher's word, pictures in the church, and biblical tales. Those who could lay claim to some education participated through popular works and to varying degrees in the teaching of the schools.[30]

In the schools, the world was accepted as created by God yet understandable as a coherent system based on metaphysical notions of matter and form, on four causes (among which the final cause loomed large), on the four elements of air, earth, fire, and water, themselves combinations of the four qualities of hot, cold, dry, and wet.[31] Not only was the earth, and with it man, in the center of this universe, but what happened was understandable, because much of it was expressed in terms of human experience. To say with Aristotle that things were heavy because they tended downward to their natural place, and light because their tendency was upward made such phenomena more human than a mathematical law of gravity, which does not say what gravity is.[32]

In interpreting prescribed texts, professors could point out contradictions and could deviate from their authorities. Thus some medieval schoolmen had changed Aristotle's doctrine of violent motion and had found new formulations for accelerated motion.[33] New interpretations and new observations might meet

Septima etas müdi   CCLXIIII
Jmago mortis

Orte nibil melius.vita nil peius iniqua
Opma mors boim.reqes eterna laborü
Tu senile iugum domino volente relaxas
Uinctorüq graues adimis ceruice catbenas
Exilumq leuas.τ carceris bostia frangis
Eripis indignis.iusti bona pribus equans
Atqz unmota manes.nulla exorabilis arte
A primo prefixa die.tu cuncta quieto
Ferre iubes animo.promisso fine laborum
Te fine supplicium.v.t.a est carcer perennis

Dance of the Dead, from Hartmann Schedel, *Liber chronicarum* (Nuremberg, 1493). Compare these skeletons with the drawing from Vesalius. (Permission of Harvard College Library.)

with resistance, but they were not barred; nor, on the other hand, were they expected, let alone demanded.

Works on education allow some insight into the role allotted to mathematics and natural philosophy among the cultivated class. Children, said Leon Battista Alberti, must learn to read and write. "Afterwards they should learn the abacus and, at the same time, they may also look at geometry as far as it may be useful: these two are sciences that are proper and pleasing for children's minds and no little useful in every circumstance and age."[34] That is little enough, even for children, coming as it does from an architect and theoretician of the arts, especially if compared with Alberti's enthusiastic praise of the study of literature, above all Latin literature. Alberti died a year before Copernicus was born; his thoughts may be considered influential at the time of Copernicus' youth. Juan Luis Vives, on the other hand, published his great pedagogic work *De tradendis disciplinis* in 1531. According to him, the chief aim of education was religious, moral, and civil. Within this framework, the practical uses of the astronomical sciences and of natural philosophy were stressed, and Vives was particularly interested in such matters as medicinal plants, navigation, architecture, and all crafts. They need not, however, be taught in school; rather, the taste to acquire such knowledge should be cultivated.[35] But Vives also warned that knowledge gained of nature "can only be reckoned as probable and not assumed as absolutely true." Men, like Pliny and Aristotle, who insist on sensual or logically indisputable causes for everything "become incredulous of the discoveries of others, and unbelieving in matters of religion."[36]

Vives belonged to the circle of Sir Thomas More and Erasmus of Rotterdam. In More's *Utopia*, "Science" does not play any conspicuous role. Erasmus, on the other hand, in a short letter that was prefaced to Agricola's *Bermannus*, allows a glimpse of his attitude. The book pleased him, and the veins of silver and gold almost induced in him a desire for such things. "If only," he exclaims, "our minds carried us heavenward with the same zeal with which we search the earth." There is nothing wrong with the mining industry, but "only the vein of sacred books can truly enrich man."[37]

Some 300 years later, Dr. Arnold, the headmaster of Rugby, wrote to a friend:

Rather than have physical science the principal thing in my son's mind, I would gladly have him think that the sun went round the earth, and that the stars were so many spangles set in the bright blue firmament. Surely the one thing needed for a Christian and an Englishman to study is Christian and moral and political philosophy.[38]

In the days of Dr. Arnold, science did exist and it posed a serious threat to fundamentalist beliefs and to the primacy of a classical and moral education. In the days of Vives and Erasmus no such threat existed—the natural knowledge of the time was itself full of miraculous stories. The various branches of the pseudosciences were controversial, and there were those who rejected the casting of horoscopes and any form of black magic; but there was near unanimity in the admission of white magic, the knowledge of and command over the occult forces of Nature. How could it be otherwise, when all religions firmly asserted the historical reality of miracles that were not verifiable? Their acceptance presupposed a habit of mind that did not insist on verification of alleged facts of experience, if vouched for by the authority of tradition or of a venerable name.

Be that as it may, tradition and authority, we shall concede, had their place in theology and the law, both of which relied on authorities of the past: the Bible, the Fathers and Councils of the Church; the code of Justinian, and the Canon law. Here, and perhaps in metaphysics, was a place for interpretation by argument. But it ought to have been clear, we may think, that such methods and beliefs were futile in the explanation of Nature, which had to be discovered rather than interpreted.

In our impatience with authority and tradition in science we may well ask what happened when they were seriously challenged. For such challenge and pressure for new discoveries were not altogether lacking, as our discussion of the great magi suggested and for which Paracelsus is the prime example. Paracelsus condemned the medicine and natural philosophy of the schools; he appealed to experience in the light of nature; to the Aristotelian elements he opposed the chemical principles of salt, mercury, and sulfur; and he urged the finding of cures for allegedly incurable diseases. His declared motives for asking for the new were religious and social. God had not created physicians to let people die. It was their duty to find cures, rather than to grow rich and to parade the insignia of their doctoral rank. Without trust in God, the search would be futile.[39]

Paracelsus met with violent opposition, which, toward the end of his life (he died in 1541), drove him to write his *Seven Defensiones*.[40] Of these, characteristically enough, three are defenses of his discovery of a *new* medicine, of *new* diseases and of *new* receipts. Much of the opposition to him was shortsighted; but if, for a moment, we imagine that the opposition had failed, the immediate result could hardly be envisaged as anything but chaos. Paracelsus's writings are very hard to understand, his principles of research lack methodological clarity, and the curative power of his remedies is far from certain. To be sure, authority covered much sham, but it also was intended to protect against unscrupulous empiricism and experimentation. In a popular German therapeutic work of 1524, the author recommended the authority of Hippocrates with these words:

> For many years the art of medicine lay neglected until holy Hippocrates . . . saw the light of the world. And rightly is he called 'holy', for God Himself was without doubt in him, so that he gave the human race such fruitful teaching, and none can say that this same Hippocrates erred in any single thing in medicine. He was a pious, virtuous man, not idle, and always wrote for man's well-being.[41]

Generally speaking, experience needed authority; it could not be relied upon outside of a few fields like anatomy and botany. For the rest, moderate antagonism to authority and tradition did exist independently of Paracelsus and before the end of the age of Copernicus. For example, Girolamo Fracastoro wrote a short treatise in which, very apologetically, he dared to take issue with Galen.[42]

Quite possibly Fracastoro and Copernicus were fellow students in Padua, when Copernicus studied medicine there. They had more in common than medicine, for Fracastoro also wrote on astronomy. The versatility of the learned corresponded to the versatility of those who expressed themselves in drawing, viz., architects, painters, and sculptors;[43] it corresponded to a lack of specialization that united smiths or gunfounders and clockmakers.[44] Booklearning was not an adequate method for experimental science, but it allowed familiarity with many branches of study. Many outstanding philosophers, Marsilio Ficino, Pomponazzi, Achillini, and Zimara, were also physicians, or at least trained in medicine. Others, like Cardanus and the older Scaliger, are hard to pigeonhole, and Copernicus himself held the degree of Doctor of Canon Law. Versatility, of which Fracastoro was an outstanding example, was as much a matter of lacking need for specialization as a manifestation of the urge to universality.

Today, the fame of Fracastoro rests on his study of contagious diseases and on his poem *Syphilis sive De morbo Gallico*, because it coined the name syphilis for the disease then known to a majority as the French Pox, and as the *Mal de Naples* to the French. In his own days, this poem was highly esteemed for the beauty of its Latin hexameters. Fracastoro also wrote on sympathy and antipathy between things, on intellectual comprehension (which he considered part of natural philosophy), and on the soul.

Fracastoro's astronomical treatise, which preceded Copernicus' *De revolutionibus* by five years, may here be cited not for his theory of homocentric cycles but for a sentiment about the usefulness of astronomical study. After stating the many practical uses of astronomy Fracastoro wrote:

Besides, this contemplation seems in no small degree conducive to putting our minds at rest, first, because we never recognize our own and our life's insignificance as much as when we admire the greatness and perpetuity of their immortal bodies, also because lifting our mind from these terrestrial affairs we become accustomed to those that are divine. From here first came knowledge of the gods, love for them and veneration, and admiration of eternity.[45]

These thoughts were not altogether original. The Roman philosopher, Seneca, had voiced similar sentiments some 1500 years before.[46] The earth might be in the center of the world, but together with man it was believed to be of inferior material and insignificant if compared with the divine majesty of the stars. Copernicus moved man and earth out of the center, but he made them the equals of the other heavenly bodies.

How man's image of himself changed with such different cosmological views is an intriguing question. Which of the two was more likely to make him feel proud or humble? Whatever the answer may be, to us the whole subject seems to lie outside scientific astronomy. The effects of science on man are extremely important, but they are only effects, they do not belong to the science itself. Science pursues truth in its own way; its verified results are valid regardless of human feelings, social application, moral and esthetic values, and religious convictions.

Such a detachment may have been in the minds of a few individuals. Here we are not concerned with individuals in social isolation but in their interactions. In such a social sense, I believe, this kind of detachment did not yet exist. Even if the astronomer, the anatomist, or the philosopher was allowed to pursue his research in his own way, it was on condition that in the end any conflict with religion was resolved in favor of the latter.[47] When Osiander, in his preface to the reader of Copernicus' work, claimed that divine revelation alone contained the truth, which neither the pragmatic hypotheses of astronomers nor the probabilistic speculations of philosophers could offer, he misrepresented Copernicus.[48] As Professor Hall has told us in the preceding article, Copernicus' achievement to no small degree consisted in upholding the reality of the heliocentric system. But calling this an achievement underlines the exceptional nature of such independence in controversial matters. Moreover, though yielding to religious truth was the most obvious obstacle to the detachment of science from nonscientific assumptions and values, it was not the only one, and here Copernicus himself can be quoted.

He too refers to the smallness of the earth as compared with the quasi-infinite size of the universe, and he uses this argument against the view that this vast world, rather than the small earth, revolved in 24 hours. Immobility is held to be more noble and more divine than change and instability, which therefore is more becoming to the earth,[49] and he praises the central position of the sun with words that appeal to old associations linking the sun with power and dignity.

In this most beautiful temple, who would care to put this lamp [the sun] in any other or better place than whence it can illuminate the whole at the same time? Thus some people not inaptly call [it] the lamp of the world, others [its] mind, others [its] governor. Trimegistus [calls it] the visible god, Sophocles's Electra [refers to it] as seeing everything. Indeed, residing on a royal throne, as it were, the sun rules the revolving family of the stars.[50]

The fact that we can easily disregard such passages does not allow us to think of them as irrelevant for the conviction of Copernicus that the heliocentric world was real and no mere hypothesis. Similarly, the philosophical postures that Vesalius (or his artist) gave to the skeleton, and the landscape within which he placed his muscle-men hint at the feelings and thoughts associated with anatomy. In the popular dance of death, the skeleton appears as the symbol of death. Anatomically much more exact, the symbolism is still there.

To us, the landscape appears irrelevant for anatomy; the beauty of an anatomical illustration lies in its exactness and clearness. Yet for Vesalius' artist this functional beauty, if I

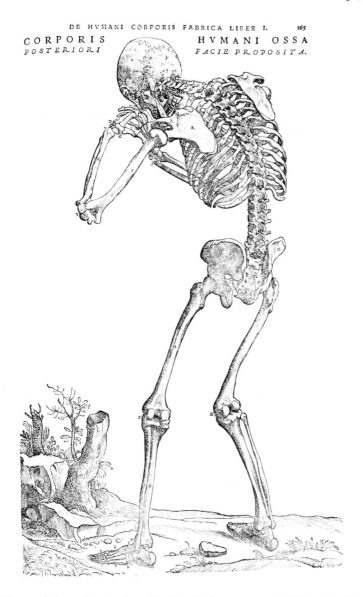

DE HVMANI CORPORIS FABRICA LIBER I.   165

CORPORIS
POSTERIORI

HVMANI OSSA
FACIE PROPOSITA.

Human skeleton, from Andreas Vesalius, *De humani corporis fabrica* (Basel, 1543). This great work was published in the same year as Copernicus' *De revolutionibus*. (Permission of Harvard College Library.)

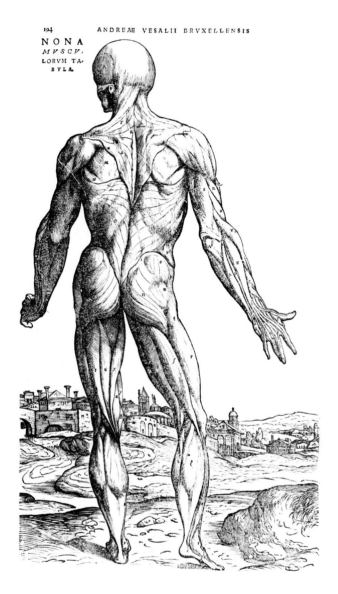

Muscle man in Italian landscape, from Andreas Vesalius, *De humani corporis fabrica* (Basel, 1543). (Permission of Harvard College Library.)

may say so, does not suffice; even a flayed human figure has to be given a place in the world, rather than on a bare page. This distinguishes the anatomical picture from the schematic drawings in the *Fabrica*, and the incorporation of such pictures is no less remarkable than the writing on syphilis in beautiful verse.

As a last example, a chapter can be cited from the *Narratio prima* of 1540, in which Rheticus gave his preliminary account of Copernicus' astronomy.[51] The chapter is entitled, "The Kingdoms of the World Change with the Motion of the Eccentric," the eccentric being the device by which the varying distances of the earth from the sun were explained.[52] In this chapter, Rheticus establishes a relationship between eccentricity and Rome's becoming a monarchy, its decline "as though aging," and its fall, the establishment of Islam, and the extension of the latter's realm (i.e., by the Turkish empire).

In our time it is at its pinnacle from which equally swiftly, God willing, it will fall with a mighty crash. We look forward to the coming of our Lord Jesus Christ when the center of the eccentric reaches the other boundary of mean value, for it was in that position at the creation of the world. This calculation does not differ much from the saying of Elijah, who prophesied under divine inspiration that the world would endure only 6,000 years, during which time nearly two revolutions are completed.[53]

Such connections of human history and heavenly events were common enough, and they usually went back to the book of Daniel and the Apocalypse of John. The idea of Heaven's having an age was forcefully expressed by Paracelsus: "Now the sky too was a child, it too had its beginning and is predestined to its end, like man, and death is in it and around it."[54]

Are such ideas to be put together with astrology, that exercised so profound an influence on high and low, on rich and poor? Whatever the answer may be, I wish to substitute a more general thought. To the generation that believed in the Bible, God had created the whole world, heavens, earth, plants, animals, and man within six days. Man and the sky have a common creation; should they not also have a common history from creation to the day of judgment?[55] Once such a notion is accepted, parallels between macrocosm and microcosm, between heavenly events and human ones, are not intrinsically fantastic. Constellations in the sky may take on the nature of portents and thus allow prognostications, even if causative power is denied to the stars. Such basic convictions rooted in a religion common to all can be much stronger than speculative theories. They make it difficult to believe that stars and men go their separate ways, that the world of nature is without meaning, without beauty, and without justice.[56]

We have spoken about detaching nature from human values; we have not yet spoken about the psychological detachment that allows man to study the world free from passion. In the age of Copernicus, people by and large more easily gave way to their impulses of anger, aggressiveness, and immediate joy than we do, who are compelled to exert constant control over ourselves and rational foresight.[57] Did those who studied nature and the behavior of man stand out by their dispassionate character?

There is the story of Leonardo da Vinci and the centenarian in the Hospital of Santa Maria Nuova in Florence.

This old man [Leonardo relates], a few hours before his death, told me that he was conscious of no bodily failure other than feebleness. And thus sitting on a bed . . . without any worrying movement or sign, he passed from this life. And I made an anatomy to see the cause of a death so sweet.[58]

The story has more than anecdotal value when we realize that anatomists who stole bodies from the gallows might have to dissect them clandestinely under conditions which would make their modern colleagues shudder.[59]

The other example concerns Guicciardini, the great Florentine historian and statesman and friend of Macchiavelli. In his *Ricordi*, Guicciardini remarks that people are quite wrong who believe that military victory depends on the justice or injustice of a cause. The belief in justice of one's cause can contribute to victory, because it makes people bold and obstinate. "Hence having a just cause may indirectly help, but it is false to say that it does so directly." [60]

So much these two examples show: There were men in the age of Copernicus who could view the life and the behavior of man dispassionately, even to a frightening degree as demonstrated by Machiavelli. It may be hard for us to realize that the desire to see things as they are and yet to see them in a framework in which science and pseudoscience, nature and values are not clearly separated can coexist. This difficulty of ours, I think, lays bare some of *our* assumptions about science and, implicitly, about the significance of scientific discovery.

We are celebrating the 500th anniversary of the birth of Copernicus; i.e., we are celebrating the man and not only his work. Therefore, it seemed appropriate to me to speak of his age in the narrow limitation of his birth and death. To bring this age into clearer focus, I have, as far as possible, refrained from using such broader concepts as renaissance and humanism, whose importance is beyond doubt, but which would blur the limits. In consequence, I have been able to offer no more than an impression, which undoubtedly could be supplemented, perhaps even supplanted, by different impressions based on a different and wider selection of examples and witnesses. The impression I offered was not that of a neat historical entity generated by the preceding age and itself generating the next. Copernicus' book was a product of his own age, but its fate was determined by old and new circumstances and by people of whom some had survived him, others were young when he died, still others were born after him.

How they dealt with Copernicus' doctrine and why they did so is their own story.[61] By narrowing our view, we may, however, have come a little closer to "the untidiness of human affairs," [62] which marks reality and which ruled over the relationship of science and society then, as it does now.

Notes

[1] On the interrelationship of the discovery of America and European intellectual and economic life, see J. H. Elliott, *The Old World and the New: 1492-1650* (Cambridge, 1970).

[2] The surprising rise of Spain has been studied in some detail by J. H. Elliott, *Imperial Spain: 1469-1716* (New York, 1963).

[3] Norbert Elias, *Über den Prozess der Zivilisation: Soziogenetische und psychogenetische Untersuchungen,* second edition (Bonn and Munich, 1969), has investigated the development of behavior within the political and social transformations of Europe since the Middle Ages.

[4] Horst De la Croix, "The Literature on Fortification in Renaissance Italy," *Technology and Culture,* vol. 4 (1963), p. 31 (with regard to Italy): "Not only military men, but humanistic scholars, architects, and artists began to devote their time and energy to an intense study of the problems raised by the apparently irresistible power of the cannon."

[5] Niccolò Tartaglia, *Nova Scientia,* quoted from Stillman Drake and I. E. Drabkin, *Mechanics in Sixteenth-Century Italy: Selections from Tartaglia, Benedetti, Guido Ubaldo, and Galileo* (Madison, 1969), pp. 63-64.

[6] Ibid., p. 72 ff.

[7] Ibid., pp. 68-69.

[8] *Der deutsche Kalender des Johannes Regiomontan, Nürnberg, um 1474, With* an introduction by Ernst Zinner (*Veröffentlichungen der Gesellschaft für Typenkunde des XV. Jahrhunderts,* "Wiegendruckgesellschaft," series B, vol. 1), (Leipzig, 1937), f. 24ʳ.

[9] As Zinner (ibid., p. 9) points out, in writing the German edition Regiomontanus had to cope with a language as yet quite unadapted to mathematical and astronomical discourse.

[10] For Agricola and his works see now *Georgius Agricola: Ausgewählte Werke* (Berlin, 1955—). In particular, Volume 1: Helmut Wilsdorf, *Georg Agricola und seine Zeit,* is important for the biography of Agricola, his forerunners in mineralogy and mining, and the conditions of his time. Volume 2: *Bermannus oder über den Bergbau: Ein Dialog,* contains a German translation of the work by Wilsdorf. For the Latin text I have used the edition in *De re metallica* (and other works) (Basel, 1657), pp. 679-701.

[11] *Bermannus,* p. 70 (German) and p. 683a (Latin).

[12] Ibid., p. 70 (German) and p. 683a (Latin).

[13] Ibid., p. 77 (German) and p. 684b (Latin). According to Wilsdorf, *Georg Agricola und seine Zeit* [note 10], pp. 163-164, there were about 900 pits and tunnels in Joachimsthal in Agricola's time, whereas by 1579 the number of tunnels had fallen to 13.

[14] *Bermannus,* pp. 107, 110, 112 (German) and 690b, 691a, 691b (Latin).

[15] For a general survey of technology in our period see A. Rupert Hall, "Early

Modern Technology, to 1600," *Technology in Western Civilization*, edited by Melvin Kranzberg and Carroll W. Pursell, Jr. (New York, 1967), vol. 1, pp. 79-103.

[16] Leonardo Olschki, *Geschichte der neusprachlichen wissenschaftlichen Literatur* (Leipzig, 1919-1922, and Halle, 1927), is fundamental for the study of the artist's role in the Renaissance. See also Erwin Panofsky, *The Codex Huygens and Leonardo da Vinci's Art Theory* (London, 1940), p. 91.

[17] Erwin Panofsky, *The Life and Art of Albrecht Dürer* (Princeton, 1955), p. 243: "The Renaissance . . . established and unanimously accepted what seems to be the most obvious, and actually is the most problematic dogma of aesthetic theory: the dogma that the work of art is the direct and faithful representation of the natural object."

[18] Ladislao Reti, "Francesco di Giorgio Martini's Treatise on Engineering and Its Plagiarists," *Technology and Culture*, vol. 4 (1963), pp. 287-298.

[19] For details see Olschki, op. cit. [note 16] vol. 1, pp. 30-46.

[20] Panofsky, op. cit. [note 17], p. 270.

[21] The methodological importance of Vesalius lies in his emphasis on learning human anatomy from the dissection of human bodies; see Owsei Temkin, *Galenism: Rise and Decline of a Medical Philosophy* (Ithaca, N.Y., 1973). Characteristic for the realistic attitude of the botanists is a picture in Leonhart Fuchs' *De historia stirpium commentarii insignes* (Basel, 1542), showing two artists at work.

[22] See the English translation of the Preface to the *Fabrica* by C. D. O'Malley, *Andreas Vesalius of Brussels, 1514-1564* (Berkeley-Los Angeles, 1964), pp. 317-324.

[23] Henricus Cornelius Agrippa ab Nettesheym, *De occulta philosophia libri tres* (n.p., 1533), "Ad lectorem," f. aaii[r]. The English translation is quoted (with modification) from *Three Books of Occult Philosophy or Magic by Henry Cornelius Agrippa, Book One: Natural Magic*, edited by Willis F. Whitehead (1892) (reprinted New York, 1971), p. 26. For the relationship of Agrippa's magic work to his religious endeavors see Charles G. Nauert, Jr., *Agrippa and the Crisis of Renaissance Thought* (Urbana, 1965) [*Illinois Studies in the Social Sciences*, vol. 55]. For Agrippa's place in the Hermetic tradition see Frances A. Yates, *Giordano Bruno and the Hermetic Tradition* (Chicago, 1964). Wayne Shumaker, *The Occult Sciences in the Renaissance: A Study in Intellectual Patterns* (Berkeley-Los Angeles, 1972), devotes a lengthy section (pp. 134-156) to the analysis of Agrippa's work. Shumaker's book is the most comprehensive, recent discussion of all forms of "occult sciences" in the Renaissance.

[24] George Boas, "Philosophies of Science in Florentine Platonism" in *Art, Science and History in the Renaissance*, edited by Charles Singleton (Baltimore, 1967), p. 241. See also Frances A. Yates, "The Hermetic Tradition in Renaissance Science," ibid., p. 255.

[25] Cf. Hugh Kearney, *Science and Change 1500-1700* (New York, 1971) (World University Library), pp. 37 ff. and 96 ff.; and W.P.D. Wightman, *Science in a Renaissance Society* (London, 1972), ch. 11.

[26] *Bermannus* [note 10], p. 88 (German) and p. 686b (Latin).

[27] In "Towards a New History of the New Science," *Times Literary Supplement* 15 September 1972, p. 1058, the anonymous reviewer, noting the absence, in 1556, of Copernicus' *De revolutionibus* from any Oxford college library, remarked on "the caution which it is necessary to use when accepting the word 'science' in an anachronistic sense." See also Wightman, op. cit. [note 25], p. 19.

[28] For the debate on "method" see John Herman Randall, Jr., *The School of Padua and the Emergence of Modern Science* (Padua, 1961), and Neal W. Gilbert, *Renaissance Concepts of Method* (New York, 1960). Copernicus himself, in the preface to Pope Paul III (*De revolutionibus orbium coelestium, libri VI* (Nuremberg, 1543), f. iii^v, wrote of earlier astronomers: "*Itaque in processu demonstrationis, quam μέθοδον vocant, vel praeteriisse aliquid necessariorum, vel alienum quid, et ad rem minime pertinens, admisisse inveniuntur. Id quod illis minime accidisset, si certa principia sequuti essent. Nam si assumptae illorum hypotheses non essent fallaces, omnia quae ex illis sequuntur, verificarentur proculdubio.*" ["Hence in the course of the demonstration, which they call 'method,' they are found either to have omitted something indispensible or to have introduced something extraneous and wholly irrelevant. This would not have happened if they had followed established principles. For if the hypotheses they assumed were not false, everything that followed would have been verified beyond any doubt."] As Edward Grant, "Late Medieval Thought, Copernicus, and the Scientific Revolution," *Journal of the History of Ideas*, vol. 23 (1962), p. 213, has shown, Copernicus used the argument to demand of hypotheses that they be true; but it did not lead him to his own discovery.

[29] *Margarita philosophica nova* (Strasbourg, 1515), f. A iv^r: "Philosophiae particio." Theoretical philosophy has two parts: "*realis*" and "*rationalis*." The former is subdivided into "*Metaphysicam,*" "*Mathematicam*" (consisting of the traditional quadrivium, i.e., arithmetic, geometry, music, and astronomy), and "*Physicam sive naturalem sub qua et medicina theorica continetur et traditur in libris Phisicorum, De caelo et mundo, De generatione et corruptione....*" "*Philosophia rationalis*" is represented by the *trivium* (grammar, rhetoric, and logic). Medicine as such, i.e., without what we would call the basic sciences, belongs to the mechanical arts (*Philosophia factiva*) together with agriculture, navigation, etc.

[30] Karl Sudhoff, *Deutsche medizinische Inkunabeln* (Leipzig, 1908) [*Studien zur Geschichte der Medizin,* fasc. 2-3], offers examples of this literature for one country.

[31] Regarding discrepancies between biblical authority and philosophical speculation, especially in such matters as creation of the world versus its eternity, and the immortality of the soul, see below, note 47.

[32] Aristotle, *Physics* 8. 4; 235b14.

[33] For a recent survey of these developments see Edward Grant, *Physical Science in the Middle Ages* (New York, 1971).

[34] Leon Battista Alberti, *I libri della famiglia*, edited by Ruggiero Romano and Alberto Tenenti (Turin, 1969), p. 86. Eugenio Garin, *Scienza e vita civile nel rinascimento italiano* ([Bari], 1965), pp. xiv-xvi, considers Alberti an important example of the sharing of scientific interests by humanists.

[35] Joannes Ludovicus Vives, *De tradendis disciplinis* 4. 6. In *Opera omnia*, edited by Gregorius Majansius (Valencia, 1785), vol. 6, p. 373 ff. [reprinted London, n.d.]. English translation by Foster Watson, *Vives: On Education* (Cambridge, 1913), p. 208 ff.

[36] *De tradendis disciplinis* 4. 1; Watson's translation p. 166 ff. with slight change. Cf. *Opera omnia* [note 35], vol. 6, p. 347. Vives is hostile to all immersion in fruitless theoretical studies of nature; we should direct all our studies "*ad vitae necessitates, ad usum aliquem corporis aut animi, ad cultum et incrementa pietatis.*" ["Toward the necessities of life, toward some use of the body or soul, toward the cultivation and increase of piety."] (ibid.)

[37] Erasmus, letter to Andreas and Christoph of Koenneritz (Agricola, ed. 1657 [note 10], p. 679): "*Visus sum mihi valles illas et colles, et foedinas et machinas non legere, sed spectare. Nec multum abfuit, quin ex tot venis argentariis et aurariis conceperim aliquam ejusmodi rerum cupiditatem. Utinam animis eo studio feramur in coelum quo scrutamur terram, non quod improbem hanc industriam, nobis enim terra gignit quicquid gignit, sed quod hae venae quantum vis faecundae, beatum hominem adeo facere non possunt, ut non paucos operae et impendii paenituerit, sola divinarum literarum vena vere locupletet hominem.*" ["It seemed to me that I saw those vales and hills, those mines and machines, rather than read about them. So many veins of silver and gold all but engendered in me a desire for things of that sort. If only our mind carried us heavenward with the same zeal with which we search the earth. Not that I disapprove of this activity— for whatever the earth produces, it produces for us—but because however rich these veins may be, they cannot make man entirely happy: some few have regretted the labor and the cost. Only the vein of sacred books can truly enrich man."] The German translation of *Bermannus* [note 10] carries the letter on pp. 59-60. For the history of the letter see Wilsdorf, op. cit. [note 10], pp. 187-188.

[38] Quoted from Lytton Strachey, *Eminent Victorians* (Garden City, N. Y.: Garden City Publishing Co., n.d.), p. 220.

[39] The idea that only God will make the work succeed is often expressed by Paracelsus, e.g., *Seven Defensiones*, translated by C. L. Temkin (in *Four Treatises of Theophrastus von Hohenheim Called Parcelsus*, edited by Henry E. Sigerist [Baltimore, 1941], p. 30): "Wherefore it follows from this that to those who walk in the way of God, perfect works and fruits grow of their talents which God gave them." This conforms to the Christian warning against the pride and haughtiness of those who ascribe their successes to their own prudence and virtue; cf. St. Ambrose, *Cain and Abel* 1. 7. 25 (*The Fathers of the Church: A New Translation*) (New York, 1961), vol. 42, p. 384.

[40] See note 39.

[41] Laurentius Phries, *Spiegel der Artzney* . . . gebessert . . . durch Othonem Brunfels (Strasbourg, 1529), f. viii^v.

[42] Hieronymus Fracastorius, *De causis criticorum dierum libellus*. In *Opera omnia*, second edition (Venice, 1574), folios 48v, 49v, 50v.

[43] Giorgio Vasari, *Le vite de' più eccellenti pittori scultori e architettori* (Nelle redazione del 1550 e 1568), edited by Rosanna Bettarini, commentary by Paola Barochi (Florence), vol. 1, p. 31: "*Introduzzione di Messer Giorgio Vesari pittore Aretino alle tre arti del disegno cioè architettura, pittura e scoltura.*"

[44] See Carlo M. Cipolla, *Clocks and Culture 1300-1700* (New York, 1967), p. 50 ff.

[45] *Homocentricorum, sive de stellis, liber unus;* ch. 1, in Fracastorius, *Opera omnia* [note 42] f. 2v. For Fracastoro's homocentric theory see J. L. E. Dreyer, *A History of Astronomy from Thales to Kepler,* second edition (New York, 1953), p. 296 ff.

[46] Seneca, *Naturales quaestiones,* Preface to book 1. Ibid., 7. 2.3 (vol. 2, p. 230 of the Loeb Classical Library), Seneca queries "whether the world goes round while the earth stands still, or whether the earth turns around while the world stands still." Although this may refer to the daily rotations only, Seneca yet concludes: "It is a matter worth contemplating, so that we may know where in the order of things we are: whether we are allotted the laziest seat or the fastest, whether god moves everything around us or us [around everything]."

[47] Grant, op. cit. [note 33], p. 25 ff. and Paul Oskar Kristeller, *La tradizione aristotelica nel Rinascimento* (Padua, 1962), p. 21, seem to disagree on the medieval meaning of "double truth." For the age of Copernicus, however, it seems conceded that the theory meant "that something can be more probable according to reason and Aristotle, while the contrary must be accepted as true on the basis of faith." (Kristeller, op. cit., p. 21.) Granting to the theory its role in the emancipation of philosophy and science from theology, Kristeller adds (p. 21 ff.), "I don't believe that the theory of the double truth as such was a conscious expression of free thought . . . but it certainly paved the way for the free thinkers of a later epoch, especially for those of the 18th century, who abandoned theology and faith and took advantage of a tradition which had established purely rational research as an independent enterprise."

[48] This seems now generally agreed upon.

[49] Copernicus, *D revolutionibus* [note 28], book 1, ch. 6; f. 6v and ch. 8, f. 7r.

[50] Ibid. 1. 10; f. 9v. This passage has often been quoted; e.g., by Wightman, op. cit. [note 25], p. 121, and Kearney, op. cit. [note 25], p. 99 ff., usually to establish the link between Copernicus and Hermetic philosophy.

[51] Edward Rosen, translator, *Three Copernican Treatises: The Commentariolus of Copernicus, The Letter against Werner, The Narratio prima of Rheticus* (New York, 1939), pp. 121-127.

[52] For details see Rosen's introduction, op. cit. [note 51], p. 34 ff.

[53] Rosen, op. cit. [note 51], p. 122.

[54] *Seven Defensiones* [note 39], p. 20.

[55] The Revelation of St. John 21:1-2: "And I saw a new heaven and a new earth; for the first heaven and the first earth were passed away; and there was no more sea. And I John saw the holy city new Jerusalem, coming down from God out of heaven, ..." The new world and the Kingdom of God together follow the destruction of the old world.

[56] On beauty and goodness in science of the time see Boas, op. cit. [note 24], p. 250.

[57] Cf. the work by Norbert Elias, op. cit. [note 3].

[58] Quoted from Martin Kemp, "Dissection and Divinity in Leonardo's Late Anatomies," *Journal of the Warburg and Courtauld Institutes*, vol. 35 (1972), p. 202, n. 11.

[59] See Andreas Vesalius, *De humani corporis fabrica libri VII* (Basel, 1543), p. 161 ff.

[60] *Ricordi*, no. 147. Quoted from *Francesco Guicciardini, Selected Writings*, edited by Cecil Grayson, translated by Margaret Grayson (London, 1965), p. 38. Guicciardini does not doubt God's justice, but he thinks that "His counsels are so profound that they are rightly called *abyssus multa*." (ibid., no. 92, p. 26).

[61] For this story see Alexandre Koyré, *From the Closed World to the Infinite Universe* (Baltimore, 1957); and Thomas S. Kuhn, *The Copernican Revolution: Planetary Astronomy in the Development of Western Thought*, reprint (New York, 1959), ch. 6 and 7.

[62] Unfortunately, I am unable to trace the source to which I owe this useful expression for one of the main obstacles to historical simplifications and generalizations.

REFORMATION AND REVOLUTION: COPERNICUS' DISCOVERY IN AN ERA OF CHANGE

Heiko A. Oberman

This address is not to be just an interesting commemoration of the historical past. Copernicus has become more than a private scholar who made a scientific discovery. Copernicus has become a symbol if not a syndrome; and it is not easy to define exactly what this symbol stands for, so varied are the reactions to his name and the associations it evokes. The nerves of Western man are hit, titillated, or hurt, and sometimes all of these at once. By no means without precedent, but certainly most intensively, the community of scholars and, with a remarkable intuition for essentials, society at large is probing the ultimate questions of man and matter, of time and space.

The inability to present the Copernican Revolution in a more or less objectively descriptive fashion, myth-proof as it were, is already obvious from the first series of articles and television programs so far presented to the public—and many more are to follow in this commemoration year! But the serious scholarly

---

Discussions of this paper begin on pages 511 and 386.

tradition on which these popularizations had to rely gave ample occasion, reason, and cause for the spread of myth. With ill-hidden ideological passion the name of Copernicus has been used to propagate the values of the French or Russian revolutions as his legitimate heirs. Replacing the Aristotelian hierarchy of multiple spheres, Copernicus then would mark the end of feudalism and emerge as the herald of our modern society. His name also suffices to connect the Christian faith with the dark Middle Ages. Pre-Copernican man is seen as caught in the blinding spiritual captivity of the "Ptolemaic Church," from which this astronomical giant liberated us to lead us into the promised land of modern times.

What is at stake in this complex issue—indeed a central issue underlying our Western Copernicus complex—is the question of what the access-route to knowledge is and, concomitantly, what the universities on the tightrope—tottering between impatient relevance and vain curiosity[1]—can do and should do. It concerns the question of theory and practice, of reason and test, speculation and experience. It is the classical clash between Plato and Aristotle, today intermittently illuminated by tensions between the German and the Anglo-American tradition of scholarship and research underlying in parallel but different ways the student revolution of our times.

Finally, the Copernican Revolution touches upon, and is rooted in, man's new relation to nature, suggested by the development from prehistorical animistic veneration to the classical adoration of nature, and via Christian admiration to the post-Christian administration of nature—therefore implying man's own changing role.

This year's celebrations may appear as a feast for fools: after all, Copernicus' heliocentric cosmology places man off center and unmasks him as cosmically eccentric. Yet we have to reconcile this with another phenomenon, on the level of anthropology, where we see a geocentricity reemerging in a sublimated form as anthropocentricity: Man in a succession of stages developed from the microcosm and image of God into the *Homo faber*[2] and partner of God, to end up, finally, as the *Homo manipulator,* God in his own realm. At that very point, what used to be the mysterious dwelling place of man and dewed path for the feet of God becomes the secular "environment"—the mechanical context of Man's survival; that is, the contemporary sore point where cosmological, behavioral, and environmental studies converge.

This short "tour d'horizon" suggests the range of concerns and apperceptions with which I have approached the given theme. If done well this lecture will be a festive historical commemoration but at the same time something of an acupuncture of nerve centers, without the Chinese promise that it will not hurt.

We begin by looking into the first encounter between the two 16th-century reform movements—in theology and in cosmology—for a time suspended in a precious but precarious balance between partnership and rivalry. In a second part we gain historical perspective and distance by dealing with the preceding late medieval phase in which both the modern sciences and the modern consciences prove to evolve simultaneously, in terms awaiting translation to reveal their effect on modern man.

THE ENCOUNTER BETWEEN COSMOLOGY AND THEOLOGY

In his play *Galileo,* Bertolt Brecht has the old cardinal say to Galileo: [3]

So you have degraded the earth despite the fact that you live by her and receive everything from her. I won't have it! I won't have it! I won't be a nobody on an inconsequential star briefly twirling hither and thither. I tread the earth, and the earth is firm beneath my feet, and there is no motion to the earth, and the earth is the center of all things, and I am the center of the earth, and the eye of the Creator is upon me. About me revolve, affixed to their crystal shells, the lesser lights of the stars and the great light of the sun, created to give light upon me that God might see me—Man, God's greatest effort, the center of creation. "In the image of God created He him." [4]

"OLD CARDINAL: . . . I won't have it. I won't be a nobody on an inconsequential star briefly twirling hither and thither . . . *(His strength fails him.)*" Galileo and the old cardinal from scene five of the Harvard Dramatic Club production of Brecht's *Galileo*, Loeb Drama Center, Cambridge, 19 May 1973.

The scene here presented by Bertolt Brecht is as moving as it is misleading. Granted, there is some truth in seeing in Galileo's plight the clash of science and faith; therefore, we cannot avoid asking whether the same applies in the case of Copernicus—whether, just as the Roman Catholic Church forced Galileo to recant, so, some twenty years after Luther's appeal to his conscience at Worms, the Reformation did not unmask itself as an intolerant, repressive, and anticonscientious movement, which tried to suppress, and for a time succeeded in subverting, the Copernican Revolution.

It is to be said with all possible clarity, however, that pre-Copernican cosmology did not posit the earth at the static center as a place of glory but as a place of inertia, the farthest removed from divine movement so perfectly reflected in the circular movement of the stars. Man, not his earth, held the cosmic place of honor, reaching in the summit of his soul *(apex mentis)* the greatest proximity to God. As far as I can see, we owe it to the mystical tradition that "center" and "summit" could become interchangeable and equivalent in dignity [5] —as can still be noticed in the parallel mixture of spacial and anthropological components in the words "depth" or "profundity."

The resistance against Copernicus may have had other causes than the normal healthy resistance in intellectual man to novelty; it might have been furthered by a mystical sense of the cohesion of man and his cosmic environment. This resistance, however, cannot be explained in terms of hurt pride as the defence mechanism of Ptolemaic-Medieval man. To the contrary, Copernicus gave the earth a cosmic dignity in keeping with the ontological rank of man, its divine inhabitant. It is thus all the more important to analyze the first reactions to Copernicus from close quarters. In order to test the traditional story of Copernicus'-lone-battle-against-the-mighty-Church we have to listen to Luther's oft-quoted *Tischrede (Tabletalk)* and weigh more extensively the arguments in the famous case of Osiander's "fraud." If this story can be substantiated, Copernicus would have a valid claim on the gratitude of all those who see in the emancipation from Christian faith the basis for the cultural progress of Western man; more gratitude, at least, than is owed to Galileo, whose similar claim is convincingly rejected by Friedrich von Weizsäcker [6] and, for very different reasons, by Bertolt Brecht himself.[7] Subjectively the case is clear: Copernicus felt intimidated by the anticipation of the charge of innovation; the very fact and the carefully worded content of his letter of dedication to Pope Paul III make this abundantly clear (see Appendix I). This is the element of truth in Arthur Koestler's *Sleepwalkers*[8] where he casts Copernicus as a fearful and submissive weakling. Objectively seen, however, Copernicus' expectation of curt, if uninformed, rejection had already been proven to be well founded. Luther spontaneously exclaims on hearing the advance rumor:

Nowadays people try to show their genius by producing new deviating ideas; this man subverts the whole field of astronomy. Even when that whole field stands topsy turvy I believe Holy Scripture. After all Joshua (10:13) commanded the sun to stand still and not the earth.[9]

Calvin—who without documentation and basis in fact is held by some recent scholarship to have been a critic of Copernicus [10] —seems to present an alternative to Luther by introducing another relation between revelation (in Scripture) and (experienced) reality. After all, in his commentary on *Genesis* Calvin points out [11] that the story of creation does not compete with the "great art of astronomy," but accommodates to and speaks in terms of the unlettered *idiota*, the common man.[12] Exactly the same argument as we find a hundred years later with Kepler, when this admirer of Copernicus reconciles Joshua with his new cosmology.[13]

Since we touch here, in this difference between Luther and Calvin, upon one of the main phenomena of change in the Copernican era, we want to take a closer look. For those who know this period, Luther's reaction is predictable: he does not give his considered opinion of the Copernican thesis; he sees him merely as an instance of the sickness of the times, *"vana curiositas."* Luther stands in a late medieval tradition *"contra curiositatem,"* which is pre- and supra-confessional, as can be seen from the identical views of Gerson and Erasmus. More generally a characteristic of the *via moderna* and *devotio moderna*, this acute aversion to *curiositas* is the awareness of the danger of one-sided intellectualism. At its worst this "modern" attitude is pietistic and anti-intellectualistic, reeling back from secular scholarship as a threat to the sacrality of the inner life. At its best—and closer to its historical origins and main thrust—it calls for a reform of the universities to discard intellectual games, far removed from experienced reality. As we shall see, this very thrust of seeming obscurantism proves to be the great wedge that is to provide Copernicus with the metaphysical antidote and the intellectual antecedents presupposed in his discovery.

The appropriate slogan for this campaign "*contra curiosita-tem*" we find in the *Adagia* of Erasmus: "*Quae supra nos nihil ad nos*" [14] ["The things above us do not concern us"]. Erasmus found it as a *dictum socraticum* (a Socratic saying) in the Church-father Lactantius († 320) and he knows that its main thrust is directed against cosmic speculation as "the curious investigation of things celestial and the secrets of nature." [15] After Lactantius, Augustine had dedicated an excursus in his *Confessiones* to the dangers of curiosity,[16] but left the deepest impression in a more direct parallel to Lactantius in his *Enchiridion*,[17] confronting the Greek metaphysical-cosmological speculation by arguing that to know the cosmic forces, the *causas motionum*, does not bring happiness; what we should know are the causes of good and evil. Hence not metaphysics, but ethics fully deserve our dedication and pursuit. This Augustinian legacy of the contrast and even mutual exclusiveness of metaphysics and ethics, of cosmology and theology had been submerged [18] and was lying dormant throughout the era of the successful Aristotelian band wagon, till in the 14th century human experience in physics and theology started to pull at the dogmatic Aristotelian chains. It is this pull-ing that is expressed in the campaign "*contra vanam curiosi-tatem.*"

It would be a fatal mistake to see in this campaign the high tide of medieval obscurantism thwarting the emergence of mod-ern science. On the contrary, "*contra vanam curiositatem*" is best translated as "against distorting intellectualism" and marks the revolt that not only paved the way, but also provided method and models, for the coming era of science.

Before reform and revolt grew into religious reformation and scientific revolution as two distinguishable movements, we see how a man like Gerson can hold in one hand the threads of re-newal in both fields. The common impetus is the call for experi-ence as the best antidote against curiosity. On the one hand the reform of religion, Church, and theology urged a return to mystical piety (Bonaventure!) and thus stressed experience as the hallmark of the true Christian. On the other hand, the renewal of the sciences called for a revolt against metaphysics and thus based the new physics on experiment and experience. Less than a century and a half later, the common impetus was severely tested when the experience of faith and the experience of science were in the process of turning against each other as alternative bridges to the future. It is here that the voice of John Calvin carries particular weight. Calvin applies the slogan "*Quae supra nos nihil ad nos*" not to the reader of Scripture, but to the intention of Moses as the author of Scripture ("*nihil attinet supra coelos volare*"; "It accomplishes nothing to fly above the heavens"). The discrepancy between the story of creation and the secured data of astronomy are not to be solved by condemning astronomy as the obscurantists (*phrenetici*) do, who arrogantly reject everything unknown to them. Nor should the data of astronomy be taken as proof that Moses erred. Moses was not a teacher of astronomy, but a theologian, hence concerned with the glory of God, which, contrary to vain curiosity, is most useful to man. In his field, the astronomer does exactly the same: his field is not only exciting, but also most useful, providing access to the breathtaking wisdom of God: "*nam astrologia non modo iucunda est cognitu, sed apprime quoque utilis: negari non potest quin admirabilem Dei sapientiam explicet ars illa.*"[19] ["for astronomy is not only nice to know, but also very useful: it cannot be denied that this study reveals the marvelous wisdom of God."]

In words almost identical with Kepler's in his *Astronomia nova*,[20] Calvin sees theology no longer in competition with astronomy or as penetration of the heavens to be rejected with the charge of vain curiosity and audacious preoccupation with the things "*supra nos.*" The sky above us is no longer the realm beyond us, beyond our ken, *supra nos.* Calvin's solution is not the obscurantist rejection of astronomy, nor does he go along with the adherents of the doctrine of "double truth"—what is true in theology is not true in philosophy. Rather one has to respect the

limits. To stay within one's personal limits—the medieval definition of humility and the alternative to proud curiosity [21]—means
now to stay within the limits of one's field of competence. The
medieval differentiation between the university faculties, programmatically transcended in the preceding stages of the Renaissance, is here recaptured by Calvin to defend and respect the
different methods of illuminating the common object, the glory
of God.

It is here that I find the historical basis for the early latitude
in Calvinism to favor or reject the Copernican vision. This
stance helps to explain as well why, against all expectations, the
relationship between Puritanism and science is to be a most
intensive and fruitful union.[22] After all, before the restoration,
the Puritans "were the main support of the new science." [23]

The so-called fraud of Osiander, who in his introduction to
De revolutionibus tried to pass Copernicus' heliocentricity thesis
off as "hypotheses," [24] was intended to raise the toleration level
in the scholarly world. It may explain Melanchthon's shift from
early condemnation to cautious support;[25] it could, of course, not
provide a more lasting basis for welding together the new science
and religion. Such a basis could only be found in the conviction
formulated by Calvin that Scripture is not a supernaturally revealed book of nature, so that religious experience and scientific
experience can go hand in hand.

Only after Darwinism as the scientific "arm" of Cartesianism
programmatically separated these two hands, was the threat of
Descartes to Christian faith met with obscurantist fanaticism.[26]
One of these means was to hold the book of *Genesis* against the
book of Darwin and to match God's Adam with Darwin's ape,
hence falling back into a pre-Calvin stage of unenlightened obscurantism, all the more offensive since science had made such
remarkable progress in the meantime. It is a serious mistake,
however—and very often made—to read the reactions against
Copernicus in the light of the anti-Darwin crusade. Even the
17th-century stir over Galileo is a misleading paradigm. Apart

from the overcautious suspicion of vain curiosity, which all that emerged from academic circles had to face in late medieval society, the opposition to Copernicanism was due rather to weaknesses and obscurities immanent in the Copernican system itself, as well as to his assumptions (i.e., hypotheses!), which were not to be substantiated until the time of Kepler, Galileo, and Newton.

It is safe to say that even had there been no religious scruples whatever against the Copernican astronomy, sensible men all over Europe, especially the most empirically minded, would have pronounced it a wild appeal to accept the premature fruits of an uncontrolled imagination, in preference to the solid inductions, built up gradually through the ages, of man's confirmed sense experience. . . . Contemporary empiricists, had they lived in the sixteenth century, would have been [the] first to scoff out of court the new philosophy of the universe.[27]

OSIANDER'S UNAUTHORIZED PREFACE

The unauthorized introduction or preface to *De revolutionibus* by the astronomer and first Lutheran minister in Nuremberg, Andreas Osiander, has been characterized by Bishop Tideman Giese as a "fraud"[28] and has, ever since, drawn a major portion of research energy away from the real subject, Copernicus. Some of the charges against Osiander can be easily disposed of. There is no sly effort on his part to suggest that his preface is actually written by Copernicus. Content and style—he speaks about the author in the third person—clearly pointed to a third person, often a friend of the author who introduced the book to the reader, as was often the case in this *genre* during the 16th century.[29]

One more aspect of Osiander's subjective honesty: The basic structure of his preface can be found in a letter sent two years earlier to Copernicus and to the first Copernican and original editor, Georg Joachim Rheticus. Here we already find the proposal to placate and then win the Aristotelians and theologians by emphasizing that the Copernican theory is based on a series

Andreas Osiander, author of the anonymous introduction to *De revolutionibus;* copper engraving by Bathasar Jenichen, 1565. (Courtesy of Nürnberg Stadbibliothek reproduced from Gottfried Seebass, *Das reformatorische Werke des Andreas Osiander,* Nuremberg, 1967.)

of assumptions (hypotheses) and hence cannot claim ultimate truth. Since several hypotheses can be offered to explain one and the same phenomenon, it should be regarded as belonging to scholarly freedom (*"Freiheit in Forschung und Lehre"*) that more convincing hypotheses can be always advanced: "In that way the potential opponents will be lured away from massive criticism to more intensive research; and, through newly gained respect and a lack of counter arguments, be moved to fairness and ultimately to acceptance."[30]

The word "hypotheses" should not be as offensive to us as it was to Kepler and many a Copernican scholar since.[31] In one of the most concise, but also most accurate, treatments of the Copernican discovery, Edward Rosen[32] has established Copernicus' own use of the term in his main works. And the first believer, Rheticus, describes the achievement of his beloved master as *"renovare hypotheses."*[33] In his own dedicatory letter to Pope Paul III, Copernicus describes the genesis of his breakthrough and provides us with a number of significant parallels with Osiander's preface. But more importantly, the letter lays the basis for our effort to place the Copernican Revolution in an era of change.

For Copernicus the point of departure was that the hypotheses of preceding astronomy—the theoretical explanations of the postulated mathematical astral movements—did not jibe with observed reality; i.e., the actual forecasts of future movements of sun and moon on the basis of the assumption of concentric circles did not prove true. Above all—and now comes the explicit goal that Copernicus had set for himself—earlier assumed explanations did not lead to the discovery of the true shape of the universe (*forma mundi*), or to the symmetry of its structure (*partium eius certam symmetriam*). What had been lacking was a blueprint explaining the inner workings of the universe (*ratio motuum machinae mundi*), the world machine which, after all, the greatest and the most orderly machinist has produced for us (*propter nos*), because of us:

Encouraged by the witness of classical authors I too started to think in terms of a moving earth. And even though this seemed an absurd view [*opinio*],[34] I felt that I had the same freedom to advance hypotheses to explain astral phenomena as others before me; and hence that it would be permissible to find out whether on the assumption of global motion [*ut experirer an positio terrae aliquo motu*] a more reliable explanation could not be advanced for the revolution of the celestial spheres.

Up to this point there is a striking double parallel with Osiander. First, the appeal to the freedom of scholarly investigation in a time of emancipation from the homogenizing weight of tradition. This is the very juncture at which the ideal of self-directed research frees itself from the pious shackles of metaphysical orthodoxy. This is an implicit plea against "vain curiosity" and for respecting the limits of each discipline. For Copernicus the piety of the Church Father Lactantius leads to obscurantism—quite audaciously put in a letter to the Pope! After all, Lactantius, who handed down to posterity the slogan *"Quae supra nos, nihil ad nos"* came to ridicule those who discovered the rotundity of the earth. Secondly, there is the common description of the Copernican research-process in terms of "hypotheses." This is as far as the parallel goes. What Copernicus now discovers remains for Osiander on the level of *opinio*, that is, "assumption," or "hypothesis," without an ultimate claim to a true explanation of cosmological causality, of what makes the universe tick. For Osiander that was the sole domain of God and of those with whom He cares to share His wisdom.[35]

Copernicus, on the other hand, left the level of assumption behind at the moment when he made his breakthrough: at that moment—namely, when his hypothesis of the movement of the earth was hardened, as he claimed, by experience and confirmed by observations (*multa et longa observatione tandem repperi*); sense-data suddenly fell into place, and above all, showed a universal pattern,[36] a true cosmos, where blueprint and global machine fit perfectly together.

We are now in a better position to assess the charge of fraud against Osiander. Osiander is a misleading guide to the world of Copernicus. Without questioning the former's good intentions, the worlds of Osiander and of Copernicus are as a matter of fact light-years apart. But we would miss the true nature of science's advance, if we cede traditional scholarship the point that these two worlds are to be designated as "medieval" and "modern." With great erudition, as well as with dizzying rhetorical magnetism, Hans Blumenberg has advanced the thesis that Osiander embodies a basic nominalist position. As Blumenberg sees it, the nominalist stands in an alien, unreachable universe, which is metaphysically systematized as "astronomical resignation."[37] Out of this heteronomous world, the great humanist Copernicus, as it is claimed, freed us to relate man to his *Umwelt* within which he is to gain his conscious autonomy.[38]

In a last section, I shall attempt to show that where the Copernican Revolution is a cause of celebration for modern man, it presupposes and is based on a nominalist platform—and that when we let ourselves be waylaid and lured away from this platform, we are bound to regress into something worse, to confuse again astrology and cosmology, *Weltbild* and *Weltanschauung*. Neither Protestantism nor Roman Catholicism, much less Blumenberg's philosophical humanism, gave birth to modern science. For that we have to look at a preceding stage, a true fertile crescent.

THE NOMINALIST'S BACKGROUND TO THE COPERNICAN REVOLUTION

"Nous avons du ciel trop peu d'expérience."[39] That is an exclamation, a *cri de coeur* of the leading nominalist philosopher in the generation between Ockham and Gerson (†1429), Nicole Oresme (†1382), 150 years before Copernicus, younger contemporary of Thomas Bradwardine (†1348) and disciple of Jean Buridan (†1348). We do not quote Oresme here because we believe that he influenced Copernicus directly, though Copernicus

had in his library, besides some Bradwardine, at least one nominalist source, the *Quaestiones* of Pierre d'Ailly (†1420).[40] Copernicus, however, probably did not know French; and it is in beautiful, indeed creative French, that Oresme made available the works of Aristotle—in translation, commentary and critique. In France we find Oresme quoted by d'Ailly and Gerson, and his name was soon respected both in Germany and in Italy. Moreover nominalism is such a powerful and all pervasive movement that we cannot ignore Oresme, one of its pacesetting spokesmen, if we want to catch at least a glimpse of its originality and constructive revolt.[41]

It is by no means a novelty to introduce Oresme's name in our attempt to understand the significance of Copernicus. Since Pierre Duhem, modern scholarship has been very aware of this point; among others, Lynn Thorndike,[42] Anneliese Maier, and Marshall Clagett have furthered our knowledge of Oresme significantly. Yet the high claims of Duhem for Oresme's role as precursor and even as preemptor of later discoveries has now generally led to an overcautious reaction. With their usual nod to Anneliese Maier's impressive phalanx of manuscript-based evidence,[43] scholars invariably tend to come to the conclusion that Oresme may have had some theoretical insights but remained "Aristotelian" and offered mere speculative possibilities without relation to fact and reality.[44] It is, however, misleading to speak here of "mere speculation" for we then miss, I believe, the decisive access-route to the phenomenon of modernity. Hence we should be prepared to listen more patiently to the sources.

The systematic application of the theological distinction *potentia absoluta* (absolute power)—"what God could have done without contradiction"—and *potentia ordinata* (God's commitment)—"what God *de facto* did," or, as Oresme puts its "*selon verite*," actually "revealed, decided to do or ordained"—functioned, in line with the condemnations of Averroism in 1277, to place God beyond the fangs of necessity in thought or action. In other words, the transcendence of God is what really concerned the nominalist here. The distinction—and this we have not seen before—works itself out in two different ways: in theology and physics, which includes, of course, astronomy. In theology, the distinction shows the irrelevance and irreverence of speculative theology and man's absolute dependence on God's own revelation. Speculation makes us leave reality behind and orbit in the infinite realm of the *potentia absoluta,* disoriented and lost amongst the infinite number of possibilities God *could* have decided to realize. To penetrate this realm of the *Deus absconditus,* is *vana curiositas,* to fathom the thoughts of God is vain curiosity, whereas it is the task of religion and faith to base itself on God's own revelation, the *potentia ordinata.* Together with the humanist quest for authentic sources (*fontes*), the insistence on nothing but God's commitment, the *sola potentia ordinata,* may evolve into a *sola scriptura,* the Reformation principle "Scripture alone." As history can document, however, nominalism has left its profound impact not only on Luther, but also on Erasmus and the decrees of the Council of Trent.

In both theology and physics the distinction between possibility and reality helped to free man from the smothering embrace of metaphysics. Yet in physics the same distinction works itself out in a different way. Here the main shift from preceding tradition is that the investigation of final causality is recast in terms of efficient causality.[45] This means that the *Weltbild,* the experienced world, is set free from the fangs of a *Weltanschauung,* the postulated world. Simultaneously the unmoved Mover thus cedes his place to the inscrutable Lawgiver. Here the *potentia ordinata* stands for the realm of nature, the "present order," or as Oresme puts it, *"le cours de nature."*[46] Whereas in theology the established order (e.g., of the Church) is at the same time the revealed order (through Holy Scripture, and/or Tradition), in the realm of physics the established order is the order of the established laws of nature,[47] still to be investigated and freed from the Babylonian captivity of metaphysical a priori.

Nicole Oresme in his study with an armillary sphere. (Bibliothèque Nationale, Paris, MS Fr 565 Fº lʳ, permission of the keeper of manuscripts.)

In this climate there emerge before our eyes the beginnings of the new science. I see the first contours of this science in a double thrust:

1. The conscious and intellectually ascetic reduction and concentration on *experientia* both as collective experience entered in the historical record of mankind; and as sense or "test tube" experience (*cognitio intuitiva*) that allows for general conclusions and the discovery of laws; and such only by induction.[48]

2. The discovery of the scientific role of imagination[49] allows for mental experiments. Where facts are not in the reach of experience, we grope for the facts with our imagination, the realm of the *potentia absoluta*, the *terra incognita*, the unknown realm, of logical possibilities. In the field of theology this would be "vain curiosity"; in the field of natural philosophy this is research, *investigation*. This is the breeding ground of the so-called hypotheses that are completely misunderstood as "mere speculation": hypotheses are at once the feast of free research unhampered by *a priori*, unassailable assumptions and the forecast of possibilities based on experiments, the formulations of scientific expectations. The nominalist scientific revolution cannot be sufficiently measured when one merely looks at the research results, even though these are most impressive; but nominalism brings about a revolution in research methodology, which is strictly oriented to experiment and experience.

In the field of astronomy the nominalist hunger for reality is all the more acute, since the heavens are so far removed from collective (the incomplete lists of observations!) and individual experience. Hence the *cri de coeur* of Oresme: "... *nous avons du ciel trop peu d'expérience.*" While this very hunger will lead to the development of such instruments as the telescope (the needed extensions of the human senses), in the meantime imagination has to fill the gap left by actual experiments, in a conscious suspension to final judgment. All in all mental hypotheses reach out to reality and expect to be verified by it.

When the distinction is allowed between *microcosm* (for man), *macrocosm* (for the universe) and *metacosm* (for the realm of God)[50] we may say that (1) nominalism has discovered "space" by transforming the metacosm from the habitat of God into the infinite extension of the macrocosm,[51] while the omnipresence of God binds Him no longer to circumscribed space, hence placing His presence squarely in the macrocosm—an aspect pursued in Luther's theology and particularly in his doctrine of the Eucharist.[52] (2) In concentrating on the macrocosm as *machina mundi* or the reliable "horloge" set by God,[53] under the exclusion of the metacosm, the demarcation line between God and nature is clearly marked and hence space is demythologized and dedivinized.[54] The thrust of this development is better expressed in the designation "naturalization of the universe" than in the more depreciating two-dimensional "mechanization of the universe." (3) The demarcation line drawn by God himself between his own being and his creation terminates the centuries-long argument that the very existence of God requires celestial movement including the orbiting of the sun.[55]

The example from the book of Joshua (10:13) that was going to be used as a biblical argument against Copernicus to prove that the sun moves—"Sun stand still"—is adduced by Oresme to show that creation is not a necessary function of the Highest Being, but the result of a voluntary decision of the Highest Person.[56] It is important to note that for Oresme Scripture admits the possibility that the earth moves—"*qui commovet terram de loco suo*" (Job 9:6)—so that henceforth the investigator is forced to offer a physical instead of a theological solution.[57] Therefore for Oresme, as well as for those who stand in his tradition, the issue of the heavenly movements—of the orbits of the sun, the moon, the moving stars and the earth—is no longer to be solved in terms of a deductive speculative cosmo*logy*, but in terms of an experiential inductive cosmo*nomy*, with the aid of imagination but without claim on scientific accuracy until the mental experiments are confirmed by experience.

It is impressive to see how far Oresme has come in opening

up the realm of imagination and of theoretical astrophysics—the impetus theory, the three-point requirement in perspectives, the diurnal rotation of the earth—thus finding significant pieces in the puzzle that would reveal to Copernicus the vision of heliocentricity. But again, outlasting by far the significance in material progress, we emphasize the advance in scientific attitude, an attitude that is not tacked on but integrally related to the new religious and theological attitude: *vain* curiosity is the effort to penetrate the unknown realm of God omnipotent (*qui supra nos; potentia absoluta*); *true* and *valid* curiosity is concerned with the whole *machina mundi,* which includes earth *and* heaven (*quae supra nos; l'ordre selon nature*). Programmatically God and the heavens are separated: the wise Greek Thales, once the laughing stock and object of jokes about the ivory tower of speculative Platonism, may still stumble into his pitfall, but now because of his proud penetration of divine mystery, no longer because of his astronomical curiosity. In our modern parlance, the mysteries of the heavens have been "declassified."

THE COPERNICAN MANIFESTO

I have presented a sketch of Oresme because I sense here a remarkable proximity to the birth of the modern theory of research in the natural sciences. With much truth, yet with little humility and hence in a strikingly post-medieval way, Oresme concludes[58] his *Livre du ciel et du monde* with the words: "I dare say and insist that there is no human being who has seen a better book on natural philosophy than this one."[59]

I have not dealt with Oresme, the Parisian master, to reopen the issue of the forerunners of Copernicus, though it may have become clear that I do not support the theory of "spontaneous generation."[60] The point is rather that we gain a revealing perspective on Copernicus, and this evaluation necessarily includes his two unfortunate editors. To begin with there is Georg Joachim Rheticus, who had cause to feel slighted by his beloved master Copernicus when the latter decided not to mention this Protestant disciple in his dedicatory letter to Pope Paul III. In a letter to Peter Ramus, dated Cracow 1568, Rheticus describes his future program as the task of liberating astronomy from hypotheses; henceforth astronomy was to be, as he insists, solely based on observation (*solid observationibus*);[61] in the field of physics modern research should be freed from the shackles of tradition and be allowed a direct approach, based only on the analysis of the phenomena of nature (*ex sola naturae contemplatione*). We find here, some two centuries after Oresme, a reformulation of the nominalistic antimetaphysical program that envisoned the replacement of metaphysical *a priori* assumptions by experiment and experience. In discarding classical sources as a hindrance to progress, Rheticus proves that he has outgrown the scientifically unproductive phase of the Renaissance which, with its sun-symbolism and *magi*, blinded many a scholar until our own day.[62]

As far as Osiander is concerned we are, I think, now in a position to do justice to his vision of reality and to see the element of truth in his Preface—better, I believe, than either those who are understandably irritated by his face-saving (but not faith-saving!) devices or those who have opted for the *via antiqua* and reject the nominalist stance on principle. However harsh it may sound, astronomy cannot reveal the "true causes" of astral phenomena insofar as final causality lies beyond its purview. It can deal with efficient causality—what is called in German "system-immanente Faktoren." In our terms, however, it can deal with cosmonomy in contrast with cosmology. But even here astronomy and science in general provide "hypotheses" whose validity cannot be established without experiment and experience, which, most literally, were not yet "in sight" in Osiander's day. Whatever we may think of his claims for the Christian faith, I for one am prepared to grant that the goal of the natural sciences is validity in the sense of *accuracy*, whereas that of the humanities, particularly of philosophy and theology, is validity in the sense of *truth*. Where this distinction is lost, a mechanized and not a naturalized world view has emerged out of the process of nominalist demythologization.

And now finally the case of Copernicus himself. In the first place, heliocentricity is a significant advance and breakthrough in the accurate charting of the universe. Before Copernicus, the theological, philosophical, and physical possibility of the daily and yearly (dual) movement of the earth had been probed and approved—but indeed only as possible. Hindrances in all three fields had been cleared to make Copernicus' theory conceivable, a "Denkmoglichkeit."[63] But it was Copernicus who formulated heliocentricity with clarity and audacity, particularly when the limited basis of facts established by experience are taken into consideration. Yet, on that very point Copernicus, though materially in advance, is formally a step backwards in comparison to nominalistic research standards.[64] Copernicus presented a system mathematically equivalent with that of Ptolemy and based on the Aristotelian, pre-Newtonian hypothesis of the circular movement of the planets without the substantial addition of new observations (experience). In describing the road to his discovery Copernicus mentions heliocentricity as an initial assumption which then, however, becomes conviction and certainty (*repperi!*), a claim improperly ignored by Osiander. Until hard proof had been ascertained by Galileo, Kepler, and Newton, Copernicus asked from his readers a faith in his intuition (*fides implicita*); from such faith the nominalists had wanted to free science in their crusade against metaphysics, against arguments drawn from a dimension of faith beyond the test of experience. Copernicus' discovery would not have been less, but more "modern," if he had highlighted the gap between his heliocentric "imagination"—as Oresme would have termed it—and the compound of experiment and experience interpreted by it. Such a procedure *might* have made Osiander's "fraud" redundant; it most *certainly* would have made early Copernicanism more difficult to combat. Whatever the differences in goals and methods, common to the natural sciences and the humanities is the accurate description of the credibility gap between conceived and sensed reality as a precondition for every advance in our different accesses to reality.

The most significant and lasting aspect of the Copernican discovery is that Copernicus crowned an era hungry for reality, groping for answers, and seeking to initiate change. By the very fact that the earth is launched as a planet into space, the macrocosm is drawn into the orbit of man: heliocentricity is the extension of creation in space and infinity. This projection into space is the part of the Copernican Revolution that has not yet been "received" and absorbed by modern man; it is the part that psychologically, i.e., effectively, is still ahead of us. At the historical beginnings of our conception of the universe the Greeks projected their *polis,* their city-state, into the skies as the model of the cosmos.[65] That was at the same time the beginning of a long process of demythologization of the divinized planets. Yet the older view proved to be virile, indomitable, time- and science-resistant: the gods jealously contested Man's access to space: "What is above you, man, is none of your business (*Quae supra vos nihil ad vos)*"! The Icarus complex—the hidden motive in the Icarus story, *space-angst*—is so fundamental a trait of man that faith, science, and superstition combined to stage the fundamental antithesis between Mother Earth and Father Cosmos. This is what was and is blocking the medieval emancipation of astral physics from cosmology and obscuring the distinction between *Weltbild* and *Weltanschauung,* between astronomy and astrology, and finally, between legitimate research and vain curiosity.

In this long drawn-out intellectual twilight Copernican heliocentricity is at the same time a manifesto proclaiming the *secular cosmos* and a call for the radical colonization of space: "*Quae circa nos tota ad nos,*" ["the cosmos around us is our immediate concern"].[66] Till this very day we modern men have not been existentially able to absorb this vision of reality, as is clear

Two U.S. spacecraft on the moon: astronaut Alan L. Bean with Surveyor III, which soft-landed 19 April 1967, and the Apollo 12 Lunar Module, which landed 20 November 1969, is in the background. (National Aeronautics and Space Administration photograph by astronaut Charles Conrad, Jr.)

from the fact that the designation "cosmopolitan" has been reduced to the tourist badge for the well traveled on this very small globe.

I would like to conclude with an allusion to Paul Tillich's book *The Courage To Be* (1952) by saying that Copernicus is properly celebrated when in the name of the survival of man (Copernicus: "*propter nos*") the dedivinization of space finds its completion in the exorcism of our residual *space-angst,* thus freeing us to face the future with the courage to be in space.

Appendix

FROM COPERNICUS' DEDICATION TO POPE PAUL III

After I had pondered at length this lack of certainty in traditional mathematics concerning the movements of the spheres of the world, I became increasingly annoyed that the philosophers, who in other respects made such a careful scrutiny of the smallest details of the world, had nothing better to offer to explain the workings of the machinery of the world—which is after all built for us by the Best and Most Orderly Workman of all. Hence I assigned myself the task of reading and rereading all the philosophers whose books I could lay my hands on, to see if anyone ever advanced the view that the movements of the spheres of the world are different from those postulated by the specialists in the field of mathematics.

As a matter of fact I first discovered in Cicero that Nicetas thought that the earth moved. Afterwards I also found in Plutarch that there were others of the same opinion. I shall quote his words here, so that they may be known to all:

"Whereas the others hold that the earth is immobile, Philolaus, the Pythagorean, claims that it moves around the fire with a nearly circular motion, not unlike the sun and the moon. Herakleides of Pontus and Ekphantus, the Pythagorean, do not assign to the earth any movement of locomotion. Instead they think in terms of a limited movement, rising and setting around its center, like a wheel."

This was reason enough so that I too began to think seriously about the mobility of the earth. And although this still seemed to me an absurd point of view, I knew that others before me had been granted the liberty of postulating whatever cycles they pleased in order to explain astral phenomena. Therefore, I thought that I too would be readily permitted to test, on the assumption that the earth has some movement, whether a more convincing explanation, less shaky than those of my predecessors, could be found for the revolutions of the celestial spheres.

Translated by the author from *Nicolai Copernici Thorunensis, De revolutionibus orbium libri sex*, vol. 2, edited by Franciscus Zeller and Carolus Zeller (Munich, 1949), pp. 5,18-6,3.

Notes

[1] The association of astral (meta)physics and irrelevance has a classical root. This association is reflected in Christian antiquity when St. Augustine applied to it the term *curiositas*. Since that time the growth—and stagnation—of intellectual European man is reflected in his attitude to "the heavens," "*quae supra nos*" (See Note 16). For the classical root cf. the wise Thales falling in a pit while watching the heavens; Werner Jaeger, *Paideia. Die Formung des griechischen Menschen*, second edition (Berlin, 1954), vol. 1, p. 211.

[2] "During the first Christian millennium, in both East and West, God at the moment of creation is represented in passive majesty, actualizing the cosmos by pure power of thought, Platonically. Then, shortly after the year 1000, a Gospel book was produced at Winchester which made a great innovation: inspired by Wisdom of Solomon 11:20, '*Omnia in mensura et numero et pondere disposuisti*,' the monastic illuminator showed the hand of God—now the master craftsman—holding scales, a carpenter's square, and a pair of compasses. This new representation spread and, probably under the influence of Proverbs 8:27, '*certe lege et gyro vallabat abyssos*,' the scales and square were eliminated leaving only the compasses—the normal medieval and renaissance symbol of the engineer—held in God's hand." Lynn White, Jr. "Cultural Climates and Technological Advance in the Middle Ages," *Viator*, vol. 2 (1971), pp. 171-201, esp. p. 189. See also the illustration on p. 368 of this volume.

[3] Bertolt Brecht, *Galileo*, English version by Charles Laughton, first publication, 1952. Scene 5.

[4] For the original presentation of this paper, the Laughton translation authorized by Brecht was not available to me in a European library, so Mr. Philip J. Rosato made an elegant translation. Generally, I am most indebted for his critical perusal of my English efforts. In the following notes, Prof. Owen Gingerich, the editor of this volume, has provided various English translations.

[5]Cf., Maximilianus Sandaeus, editor, *Theologiae mysticae clavis*, (Cologne, 1640), f. 12. The problem of Cusanus as "forerunner" of Copernicus is best presented by A. Koyré, in a fashion that deserves a full quotation: "No doubt it could be objected that a century before Copernicus, in 1440, Nicholas of Cusa in *De docta ignorantia* (II, 17) had already proclaimed 'the Earth is a noble star' (*terra est stella nobilis*), and had removed it from the centre of the Universe, declaring, moreover, that this centre has no existence, seeing that the Universe is 'an infinite sphere having its centre everywhere and its circumference nowhere'; and it could be maintained that the work in question was probably known to Copernicus, whose mind could have been influenced by it (R. Klibansky, 'Copernic et Nicolas de Cues' in *Léonard de Vinci et l'expérience scientifique du XVIe siècle*, Paris, 1953). I do not dispute it. Yet, it is nonetheless true that the metaphysically very bold concept of Nicholas of Cusa, namely that of an undefined, if not infinite, Universe, was not accepted by Copernicus, nor by anyone else before Giordano Bruno; that his cosmology, scientifically speaking, is non-existent; and if he attributed any motion to the Earth, he does not endow it with any motion around the Sun. On the whole, his astronomical notions are so vague, and often so erroneous (for example, he endows both the Moon and the Earth each with its own proper light) that Nicholas of Cusa cannot by any means be ranked among the forerunners of Copernicus (except in dynamics); nor can he claim a place in the history of astronomy." A. Koyré, *The Astronomical Revolution, Copernicus–Kepler–Borelli*, translated by R. E. W. Maddison (Paris-London-Ithaca, 1973), p. 72.

[6]"We can thus even maintain that the Inquisition desired nothing more from Galileo than that he should not say any more than he was able to prove. In this argument he was the fanatic." Weizsäcker's radicalization, and romantization of inquisitional objectives, is decisively mitigated when he introduces his views on the parallels between faith and science: "In this it was appropriate that he was the fanatic. The great advancements of science do not happen while one sticks anxiously to proof. They happen through bold propositions which themselves open the path to their own confirmation or refutation. All that I have said about the fall of bodies and about the principle of inertia is illustrated by this sentence, and we cannot doubt that Galileo himself was fully aware of this methodological situation. Science, as well as religion, needs faith, and both kinds of faith submit themselves, provided that they understand themselves, to their adequate tests: the religious faith in men's lives, the scientific in continuous research." Translated from "Kopernikus, Kepler, Galilei: Zur Entstehung der neuzeitlichen Naturwissenschaft," in Klaus Oehler and Richard Schaeffler, editors, *Einsichten, Gerhard Krüger zum 60. Geburtstag* (Frankfurt am Main, 1962), pp. 376-394, esp. p. 392. Galileo as the "glorious fanatic" is also reflected in the words of Albert Einstein. See his Foreword to a translation of Galileo's *Dialogue*: "A man who possessed the passionate will, the intelligence, and the courage to stand up as the representative of rational thinking against the host of those who, relying on the ignorance of the people and the indolence of teachers in priestly and scholarly garb, maintained and defended their positions of authority. His unusual literary gift enabled him to address the educated man of his age in such clear and convincing language as to overcome the anthropocentric and myth-ridden thinking of his contemporaries." Quoted by Stillman Drake, *Galileo Studies: Personality, Tradition, and Revolution* (Ann Arbor, 1970), p. 65. The puzzling complexity of the assessment of the significance and "human dimension"

of Galileo's achievement may be seen in the fact that exactly the anthropocentrism of Copernicus, as basis for his faith in the cosmic order, marked the path toward his discovery. See his Dedication to Pope Paul III [Appendix to this article].

[7] "In reality Galileo enriched astronomy and physics while robbing these sciences of a great part of their social meaning. With their discrediting of the Bible and the Church, they stood for a time on the front lines of all progress. It is true that a change nevertheless took place in the following centuries and they were involved in it, but it was a mere reform instead of a revolution, and the scandal—so to speak—degenerated into a dispute among experts. The Church, and with her the whole reactionary wing, could execute an orderly retreat and more or less preserve its power. As for these sciences themselves, they never again reached the high place in society held then, and no longer touched the common man.

"Galileo's crime can be considered as the 'original sin' of modern science. Out of the new astronomy, which deeply interested a new class, the bourgeoise, since it promoted the social revolutionary currents of the times, he made a severely restricted special science, which because of its 'purity,' i.e., its indifference to the means of production could develop relatively undisturbed.

"The atomic bomb, both as a technical and social phenomenon is the classical end product of scientific accomplishment and its social failure." Translated from Bertolt Brecht, *Gesammelte Werke* (Frankfurt am Main, 1967), vol. 17, p. 1108 ff.

[8] Koestler ponders a number of explanations of why Copernicus did not object to or have Osiander's Preface removed, and concludes that it is more likely he submitted to Osiander's proposal since he had already submitted his whole life long. ". . . More likely he procrastinated, as he had done all his life." Arthur Koestler, *The Sleepwalkers* (London, 1959), p. 171; second edition, 1964, p. 175.

[9] See *Martin Luther's Werke, Kritische Gesamtausgabe Tischreden* (Weimar, 1916), vol. 4, art. 4638, pp. 412-413; cf. ibid. (Weimar, 1912), vol. 1, art. 855, pp. 418-421. Basic literature: Werner Elert, *Morphologie des Luthertums*, I: *Theologie und Weltanschauung des Luthertums hauptsächlich im 16. und 17. Jahrhundert,* second edition (Munich, 1958), pp. 363-393. Heinrich Bornkamm, "Kopernikus im Urteil der Reformatoren," *Archiv für Reformationsgeschichte,* vol. 40 (1943), pp. 171-183; reprint in *Das Jahrhundert der Reformation: Gestalten und Kräfte,* second edition (Göttingen, 1966), pp. 177-185. John Dillenberger, *Protestant Thought and Natural Science: A Historical Interpretation of the Issues behind the 500-year-old Debate* (New York, 1960), pp. 28-49. Klaus Scholder, in a broad cultural and historical setting: *Ursprünge und Probleme der Bibelkritik im 17. Jahrhundert. Ein Beitrag zur Entstehung der historisch-kritischen Theologie* (Munich, 1966), pp. 57-65.

[10] See Thomas S. Kuhn, *The Copernican Revolution: Planetary Astronomy in the Development of Western Thought* (Cambridge, Mass., 1957), p. 196: "Protestant leaders like Luther, Calvin, and Melanchthon led in citing Scripture against Copernicus and in urging the repression of Copernicans. Since the Protestants never possessed the police apparatus available to the Catholic Church, their repressive measures were seldom so effective as those taken later by the Catholics, and they were more readily abandoned when the evidence for Copernicanism became overwhelming. But Protestants nevertheless provided the first effective institutionalized opposition."

Since the Protestant "attack" is interpreted as being due to its *"sola scriptura,"* the Catholic reaction to Galileo has to be explained in different terms, and much to the historian's surprise, it is presented as anti-Protestant reaction and part of Catholic reform.

R. Hooykaas has eloquently opposed the myth that Calvin mentioned and rejected Copernicus in his works " 'There is no lie so good as the precise and well-detailed one' and this one has been repeated again and again, quotation marks included, by writers on the history of science, who evidently did not make the effort to verify the statement. For fifteen years, I have pointed out in several periodicals concerned with the history of science that the 'quotation' from Calvin is imaginary and that Calvin never mentioned Copernicus; but the legend dies hard." *Religion and the Rise of Modern Science* (Edinburgh-London, 1972), p. 121.

Furthermore, Hooykaas dealt with the theological thrust of Calvin's *Genesis* commentary by pointing to another aspect: "Thus Calvin's exegetical method was based on the Reformation doctrine which held that the religious message of the Bible is accessible to everybody. The Spirit of God, as he put it, has opened a common school for all, and has therefore chosen subjects intelligible to all. Moses was ordained a teacher of the unlearned as well as of the learned; had he spoken of things generally unknown, the uneducated might have pleaded in excuse that such subjects were beyond their capacity; therefore, Moses 'rather adapted his writing to common usage'." Ibid., p. 118.

[11] See *Corpus Reformatorum,* vol. 51 (*Calvini Opera,* vol. 23), col. 20-22.

[12] Calvin, loc. cit.

[13] See K. Scholder, op. cit. [note 9], p. 68 ff. and Heinrich Karpp, "Der Beitrag Keplers und Galileis zum neuzeitlichen Schriftverständnis," *Zeitschrift für Theologie und Kirche,* vol. 67 (1970), pp. 40-55, esp. p. 46 ff.

[14] *Adagiorum chiliades,* vol. 1, p. 6,69, in *Ausgewählte Werkes,* vol. 7, edited by Theresia Payr (Darmstadt, 1972), p. 414 ff.

[15] Erasmus von Rotterdam, *Ausgewählte Werke,* vol. 7, ed. cit., pp. 414, 416. In a learned and stimulating article Eberhard Jüngel pursues the function of the "Socratic saying" in Luther's theology: *"Quae supra nos, nihil ad nos.* Eine Kurzformel der Lehre vom verborgenen Gott—im Anschluss an Luther interpretiert," *Evangelische Theologie,* vol. 32 (1972), pp. 197–240.

[16] Augustine, *Confessiones,* book 5, 3,3.

[17] Augustine, *De fide, spe et caritate sive Enchiridion,* in J. P. Migne, editor, *Patrologiae completus cursus,* vol. 40, p. 235 (= *Corpus Christianorum,* vol. 46, p. 52 ff.).

[18] We find a restrictive interpretation of the Augustinian position with Hugo of St. Victor in his *Expositio in Hierarchiam coelestem S. Dionysii Areopagitae, Patrologiae completus cursus,* vol. 175, p. 925 A.

[19] *Calvini Opera,* vol. 23, ed. cit. [note 11], col. 22.

[20] Johannes Kepler, *Astronomia nova,* in Max Caspar, editor, *Johannes Kepler Gesammelte Werke,* vol. 3 (Munich, 1937), pp. 28-34, "Introductio."

[21] Cf. Thomas de Aquino, *Contra Gentiles,* vol. 3, book 4, "De unione hypos-

tatica," chap. 55: "*. . . virtus humilitatis in hoc consistit ut aliquis infra suos terminos se contineat, ad ea quae supra se sunt non se extendens, sed superiori se subiiciat.*" ("*. . .* the virtue of humility consists in this, that someone contains himself within his limits, not extending himself to those things that are above him, but subjugating himself to the higher things.")

²²Cf. John Dillenberger, op. cit. [note 9], p. 130: "Statistical evidence points to a predominant Puritan membership in the Royal Society."

²³See R. Hooykaas, op. cit. [note 10], p. 148; cf. p. 94 ff.; pp. 135-138.

²⁴*Nikolaus Kopernikus Gesamtausgabe*, II: *De revolutionibus orbium caelestium* (Munich, 1949), pp. 403-404. "*. . . astronomus eam [hypothesim] potissimum arripiet, quae comprehensu sit quam facillima. Philosophus fortasse veri similitudinem magis requiret; neuter tamen quicquam certi comprehendet, aut tradet, nisi divinitus illi revelatum fuerit. Sinamus igitur et has novas hypotheses inter veteres nihilo verisimiliores innotescere, praesertim cum admirabiles simul et faciles sint, ingentemque thesaurum doctissimarum observationem [sic; lege: observationum] secum advehant. Neque quisquam, quod ad hypotheses attinet, quicquam certi ab astronomia expectet, cum ipsa nihil tale praestare queat, ne si in alium usum conficta pro veris arripiat, stultior ab hac disciplina discedat quam accesserit. Vale.*" *Ad lectorem de hypothesibus huius operis.* [For an English translation, see pp. 301-304 of this volume.] Note: The title page of the 1543 Nuremberg edition is enlarged by a publisher's blurb: "*Habes in hoc opere . . . motus stellarum . . . novis insuper ac admirabilibus hypothesibus ornatis.*" ("You have in this work . . . the motion of the stars . . . newly arranged with fine and admirable hypotheses.") For the bibliographical data see Gottfried Seebass, *Bibliographia Osiandrica* (Nieuwkoop, 1971), p. 130 ff.

²⁵H. Bornkamm, op. cit. [note 9], p. 182 ff.; K. Scholder, op. cit. [note 9], p. 63.

²⁶On the Cartesian dichotomy between the two experiences see J. Bots, *Tussen Descartes en Darwin: Geloof en natuur wetenschap in de 18ᵉ eeuw in Nederland, Speculum Historiale* (Assen, 1972), vol. 8., pp. 136-139 (for a German summary of this section, see p. 186 ff.).

²⁷E. A. Burtt, *The Metaphysical Foundations of Modern Physical Science*, second edition (New York, 1951), p. 25; cited by Dillenberger, op. cit. [note 9], p. 26 ff. Franz Wolf, though more restrained, presents the same argument in his 1943 commemoration address: "Furthermore, in its details the superiority of the Copernican system over Ptolemy's was not yet clearly perceived. . . . From the world of Copernicus to the distances of the spiral nebulae is but a moment in the development of scientific knowledge of the heavens." Translated from *Karlsruher Akademische Reden*, vol. 22, pp. 5-23, esp. p. 11, 1943. Cf. on Norbert Schiffers, "Die Schwäche des Kopernikus," *Fragen der Physik an die Theologie: Die Säkularisierung der Wissenschaft und das Heilsverlangen nach Freiheit* (Düsseldorf, 1968), p. 13 ff.

²⁸Karl Heinz Burmeister, *Georg Joachim Rhetikus 1514-1574*, III: *Briefwechsel* (Wiesbaden, 1968), p. 55. It seems clear that Giese refers to Osiander as responsible for putting pressure on Petreius. Giese's interpretation of Osiander's motives is, understandably, more malicious than convincing: "*. . . dolens descendendum sibi esse a pristina professione, si hic liber famam sit consecutus.*" ("*. . .* unfortunately stooping to have the opening declaration in case this book would later become famous.") In letters to Copernicus and his co-editor Rheticus,

Osiander had, as early as 20 April 1541, developed his battle plan for winning over the two expected opposition parties. See Burmeister, op. cit. [note 28], vol. 3, p. 25: *". . . Paripathetici et theologi facile placabuntur* [instead of: *placabunter*], *si audierint, eiusdem apparentis motus varias esse posse hypotheses . . ."* (". . . Aristotelians and theologians will be easily placated if they hear that the same motion as perceived can be explained by means of different hypotheses . . ."). Cf. note 30.

[29] In early copies the name of Osiander is identified; as a matter of fact, this is the way in which Kepler could name Osiander as the author of the Preface. Yet even as late as Laplace, the Preface was read as being written by Copernicus. See Koyré, op. cit. [note 5], p. 99, n. 14.

[30] See Burmeister, op. cit. [note 28], vol. 3, p. 25. For the parallel, partly identical, letter of Osiander to Copernicus dated on the same day, 20 April 1541, see *Apologia Tychonis contra Ursum, Kepleir opera omnia, ed. Ch. Frisch* (Frankfurt, 1858), vol. 1, p. 246.

[31] *Johannes Kepler Gesammelte Werke,* ed. cit. [note 20], p. 6.

[32] Edward Rosen, *Three Copernican Treatises,* second edition (New York, 1959), pp. 28-33. The appendix of annotated bibliography (pp. 201-269) proved to be invaluable.

[33] Ibid., p. 31. Cf. Rheticus' dedicatory letter to the *Narratio Prima,* quoted by Leopold Prowe, *Nicolaus Copernicus* (Berlin, 1884), vol. 2, p. 321,27.

[34] In contrast to *assertio,* which means "conviction," *opinio* means "view" in the sense of "assumption."

[35] ". . . *neuter tamen quicquid certi comprehendet aut tradet nisi divinitus illi revelatum fuerit."* (". . . neither of them will learn or teach anything certain, unless it has been divinely revealed to him.") For an accessible and emendated Latin text of Osiander's Preface, see Emanuel Hirsch, *Die Theologie des Andreas Osiander und ihre geschichtlichen Voraussetzungen* (Göttingen, 1919), App. 1, p. 290.

[36] This universal vision as the essential advance beyond Ptolemy is highlighted by Matthias Schramm in his commemoration address in Tübingen, 1 February 1973, entitled "Die Leistungen des Copernicus." The author kindly supplied me with his manuscript.

[37] "Man is not, as the Stoics would have it, disposed to examine the arrangement of the heavens, but on the contrary, his theoretical inquisitiveness puts before him the appearance of a heterogeneous and unattainable cosmology [*Weltregion*] for whose perception Nature provides him no basis. The epistemology of astronomical resignation is therewith metaphysically systematized." Translated from Hans Blumenberg, *Die kopernikanische Wende* (Frankfurt am Main, 1965), p. 64. Cf. *Die Legitimität der Neuzeit* (Frankfurt am Main, 1966), p. 346 ff.

[38] "Copernicus had not only used the humanistic rules, but with his astronomical reform he had exactly hit the genuine mentality of the humanistic movement of the fading Middle Ages and had more essentially realized it than many of those who had explicitly formulated the program of this movement." Translated from H. Blumenberg, op. cit. [note 37], p. 77.

[39] Nicole Oresme, *Le Livre du ciel et du monde*, edited by A. D. Menut and A. J. Denomy, translated by A. D. Menut (Madison, 1968), p. 356 ff (= book 2, chap. 8, f. 89d).

[40] See Ernst Zinner, *Entstehung und Ausbreitung der Coppernicanischen Lehre: Sitzungsberichte der Phy.-Med. Sozietät Erlangen* (Erlangen, 1943), vol. 74, p. 406. These *Quaestiones* are part of a genre of composite volumes described by Pierre Duhem, *Les Origines de la statique* (Paris, 1906), vol. 2, p. 59, n. 1; p. 337 ff. n. 1.

[41] In another context—with respect to the significance of neostoicism as "setting" for Descartes, Spinoza, and Calvin—the French philosopher Eric Weil observes that such "authors are credited with an originality they themselves would not have admitted, simply because we do not study what every cultured man in their times had always present in mind." See his article "Supporting the Humanities," *Daedalus*, vol. 102 (1973), pp. 27-38, esp. p. 33.

[42] On a broad (often manuscript) basis Lynn Thorndike presents Oresme's view on astrology, magic, and miracles, *A History of Magic and Experimental Science* (New York, 1934), vol. 3, pp. 398-471.

[43] Anneliese Maier, herself, has often been more ready to grant Oresme his subjective sense of exploring reality; see *An der Grenze von Scholastick und Naturwissenschaft* [*Studien zur Naturphilosophie der Spätscholastik*, vol. 3], second edition (Rome, 1952), p. 354 ff. In a characteristic formulation Anneliese Maier now ascribes to Oresme a view (earlier assigned by her to Albert of Sachsen) "in whom one can see the first hint of the equivalence principle of modern set theory," *Die Vorläufer Galileis im 14. Jahrhundert* [*Studien zur Naturphilosophie der Spätscholastik*, vol. 1], second edition (Rome, 1966), p. 309.

[44] See A. D. Menut's bibliography in N. Oresme, ed. cit. [note 39], pp. 753-762.

[45] Anneliese Maier assigns to Jean Buridan the central role: "Buridan deserves the credit for having drawn the metaphysical and methodological consequences from all this knowledge; he is the first who thought he saw that these principles suffice to explain the events of nature and that on this basis one can refrain from adopting final causes and final tendencies. And hereby he has in fact anticipated the idea that would dominate the science of the following century." Translated from *Metaphysische Hintergründe der spätscholastischen Naturphilosophie* [*Studien zur Naturphilosophie der Spätscholastik*, vol. 4] (Rome, 1955), p. 334 ff.

[46] Oresme, op. cit. [note 39], book 1, chap. 2, f. 7a; ed. cit., p. 58. See also the synonyms used by d'Ailly, as quoted by Francis Oakley, "Christian Theology and the Newtonian Science: The Rise of the Concept of the Laws of Nature," *Church History*, vol. 30 (1961), pp. 433-457, esp. p. 454 ff. n. 74.

[47] Oresme, op. cit. [note 39], book 2, chap. 2, f. 71a; ed cit., p. 288.

[48] See, however, G. W. Coopland: "Of Oresme's use of experience in the everyday sense little need be said; it is illustrated at every turn and furnishes the most attractive part of his work. It is evidently the result of wide interests and knowledge of his world, although in this connection, again, we discern that strange stopping short of closer and more searching enquiry demanded by modern standards. Of organized and controlled observation in the form of experiment we can find no trace." *Nicole Oresme and the Astrologers: A Study of his "Livre de divinacions"* (Cambridge, Mass., 1952), p. 35.

⁴⁹ For the double function of *"imaginatio"* as "not-fact" and as point of departure for inquiry, see Jean Buridan, as quoted by Pierre Duhem, *Le Système du monde* (Paris, 1916), vol. 4, p. 317 ff.; *"imaginatio"* as *"modus inveniendi loca planetarum"*, and therefore merely as *calculatio* à la Osiander, see ibid., p. 146 ff. Even the editors of Oresme's *Livre du ciel et du monde* have not always differentiated between matter and method: "Under these conditions, we may suppose that a ship could float on the surface of the sphere of air just as naturally as it would on the Seine River or on the surface of the sea (199d). This final 'ymagination' in Oresme's long critique of Aristotle's *De caelo* exhibits impressively the distance that separates the science of today from that of the 14th century," ed. cit. [note 39], p. 30.

⁵⁰ Jean Gerson uses for our "metacosm," *mundus archetypus,* in *Opera omnia,* vol. 3, edited by L. E. du Pin, "Definitiones terminorum ad theologiam moralem pertinentium" (Antwerp, 1706), col. 107 B.

⁵¹ See Anneliese Maier's Addenda to the second edition of her *Die Vorläufer Galileis im 14. Jahrhundert* [*Studien zur Naturphilosophie der Spätscholastik,* vol. 1] (Rome, 1966), p. 315, and the comparison with Bradwardine's view of the *immensitas Dei,* in note 1 on the same page. Cf. John E. Murdoch: "It is of note that in the Middle Ages such speculation on the infinite centered on the older problem, in itself less scientific, of the eternity of the world. Analyses of the infinite were initially designed to resolve this more traditional problem. On the other hand, in the 14th century in many instances the problem of the possibility of an eternal world had simply become an occasion to discuss the mysteries of the infinite." Translated by John Murdoch from his " 'Rationes Mathematice,' Un aspect du rapport des mathématiques et de la philosophie au Moyen Age" in *Conférence donnée au Palais de la Découverte le 4 Novembre 1961. Histoire des Sciences* (Paris, 1962), p. 22.

⁵² The significance of this "breakthrough of God" for Luther's theology is described well by W. Elert, op. cit. [note 9], vol. 1, p. 386 ff.

⁵³ For Oresme's fascination with the image of the clock, see also Lynn Thorndike, op. cit. [note 42], vol. 3, p. 411, n. 1. For Jean Buridan, Oresme's teacher also in this respect, see *Quaestiones super libris quattuor de caelo et mundo,* edited by Ernest A. Moody (Cambridge, Mass., 1942), book 2, qu. 22, pp. 226-233.

⁵⁴ The wide spread of popular astrology is one of the many indications that the Arabian "myth screen" had not been sufficiently effective. See here Manfred Ullmann: "The interpretive possibilities of planetary positions depended on setting the planets equivalent with the gods, a procedure already established in the 6th century B.C., next by the Pythagoreans, and then generally in colloquial speech. All the attributes, capabilities, and actions of the gods that had been found deposited in their myths were now associated with the planet in question, and thus it became possible also to interpret the constellations. For the Arabs and Moslems the names of the planets lost in translation their character as gods' names. But the Arabs undertook the complicated articulation of the interpretive possibilities, which without the ancient mythological background necessarily remained a purely mechanical, inexplicable system." Translated from *Die Natur- und Geheimwissenschaften im Islam* (Leiden, 1972), p. 348.

⁵⁵ Conversely developments in the field of physics show effects on theology.

After the Thomistic ontological relation between grace and movement, the new impetus-doctrine transforms "motion" and personalizes the concept of grace. We have pursued the history of theology and the history of the medieval sciences so long in separate departments that we stand only at the very beginning of seeing the interactions between shifts in these fields.

[56] "Thus, it is clear from what we have said that it does not follow that, if God is, the heavens are; consequently, it does not follow that the heavens move. For, in truth, all these things depend freely upon the will of God without any necessity that He cause or produce such things or that He should cause or produce them eternally, as we explained more fully at the end of Chapter Thirty-four of Book I [see f. 58b]. Moreover, it does not follow that, if the heavens exist, they must move; for, as stated, God moves them or makes them move quite voluntarily. He demonstrated this action at the time of Joshua when the sun stood still for the duration of an entire day, as the Scripture states: Was not . . . one day made as two? It is probable that the daily motion of the whole heavens and that of the planets stopped, and not only the sun. In relating this event, the Prophet said: The sun and the moon stood still in their habitation, etc." N. Oresme, op. cit. [note 39], book 2, ch. 8, f. 92b; ed. cit., p. 365.

[57] "Therefore, assuming that the earth moves with or contrariwise to the heavens, it does not follow from this /(92d) that celestial motion would stop, and so, in and of itself, this circular motion of the heavens does not demand that the earth remain motionless at the center. It is, indeed, not impossible that the whole earth moves with a different motion or in another way. In Job 9, we read: Who shaketh the earth out of her place . . ." N. Oresme, op. cit. [note 39], book 2, ch. 8, f. 92d; ed. cit., p. 367.

[58] Hardly more cautious is the explicit of a physics commentary, *Quaestiones*, assigned to Buridan: "*Tu melius scribe, qui dixeris hoc fore vile / Si melius fuerit, plus tibi laudis erit*" ("Write it better, in case you say this is worthless; if you do better, it will be more praise to you."). Quoted by Pierre Duhem, op. cit. [note 49], p. 132. The same desideratum from the inverse perspective is formulated by John Murdoch in his "Philosophy and the Enterprise of Science in the Later Middle Ages," *History of the Interaction between Science and Philosophy: Proceedings of the International Symposium on Science and Philosophy held at the Van Leer Jerusalem Foundation, Jerusalem, Israel, January 15-January 19, 1971, Honoring Professor S. Sambursky*, p. 28 (MS).

[59] With all respect for my fellow countryman, E. J. Dijksterhuis, who belongs to the pioneers in the history of science, I cannot share his view of Parisian nominalism: "The decadence into which Scholasticism has fallen is typified by the fact that in their works there is no question of any further development of the fruitful, but as yet totally unexplored, ideas contained in this theory; nevertheless, one can at least appreciate in the Parisian philosophers of this time, that they were at least able to keep the good aspects when in the same period the Italians were falling back in the field of mechanics to the Aristotelian-Averroistic errors that had in Paris long since been conquered." Translated by Arthur Loeb from *Val en worp. Een Bijdrage tot de Geschiedenis der Mechanica van Aristoteles tot Newton*, Chapter II: "Val en Worp in de Scholastiek" (Groningen, 1924), pp. 117-121, esp. p. 118.

[60] See, however, Koyré: ". . . but no-one except Copernicus produced a helio-

centric astronomy. Why? It is an idle question: because no-one before Copernicus had his genius, or his courage. Perhaps, it was because no-one between Ptolemy and Copernicus had been both an inspired astronomer and a convinced Pythagorean," op. cit. [note 5], p. 42.

[61] See Burmeister, op. cit. [note 28], vol. 3, p. 188; cf. "Near the end of his life Rheticus sought to build out of his rich experience as astronomer, physician, and alchemist, a new philosophical system whose foundation should be only nature, *ex sola naturae contemplatione*. Only in this way, he wrote in 1568 to Ramus, would his natural philosophy be grounded. As we know, Rheticus placed this demand equally on medicine, astronomy, and astrology." Translated from Burmeister, *Georg Joachim Rhetikus 1514-1574*, I: *Humanist und Wegbereiter der modernen Naturwissenschaften* (Wiesbaden, 1967), p. 173.

[62] See the in-other-respects excellent study by Francis A. Yates, *Giordano Bruno and the Hermetic Tradition*, second edition (London, 1971), pp. 241-243. See also Peter J. French, *John Dee: The World of an Elizabethan Magus* (London, 1972), p. 103; "Renaissance Hermeticism prepared the way emotionally for the acceptance of Copernicus' revolutionized universal structure. In this case, then, scientific advance was spurred by the renewed interest in the magical Hermetic religion of the world." The Sun-analogies and Hermetic traditions—alluded to by Copernicus in his Preface—are certainly important characteristics of a movement we can trace from protohumanism (Richard de Bury's library, used by Bradwardine and Holcot) to Pico and Reuchlin. See further Eugenio Garin, *Portraits from the Quattrocentro*, second edition (New York, 1972), pp. 145-149, and Wayne Shumaker, *The Occult Sciences in the Renaissance: A Study in Intellectual Patterns* (Berkeley, 1972), pp. 201 ff., on the "centrality" of the sun, ibid., p. 221. For a perspective on Copernicus' discovery, this tradition does not help us a step further. Here Koyré's evaluation of the parallel case of Cusanus applies; see note 5.

[63] Thomas S. Kuhn, *The Structure of Scientific Revolutions*, second edition, (Chicago-London, 1969), p. 62: ". . . characteristic of all discoveries from which new sorts of phenomena emerge. Those characteristics include: the previous awareness of anomaly, the gradual and simultaneous emergence of both observational and conceptual recognition, and the consequent change of paradigm categories and procedures often accompanied by resistence."

[64] And as Randall has shown, the Paduan Aristotelians a century later dominated the climate of thought with which Copernicus must have become familiar during his Italian study years. Cf. John Herman Randall, *The School of Padua and the Emergence of Modern Science* (Padua, 1961), p. 24 ff.; p. 71 ff.

Note also R. Hooykaas op. cit. [note 10], pp. 35-36: "Thus Kepler and Galileo, in contrast to Plato, put forward a mathematical empiricism. This was quite evident in one of the most decisive moments in the history of science. It had been a dogma of the 'church scientific', up to the time of Kepler, that movements in the heavens could be nothing but uniform and circular. Everywhere, everybody had always held this to be true *a priori*; Platonists and Aristotelians, Idealists and Nominalists, Copernicus and Galileo had accepted this dogma and Kepler himself was thoroughly convinced of its truth.

"Yet a difference of eight minutes between observation and calculation of the orbit of the planet Mars forced him, after a struggle of several years, to abandon this dogma of circularity and to postulate a non-uniform motion in elliptical orbits.

"He submitted to given facts rather than maintaining an age old prejudice; in his mind a Christian empiricism gained the victory over platonic rationalism; a lonely man submitted to facts and broke away from a tradition of two thousand years. With full justice he could declare: 'These eight minutes paved the way for the reformation of the whole of astronomy', and it was with full justice, too, that in 1609 he gave to his book the title *New Astronomy*."

[65] Jaeger, op. cit. [note 1], p. 220.

[66] See Karl R. Popper: "Thus we live in an open universe. We could not make this discovery before there was human knowledge. But once we have made the discovery there is no reason to think that the openness depends exclusively upon the existence of human knowledge. It is much more reasonable to reject all views of a closed universe—that of a causally as well as that of a probabilistically closed universe; thus rejecting the closed universe envisaged by Laplace, as well as the one envisaged by quantum mechanics. Our universe is partly causal, partly probabilistic, and partly open: it is emergent." "Indeterminism is Not Enough," *Encounter*, vol. 40 (1973), pp. 20-26; esp. p. 26.

COPERNICUS AS A HUMANIST

Jacob Bronowski

LIFE AND CHARACTER

The facts of Nicolaus Copernicus' life are modest and easily summarized. He was born at Torun in Prussian Poland on 19 February 1473. His father, who was a merchant of some standing from Cracow, died when he was ten, and Copernicus and his brother were in effect adopted by an eminent uncle who became Bishop of Warmia in 1489. Copernicus went to the famous university of Cracow in 1491, and in 1496 he traveled to Italy and spent nine years at universities there—Bologna, Ferrara, and Padua. His studies were mostly in church law, in which he took a doctor's degree, but he also pursued his interest in astronomy and went to medical school. His uncle had him appointed a canon of the Cathedral of Frombork [Frauenburg] in 1497, and he came home once, probably to ask for further leave of absence. Just before that he was in Rome in the jubilee year of 1500, and may have given some lectures on astronomy then.

When he returned to Poland for good, Copernicus acted as his uncle's secretary and doctor. We know that he had other talents: he painted, and in 1509 he published a Latin translation from the Greek of the Epistles of a dull seventh-century Byzantine poet.

He also wrote, probably before his uncle's death in 1512, a firm outline of his astronomical system, known as the little commentary, *Commentariolus*.[1] Though it is nonmathematical, it is clear and explicit, and it seems to have become well known in manuscript.

For the rest of his life, Copernicus carried out the everyday duties of a canon at Frombork, which concerned such things as church property; he was not a cleric. In 1522 and 1526 he advised his government on currency reform, and enunciated incidentally the dictum that bad money drives out good, which is usually ascribed to the Elizabethan banker Thomas Gresham. He made some, but not many, astronomical observations, and when pressed by a friend he criticized a work by a minor German astronomer on *The Motion of the Eighth Sphere* (which was supposed to carry the fixed stars). The Medici Pope Leo x had invited astronomers to the Lateran Council in 1514 to help reform the calendar; Copernicus sent his opinion to Rome, and later wrote, "from that time on I have directed my attention to a closer study of these topics." At some uncertain date during his fifties, Copernicus began to write the magisterial account of his world system, *De revolutionibus orbium coelestium*, which contains a full mathematical treatment and tables.

In the 1530s Copernicus was urged to publish the work; his ideas were then known and respected in Rome; but he held back. Finally in 1539, when Copernicus was 66, a young mathematician from the Protestant university of Wittenberg, Georg Joachim Rheticus, came to Frombork unbidden to learn from Copernicus for himself. He wrote a fairly full *Narratio prima*, a first account, of Copernicus' work in 1540, and had a hand in getting the great book into print at last in Nuremberg in 1543.

Copernicus died on 24 May 1543; it is said that a first copy of of *De revolutionibus* was put into his paralyzed hands on his deathbed. His character remains silent for us to the end. He never married though, like others at Frombork, he had a housekeeper who was looked at askance as late as 1539. There seems to have been no warmth between him and his brother, who became the black sheep of the family. It is strange that Copernicus also has no word of thanks for Rheticus among his acknowledgments; we are left to puzzle whether he thought it tactful to be silent because Rheticus was a Protestant or a notorious homosexual. What is plain, however, is that Copernicus did not dissemble. When in doubt, he preferred to remain silent, and said nothing that he did not believe. He made no concession to any variant of the ominous doctrine of double truth, which had dogged Catholic science since the 13th century. Copernicus believed that the system of the world that he describes is real, and that the earth does both fly round the sun and turn on its own axis; and he made his belief clear whenever he wrote, even in the dedication of the book to Pope Paul III. The unsigned preface, which suggests instead that the whole scheme is merely a hypothetical device to help astronomers in their calculations, was not written by Copernicus, but by an anxious cleric in Nuremberg, Andreas Osiander—as Hieronymus Schreiber of Nuremberg noted in the copy of the book that later belonged to Kepler.

THE INSPIRATION OF THE CLASSICS

So much for what one may call the mechanics of how *De revolutionibus orbium coelestium* was brought into the world. Yet the book is not just a matter of mechanics, not even the mechanics of the solar system. For example, the phrase *orbium coelestium* in the title is most often translated as "heavenly spheres," and with some justification: Copernicus does use *orb* with that meaning in other places. Yet Copernicus (if he supplied the title) at the least must have doubted that the planets are carried by tangible spheres. That ancient picture from Aristotle did not even fit well with the astronomical system of Ptolemy (though Georg Peurbach had worked hard to reconcile them) and does not fit at all if the earth revolves. Why then did Copernicus use those emotive images and phrases from classical

Greece? Why did he, why did Rheticus, pepper their texts with quotations from Greek writers and philosophers, all the way from Sophocles to Plato?

It would be easy to say "because they were fashionable," and it would be true, but true in the shallowest way. These classical graces were fashionable because they were powerful, and they had the same power in the minds of Copernicus and Rheticus as in what we would now call literary minds. The Greek sources were golden; they had the touch of Midas; they carried from that noble past a morning glow of wisdom and purity whose sanction was immortal. When Copernicus recalls Pythagoras (or when Newton does, almost 200 years later[2]), he is not writing as a scholar directing us in a footnote to his sources. He is presenting a blessing from the first wonder of the world, a prescholastic time when great men saw very simply but to the bottom of things.

In short, Copernicus was a humanist gentleman; his mind was formed in Italy exactly in the period which Hans Baron describes, "the period from 1490 to 1520 when the Renaissance issued forth as a movement European in scope."[3] I want first to transmit the broad and even romantic sense of humanism in that description. Later I shall come to more specific components: the main philosophic component first, then a popular component, and finally the academic definition of humanism as a program for the reform of education. I want to begin with the tang and taste of the age, its almost physical sense of bursting from scholasticism and the monastery back to classical nature, which is so vivid in the Renaissance—even in its scientific pictures, for instance in the herbals.[4]

Movements that claim to go back to natural ways of thinking usually have a bias against science, and humanism was no exception. The father and fountainhead of the movement, Petrarch in the 14th century, had such an antipathy against such techniques that he even disliked medical men. Nevertheless, one of Petrarch's good friends, Giovanni de' Dondi, was a doctor and a fine mechanic into the bargain; he was nicknamed dell'Orologio because he spent sixteen years, during a busy medical and university practice, in making the beautiful astronomical clock at Padua about 1350. The clock is lost, but it has been possible from the drawing to make a copy, which is now in the Smithsonian Institution in Washington.[5] It is a mechanical model of Ptolemy's system, with seven faces on which the seven planets of antiquity (Mars, Jupiter, Saturn, the Moon, Mercury, Venus, and the Sun) run on geared sets of wheels that trace out their epicycles. We do not know when the clock was lost; it may be that Leonardo da Vinci saw it, since there is a drawing of his that looks very like the mechanism that carried Venus.[6]

Giovanni de' Dondi was a man of many parts, and that may have endeared him to Petrarch,[7] but there is also something about the clock itself that has a humanist undertone. In some delicate way, it is not a piece of machinery but a work of art, and an act of personal homage to the grand conceptions of antiquity. The dedicated love with which it was made expresses itself in a symbolic charm, which is unmistakable and irresistible. I linger on de' Dondi's clock because I want to bring home the sense of his age that the starry heavens are a work of art, and a divine inspiration for the poet and the scientist alike. Petrarch could not be expected to find that in the classical authors that he admired and reintroduced: they were Latin, and belonged to a culture that had little head or taste for science. A hundred years later, however, the movement of humanism to the Greek classics recovered a culture with a different outlook, in which science in general and astronomy in particular had been highly regarded. This is the rich setting in which Peurbach and Regiomontanus came to teach astronomy at Vienna between 1450 and 1470. Humanism in the 15th century is preoccupied with Greek as a language and with Greek ideas.

Copernicus was not at the hub of humanism in Cracow—an eastern frontier town at the edge of Europe, as we are reminded by the trumpeter who still blows the hours from the Cathedral tower to commemorate a Tartar raid. One of his objects in going

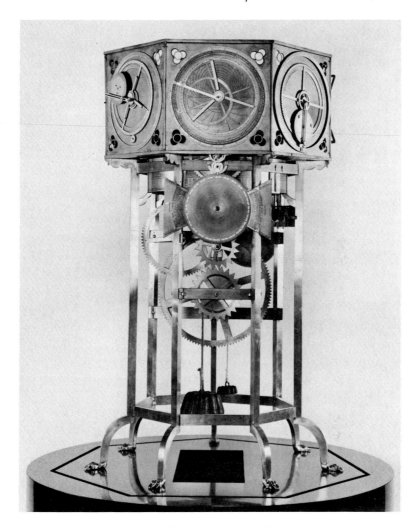

The faces of Giovanni de' Dondi's 14th-century clock traced the planetary epicycles according to the Ptolemaic theory; the front view features the solar dial. This replica stands in the National Museum of History and Technology of the Smithsonian Institution.

to Italy was certainly to learn Greek, and probably to learn Greek in a scientific context; it was just introduced, for example, at the medical school in Padua.[8] Indeed, it has been suggested by Ravetz[9] from astronomical reasoning that Copernicus had already convinced himself that the earth moves around the sun, and, therefore, that he went to Italy specifically to find support in Greek thought. The reasoning is doubtful[10] but there is no doubt that Copernicus did learn of sources that were then unfamiliar, such as Aristarchus of Samos—perhaps in the lectures at Bologna of Domenico Maria de Novara, a noted Platonist.

THE REVOLT AGAINST ARISTOTLE

At this point it is right to ask: But was not Greek science known in Europe before the Renaissance, before Petrarch even? Surely Ptolemy was known, surely Euclid and Galen were known; and surely by 1270 Thomas Aquinas had turned Aristotle into a philosophic oracle, whose Christian authority the Pope himself had to restrict sharply in 1277. Indeed, that is so: The books of these men were read and revered in Europe, having been translated into Arabic and thence into Latin during the Moorish occupation of Spain. But that roundabout and narrow channel had produced a tradition dominated by one Greek thinker, Aristotle, in all fields of science. A central thrust in humanism is the revolt against Aristotle, and this is true whether we interpret humanism broadly as I am doing, as a thirst for the whole stream of Greek knowledge, or strictly as an academic program of reform in education. For example, when the founder of the new alchemy, Paracelsus, showed his contempt for medical dogmatism by publicly burning a standard textbook in Basel in 1527, he chose the *Canon of Medicine* of Avicenna, an Arab follower of Aristotle.

Since what we would call the scientific establishment based itself on Aristotle, the up-and-coming men avid for new ideas naturally turned away to Plato. The masterly translation of Plato was begun by Marsilio Ficino in the 1460s and finally published in 1482; he had trained himself for it on the instructions of Cosimo de' Medici, who had made Florence a home for Platonists. These events had two effects on the development of science and its conjunction with humanism, one direct and one indirect.

The direct effect was to make science more mathematical, since Plato was much concerned with geometrical notions. Aristotle's insight had been into differences of quality; he saw the world as a continuum and dismissed atomism; he was a naturalist by temperament, a lover of taxonomic systems, and that was the mood of science and medicine before 1500. In contrast, Platonism brought in a more quantitive manner, in which general principles were expected to satisfy specific tests, so that the detail of nature became significant for scientists as well as artists. The new temper is evident in the arts—in the works of Michelangelo, who was a Platonist, and of Leonardo da Vinci, who was not, yet wanted his drawings of a machine or a flower not only to look right but to work right. The sense that his machinery is right, geometrically and aesthetically, is strong in Copernicus' choice of his system.[11] Aristotle has no sense of mathematics as a dynamic description; for example, he thinks of an eclipse of the moon as an "attribute" or "affection" of the moon, an inherent property, not as an effect of its motion. The idea that the world is in movement that can be pictured mathematically had to come from the Platonists. When Galileo later in his *Dialogue on the Great World Systems* explained the Copernican system, he repeatedly stressed (as did Kepler) his debt to Plato. This effect of Platonism in making science mathematical has been expounded at length by Alexandre Koyré[12] and I need not dwell on it.

NEO-PLATONIC INFLUENCE

The indirect effect of the Florentine school of Plato is more subtle and harder to trace, though I think it no less important. There is an underlying sense of mystery in Plato, and even in the

Greek fascination with mathematical relations. It was therefore foreseeable that the new Platonism in Florence and elsewhere leaned to mysticism and in time became obsessed by it. When Cosimo de' Medici's buyers in 1460 came back from Macedonia with a manuscript of the fabled Hermetic Books, which were supposed to be pre-Christian prophecies by an Egyptian magus called Hermes Trismegistus, he made Ficino put Plato aside and translate them first. Their influence was immense, for they gave to nature a numinous quality, a sacred but vibrant life, which fitted the reverend adventure of the Renaissance. They are quoted in this sense by Copernicus in a well-known passage in *De revolutionibus orbium coelestium.*

> In the middle of all sits the Sun enthroned. In this most beautiful temple could we place this luminary in any better position from which he can illuminate the whole at once? He is rightly called the Lamp, the Mind, the Ruler of the Universe; Hermes Trismegistus names him the Visible God, Sophocles's Electra calls him the All-seeing. So the Sun sits as upon a royal throne ruling his children the planets which circle round him.[13]

However, we must not let this single passage lure us to believe that Copernicus drew much on the Hermetic Books, or on Ficino's own rapturous essay on the sun; he did not. But there is no doubt that the appeal of Copernicus' system was heightened by an enthusiasm for the astrological power of the sun that came from the Hermetic Books. The fact is that until 1600 humanists knew Ficino better than Copernicus, and were quick to recognize his touch. For instance, when Giordano Bruno lectured on Copernicus at Oxford in 1583, his hearers ignored the science but were sharp to spot the quotations from Ficino.[14] This has been taken to imply that Bruno did not understand the mechanics of the Copernican system, but the contemporary account does not bear this out; on the contrary, it makes fun of Bruno for "telling vs much of *chentrum & chirculus & circumferenchia* (after the pronunciation of his Country language)."[15]

These mathematical terms do not occur in Ficino or the Hermetic Books, and they show that Bruno was indeed describing the system of Copernicus, even though his hearers thought it foolish to believe "that the earth did goe round, and the heavens did stand still; whereas in truth it was his owne head which rather did run round, & his braines did not stand still."[16]

The mystical fantasies in Neo-Platonism seem to us now merely superstitious and to obscure the science as they did for Bruno's audience at Oxford. But this is too simple a view. The Neo-Platonists were fascinated by the relations between nature and man, and they were too sophisticated to think that they could be controlled by the primitive and beastly magic that was current in the Middle Ages. They looked for more subtle influences in nature, such as the influence of the planets. Since those could not be browbeaten or controlled by man, Neo-Platonists wanted to understand nature so that they might fit their actions to the propitious moments that she presented. In this way, they moved away from the medieval concept of black or Satanic magic, which seeks to force nature out of her course, to a new concept of white or natural magic, which is content to exploit her laws by understanding them. I believe that this change in the means by which the mind hopes to master nature was an important influence of humanism on science which took place in Copernicus' lifetime. Soon after, in 1558, Giovanni Battista Porta published his book on *Natural Magick*, and by this time the phrase has come to mean simply a collection of scientific marvels.

THE POPULAR COMPONENT

Humanism in any sense is by origin an academic movement, because the sources at which it seeks its new knowledge are classical texts. But it would be unrealistic to ignore the strength that it drew from its popular appeal. Erasmus and Martin Luther were contemporaries of Copernicus, and showed before he published his book that the attack on scholastic authority needs a public that has the means to judge for itself—needs the printed book above all. Petrarch in the 14th century could find a poetic following in manuscript, but the sweep of humanism nearly 200 years after needed the backing of print; for lack of that, as Marie Boas[17] remarks, Leonardo da Vinci was forgotten (as was William Blake 300 years later). A list of the books printed in the same year (1543) as Copernicus' is fascinating: It includes the anatomical drawings of Andreas Vesalius, the first Latin translation of the mathematical works of Archimedes, and the attack on Aristotle's logic by Petrus Ramus, which did so much to change methods of reasoning and of education. A new picture of the world was forming in the public mind—a new geography that made Ptolemy old-fashioned, and a new cosmology that made him seem literal, formalist, and unimaginative. What spread the new picture through Europe as if it had wings was the printed word; for example, the most insidious advocacy of Copernicus, Galileo's vernacular *Dialogue on the Great World Systems,* could not be seized by the Inquisition in 1632 before it was sold out.

But there was also a popular element in the formation and the nature of Copernicus' picture which has been neglected and which I should stress. Consider what Martin Luther said about it even before Rheticus got into print. Here, he said in his *Table Talk* on 4 June 1539, is a "new astronomer who wants to prove that the Earth goes round, and not the heavens, the Sun and the Moon; just as if someone sitting in a moving wagon or ship were to suppose that he was at rest, and that the Earth and the trees were moving past him." Luther was an earthy man, by no means an intellectual, and of course his simile had been used long before by deeper students.[18] Yet think how surprising it is in an age when the laws of motion were unknown, and dynamics was not understood. The principle that the motions that Luther describes are equivalent is usually called Galilean relativity; but

Luther is speaking 25 years before Galileo was born. How had it come about that Copernicus could place his mind's eye in the sun and see the earth from it, and that much less mathematical minds could grasp what he was doing and see it as he did?

The question seems far-fetched today, because we have lived with perspective drawing for five hundred years, and therefore find it natural to shift our viewpoint in imagination. That was not so when Copernicus grew up; it was as novel for a man to think of moving his mind's-eye then as of moving the earth. Perspective was a new art that had been cultivated by the *Perspectivi* in Italy in the 15th century. Albrecht Dürer, who was a contemporary of Copernicus, still traveled to Bologna in 1506 "for the sake of 'art' in secret perspective which some one wants to teach me."[19] Copernicus himself was fortunate in seeing perspective at first hand as a popular art in the huge carved and colored wooden triptych in St. Mary's Church in Cracow which Veit Stosz finished about 1490.

Such simple and almost primitive church pictures had changed the perception of space in the 15th century. Before that, sacred pictures were flat and static because they represented a god's eye view: the artist shows the scene head on, not as it *looks* but as it *is*. Perspective is a different conception, mobile and human, a moment in time that the artist has caught with a glance from where his eye happens to stand. This sense of the temporal and human pervades the picture; in the Cracow triptych it comes alive in the portraits of city worthies, the everyday people that stand around the holy figures. After the coming of Luther, the Church of Rome grew alarmed at this secularization of the heavens and expressly forbade it in sacred paintings at the Council of Trent in 1563. It is a hidden element in Copernicus' view of nature; that is, he is usually accused of removing man from the center of the universe, but in fact the opposite is true—he moved him into the heavens. His system abolished the distinction between the terrestrial sphere and the crystal spheres beyond the moon, and made the heavens earthy.

The great wooden triptych by Veit Stosz in St. Mary's Church in Cracow.

THE ACADEMIC COMPONENT

The humanism that I have traced in Copernicus' outlook makes up a broad and, at the last, even a new popular mode of thought. There was, however, underlying this a narrower and more specific form of humanism to which I must give attention. This has been isolated by Paul Kristeller who shows that the central program of humanism in the universities was a reform of the curriculum.[20] In this sense, humanism was pursued in the study of a particular set of the seven liberal arts—chiefly the *trivium* of grammar, logic, and rhetoric (including such extensions as poetry, history, and moral philosophy). In time it mounted a formidable attack on the syllogistic logic of Aristotle, and replaced it by a less rigid and stereotyped choice of arguments that it called dialectic. As the association of logic with grammar, on the one hand, and with rhetoric, on the other, suggests, the aim was to uncover a more realistic *method* of inferring general results about natural phenomena, and of presenting them in nattural language. The change was brought about by younger men than Copernicus, chiefly by Luther's disciple Philipp Melanchthon and by Petrus Ramus. Even so, these pioneers failed to find an alternative in science to Aristotle's mode of reasoning from the general to the particular. That only came after 1600 when, as Lisa Jardine has shown, Francis Bacon was inspired to draw a distinction between methods of discovery and methods of exposition.[21] (What we should now call the axiomatic method was not recovered until much later, though the commentary on Euclid, in which Proclus had analyzed it, was translated into Latin in 1560.)

Nevertheless, academic humanism established in one of its disciplines results that had a far-reaching effect on science and on society together. It came in the most unlikely way from the study of Latin and Greek, which stimulated in literary scholars a feeling for exactness as passionate as the new-found Greek mathematics stimulated in scientists. As a result, they learned to analyze the grammar, the locutions, and the references in old texts precisely enough to challenge the traditional dates that had been accepted for these texts. This was more than an extension of the interest in history, which characterizes the Renaissance; it was revolutionary.[22] The discovery that texts could be accurately dated, and how to combine critical tests to do it, has a right to be ranked as a scientific discipline, a kind of literary archeology. Like the more usual archeology, it uncovered fakes; for example, it enabled Isaac Casaubon in 1614 to prove that the Hermetic Books had been forged in Christian times.

A far more damaging discovery, however, had been made before that in the 15th century. It was made by the pioneer of the method, Lorenzo Valla, an early humanist who scandalized his contemporaries by his epicurean and irreverent ways. In 1439 he electrified the Christian world by proving that a number of revered church documents had been forged, probably in Rome in the eighth century. (They were doubted about the same time by Nicholas of Cusa, a brilliant mind, who also conjectured that the earth revolves.) The most important of them was the Donation of Constantine, by which that emperor (who died in the year 337) was supposed to have granted the Popes temporal dominion in and beyond Rome. We have perhaps grown cynical now about the shuffling of treaties, and do not expect states to be scrupulous in their quest for power; but in 1439 it was catastrophic to learn that the spiritual head of the Christian world in the west was sustained by documents which had been deliberately fabricated.

THE DOCTRINE OF SINGLE TRUTH

The moment was a watershed of intellectual leadership in Europe, because it identified scholarship with tested truth. That had not been the character of academic disputation in the established tradition of scholasticism, nor was it prominent in the Aristotelian way of doing science.[23] Of course, Aristotelian and

A detail of the Veit Stosz triptych in Cracow, showing the carved hands of Mary and of an apostle.

Thomist science offered explanations for natural phenomena, but these explanations were not expected to have the precision of detail and sharpness of fit that Valla's literary archeology had shown to be possible and definitive.

It was not self-evident that the lesson of exact inference would be picked up by scientists, nor was it a foregone conclusion that it would be singled out so that it became central in their method. There is a case for saying that this was the most profound influence that humanistic scholarship had on science. It established what I will call the "doctrine of single truth."

The preoccupation with the detail of truth also created a single ethic for science. In the long run, it shifted the pursuit of science from results to methods, and the personality of the scientist from a finder to a seeker—characteristically, we call his work now *research*. In this way, science becomes an activity that demands for its collective success that all those who practice it share and adhere to certain values. As a particular case, the crucial need for mutual trust and respect has forced the scientific community to insist that ends must not govern means: There are no supreme ends—only decent and honest, namely truthful, means.

Had the church drawn the same lesson from the scandal of the Donation of Constantine, there might have been no reason for it to part company from science. Instead, Valla was long persecuted, and 150 years later Cardinal Robert Bellarmine still castigated him as *praecursor Lutheri*, one who opened the way for Luther. By then, the Copernican system had been made a religious issue; Bellarmine had a large hand in attacking it, in the trial of Bruno and in the first proceedings against Galileo; and science in the homeland of humanism was in retreat.

Copernicus was a silent man, but a committed man. He believed that his world system was true, and no ground of expediency persuaded him to say less. More is at stake here than a simple adherence to what is true; the issue is, What is the *test* of truth? Copernicus' attitude implies that the single truth exists in nature and as nature, and cannot be established or overturned by any authority other than the study of nature herself. In this direct and rational belief that nature is self-sufficient, Copernicus was a humanist pioneer, and we see in his science a base of philosophy as conscious as Isaac Newton's in a later age, and Albert Einstein's in ours.

Notes

[1] Edward Rosen, *Three Copernican Treatises*, third edition (New York, 1971).

[2] J. E. McGuire and P. M. Rattansi, "Newton and the 'Pipes of Pan'," *Notes and Records of the Royal Society of London*, vol. 21 (1966), pp. 108-143.

[3] Hans Baron, "Fifteenth-Century Civilisation and the Renaissance," in *The New Cambridge Modern History*, Volume 1: *The Renaissance 1493-1520* (Cambridge, 1957), p. 75.

[4] Marie Boas, *The Scientific Renaissance* (London, 1962).

[5] S. A. Bedini and F. R. Maddison, "Mechanical Universe, the Astrarium of Giovanni de' Dondi," *Transactions of the American Philosophical Society*, vol. 56, pt. 5 (1966).

[6] Derek J. de Solla Price, "Leonardo da Vinci and the Clock of Giovanni de Dondi," *Antiquarian Horology*, June 1958.

[7] Roberto Weiss, *The Renaissance Discovery of Classical Antiquity* (Oxford, 1969).

[8] John Herman Randall, Jr., *The School of Padua and the Emergence of Modern Science* (Padua, 1961).

[9] Jerome R. Ravetz, "The Origins of the Copernican Revolution," *Scientific American*, vol. 215, no. 4 (1966), p. 88 ff.

[10] Curtis Wilson, "Rheticus, Ravetz, and the 'Necessity' of Copernicus' Innovation," in *The Copernican Achievement*, edited by Robert Westman (Los Angeles-Berkeley, 1975).

[11] Owen Gingerich, "'Crisis' versus Aesthetic in the Copernican Revolution," in *Vistas in Astronomy*, edited by A. Beer and K. A. Strand (London, 1974), vol. 17 [in press]. J. Bronowski and Bruce Mazlish, *The Western Intellectual Tradition* (London, 1960).

[12] Alexandre Koyré, *Metaphysics and Measurement* (Cambridge, Mass., 1968).

[13] "Nicolaus Copernicus, De Revolutionibus, Preface and Book I." Translated by John F. Dobson in *Occasional Notes of the Royal Astronomical Society*, no. 10 (1947), p. 19.

[14]Frances A. Yates, *Giordano Bruno and the Hermetic Tradition* (Chicago, 1964), pp. 207-209.

[15]J. Bronowski, "Review of *Giordano Bruno and the Hermetic Tradition* by Frances A. Yates," *The New York Review of Books,* vol. 3, no. 2 (10 September, 1964), pp. 8-10.

[16]Ibid.

[17]Boas, op. cit. [note 4].

[18]Edward Grant, *Physical Science in the Middle Ages* (New York, 1971).

[19]Erwin Panofsky, *The Life and Art of Albrecht Dürer* (Princeton, 1955).

[20]Paul Oskar Kristeller, *Studies in Renaissance Thought and Letters* (Rome, 1956).

[21]Lisa Jardine, *Francis Bacon: Discovery and the Art of Discourse* (Cambridge, 1974).

[22]Weiss, op. cit. [note 7].

[23]Anneliese Maier, *An der Grenze von Scholastik und Naturwissenschaft* [*Studien zur Naturphilosophie der Spätscholastik,* vol. 3], second edition (Rome, 1952).

INTRODUCTION:
THE END OF THE COPERNICAN ERA?

Stephen Toulmin

The first section of this symposium has placed Nicolaus Copernicus in his original setting. At this point we must move forward in time and consider how the nature of science and its contribution to human life have changed in the subsequent centuries. Seen in retrospect, the introduction of Copernicus' new heliocentric, or heliostatic, picture of the planetary system, stands out as the first real modern candidate for the title of a "scientific discovery." Once we reach Copernicus (it seems) we can set those metaphysical old Greeks on one side and begin dealing with science, scientific arguments, and scientific discovery in the modern, familiar way. In this respect, Copernicus has come to symbolize the beginning of the modern era in science.

From different points of view, Gerald Holton and Werner Heisenberg will be considering the nature of scientific traditions, and the basic motives and patterns of scientific work as they have evolved during the great period of natural science that Copernicus inaugurated. In these introductory remarks, I shall myself be raising a single broader question, about the fundamental conditions—both psychological and intellectual—on which has rested the very possibility of all that scientific work that is so characteristic of the last 400 or 500 years. For this purpose, it will be necessary to recall that Nicolaus Copernicus was more than a man responsible for a particular technical transition in computational astronomy: namely, the substitution of the sun for the earth as the stationary "center" of the planetary system. In addition, we have to see him as a man who made one characteristic and essential contribution to our *general* intellectual tradition. He was the man who, after some 1200 years, staked out once again for the human mind the central claim on which any self-confident natural science depends: the claim that the human mind does have the capacity to develop consistent and comprehensive theories about the world in which we live, and can do so even on a scale that goes far beyond the direct reach of our experience.

This claim, which underlays so much in classical Greek science, has been explicitly or tacitly abandoned during the Alexandrian period. Speculation about the natural world then took second place to other religious preoccupations—and speculations about the nature of the heavens came to appear especially fruitless. Even Ptolemy himself had been content to use a variety of different astronomical constructions, which were—physically speaking—quite inconsistent with one another. As a result, his picture of the planetary system was less a life-like portrait than a botched-up picture, as though (Copernicus put it) a painter were to take the arms and legs of a figure, its body, head and hands, from quite different subjects and run them together into one portrait. Such a scientific amalgam was unacceptable. We must aim at a unified, coherent, and consistent picture, since nothing less could be "absolute and pleasing to the mind." With these words, we find Copernicus making a crucial declaration on behalf of scientific intellect—one that has played its part in the development of natural science ever since his time. From Copernicus on, therefore, one precondition of all effective scientific speculation has remained: this belief that the central goal of science is a goal that human beings can indeed hope to attain.

I say, "ever since the time of Copernicus," but I should add one gloss in passing. His robust faith in the scientific capacities of the human intellect is not something that caught on quickly or completely among his immediate successors. For that moment—indeed, throughout most of the 16th century—self-consciously "progressive" thinkers turned instinctively, not towards greater confidence, but towards greater skepticism. The Italian humanists, along with their French counterparts, such as Michel de Montaigne, were even less interested in cosmology and theoretical physics than the scholars of the medieval church had been. They turned, rather, to the newly rediscovered tragedies, histories, and biographies of the ancient writers: to Plutarch, to Sophocles, to Horace. And Montaigne in particular made it absolutely plain that in his eyes—as in those of Socrates much earlier—there was no question of human beings developing any well-founded theories about the physical character of the heavens. Our epistemological situation plainly ruled that out.

Accordingly, the intellectual example of Copernicus took as long to make itself felt among his successors as did the technical details of his astronomical innovations. In this respect, there was nothing incongruous about his position as a canon of the Roman Catholic Church. For, in spite of all the anti-Aristotelian rhetoric of Francis Bacon, the cumulative development of modern science originated not from any direct opposition to medieval thought; it arose, rather, from a secondary reaction against the secular and literary skepticism of the humanists. In this sense, it was a reaction against a reaction against the Middle Ages. To the earnest "new philosophers" of the 17th century, for instance, romantic irrationalism and a taste for "natural magic" were as foreign as they had been to their Aristotelian precursors in the medieval monastic schools. The aims of these 17th-century scientists were not literary ones: what they admired were intellectual audacity and rigor. So, for all that has been said about the scientific rejection of Aristotelianism, when we look back to the mid-17th century, we find ourselves back on the scientific road opened up by the Greeks and extended in the Middle Ages—a road on which Copernicus represents as much the last of the astronomers of Antiquity as he does the first of the Moderns.

Throughout the whole great age of modern natural science, beginning in the 16th century and continuing into the 20th, the fundamental motives underlying the work of scientists have depended on the same "realistic" spirit that originally gave force to Copernicus' own criticisms of Ptolemy. Nature has a real structure, and the mind of man is capable of discovering this structure. We cannot be satisfied with inconsistent, incoherent or merely "phenomenological" accounts of the natural world, for these will fail to give us true understanding. Without the human self-confidence expressed in this belief, the philosophers of the medieval period had never addressed themselves to a fundamental critique of the "natural philosophy" they inherited from antiquity. Beginning with Copernicus, modern science has developed out of precisely that critique. Yet, now, it seems that we must be prepared to raise afresh the question, how far that original self-confidence is still alive for us today. What is becoming of this Copernican spirit in the science of our own time? What are the conditions on which human intellect can maintain the self-confidence that Copernicus preached? And is there a danger that these conditions may—after 400 years and more—be weakening?

Let me begin by pointing out one curious feature of contemporary natural science—a feature for which I can find no exact parallel elsewhere. At the present time, one finds a quite new sense of moral ambiguity among working scientists: a novel doubt, within the scientific profession itself, about the moral justifiability of the scientists' own seemingly "pure" intellectual activities. In the past, it is true, there have often been times when painters, poets, or literary men rejected the values and achievements of science from the outside—through the satires of a Jonathan Swift, the engravings and poems of a William Blake, the plays and prefaces of a Bernard Shaw; but these doubts have rarely spread into the community of men who are pursuing scientific investigations for themselves. Nowadays, by contrast, alongside a deepening suspicion of science from the outside, we can observe a fresh lack of conviction within parts of the scientific profession itself. It is as though the psychological preconditions for science, as established by Copernicus more than 400 years ago, have at last begun to crumble away, leaving us, apparently, at the end of the intellectual era that he inaugurated.

Why should this be the case? It is easy to confuse ourselves about this point, through failing to look at the matter in a long enough perspective. Some critics, for instance, talk as though the trouble were quite simple, and merely quantitative: as though the overall quantity of scientific work being done at the present time were so different in scale as to produce—of itself— a transformation in the relation of science to society. Even now, one still hears that familiar statistic quoted, that 90 percent of all the scientists that have ever lived are alive today. Yet one only has to dig a little into the foundations of that figure—first announced by Derek Price, subsequently popularized by Robert Oppenheimer—to discover that the same thing was true 50, 100, 150, even 200 years ago. At all those previous times—in 1920, in 1870, even in 1820 or 1770—it could have been said with equal validity that 90 percent of all the scientists who had ever lived up to *that* time were alive *then*. For as Derek Price pointed out at the very start, this "90 percent doctrine" indicates only that, from around the year 1700 on, scientific activity became a novel, freely increasing social variable, which had not yet begun to approach closely to its saturation level. (For similar reasons, it is probably true that 90 percent of all the electrical carving knives ever made are in use today; but this proves nothing whatever about the social significance of electric carvers!)

Indeed, if anything quantitative about the scale of science

today is socially significant, it is rather the fact that this "90 percent doctrine" is at last *ceasing* to be true. The activities of science are at last beginning to saturate—at last beginning to run up against the limits of the resources that society is willing to provide for them; and, as a result, for the first time since 1700 it is no longer possible for science to continue expanding freely as an unchecked social variable. So I was amused the other day to see an advertisement in a news magazine that cried up the importance of contemporary science under a new heading. Triumphantly, it announced: "80 percent of all the scientists who have ever lived are alive today." In the meantime the evidence we really need about the scale of contemporary scientific work can be found by looking, not at irrelevant percentages, but at absolute figures. The most recent critical estimate I can find sets the number of fully qualified professional scientists and engineers in the United States at some 67,000; that is, at 1/30th of one percent of the population. Of these, less than one-third are working in industry; i.e., 1/100th of one percent, which hardly suggests that the current problems of science, even in the technically advanced countries, are problems of elephantiasis.

Other critics attribute current anxieties about the moral justification of science to its impact on the natural and human environment, and their views certainly appear to have a little more foundation. There may of course be some reason to foresee a possible future in which the applications of natural science to technology might have a drastic impact on the quality of human life, even on the possibility of its continuance. Yet, once again, there is an element of exaggeration in the current debate, which distracts attention from the true sources of our present ambiguities. If we leave aside the special problems created by the use of nondegradable pesticides and similar chemical agents, the remaining basic sources of environmental pollution are nearly all of them the present day products of historical processes going back to the year 1800 or before—certainly, long before any serious application of scientific ideas to industry or manufacture.

These sources are (in brief) population growth, urbanization, and ill-controlled industrialization, all of them factors whose chief origins lie outside the scientific movement. Thus, if we end up by having more children than we have the resources to support, that is not the fault of science. If cities and factories use the earth's air and waters as an open sewer, that is not the fault of science either. So it need not surprise us to find all the main themes of the contemporary ecology movement anticipated long ago, in the writings of men like Thomas Malthus, William Blake, and Anatole France. Even apart from the whole of 20th-century science, we would still be overloaded with Blake's "dark, Satanic mills"; and without the help of contemporary science these factories might well have been even more damaging to the environment than the ones we actually have. Arguably, indeed, there seems to be too little application of scientific thought and analysis to the industrial organization and practice, not too much. Industrial technology is ecologically damaging not because it is too *scientific*, but because it is too *unscientific*.

Rather than looking for the source of our contemporary scientific dilemmas in the quantitative growth of scientific work, I believe we should be looking rather at the qualitatively changing relations between the enterprise of natural science itself and the rest of human life. We shall be looking at science, not merely as an isolated intellectual enterprise that men are free to engage in or to leave alone without consideration of the broader human situation, but instead, we shall be looking at it as a human activity that is always carried on within some broader social and cultural context. As a preface to the arguments of Gerald Holton and Werner Heisenberg, I wish at this point to pose a general question, which is close to the core of their discussions. This general question arises as follows: The men who deal in a critical, innovative way with our ideas of the natural world have been known, at different places and times, by quite different names; they have been called "natural philosophers," "virtuosi," "savants," "naturalists," "scientists." If that is so, it is because the very enterprise in which they have been engaged has been viewed in different lights in different milieux. Joseph Needham has asked, "Why did Chinese science never have its indigenous Galileo?" The answer he gives will do duty for the more general question also, for it turns out that this answer is a double one. Chinese science never had its Galileo because it never had its Euclid or Heraclides or Ptolemy; and that means it never had *either* accepted scholarly traditions of critical scientific thought *or* an established intellectual profession of critically minded men to pursue and develop it. In the classical Chinese milieu, as elsewhere, the ways in which men viewed the activities of science and its agents were two faces of a single coin. The character, the vigor, even the possibility of effective scientific inquiry (that is) all depended—as they still depend—on the existence of a suitable social and cultural context.

To state my general problem explicitly, then: If the natural sciences and the men who pursue them are viewed so differently in different cultures and epochs—so that a civilization as ancient and sophisticated as that of classical China ended by providing no place for a Euclid or a Galileo—what are the factors that underlie these variations from milieu to milieu? Why classical Greece but not classical China? Why 16th-century Padua rather than 14th-century Brindisi? Why Samos in the 6th century B.C., and Pasadena in the 20th century A.D.? Why Göttingen, Königsberg, and Aberdeen of all places? How does it come about that science could be effectively pursued here rather than there, now rather than then, continuously rather than intermittently? If we had answers to all these questions, we should be well on the way to a new level of self-awareness and self-understanding in our attitudes towards science. We should have achieved a fully-fledged, well-established view of the comparative history and sociology of science, of a kind that we now lack. For the moment, all we can do is to glimpse at some of the key factors and relationships involved, and recognize certain of the key phenomena needing to be explained. I want to pick on one of these phenomena as central to my discussion here, since it may help to throw some light on our own situation—and on the needs of Science—in the present decade also.

Basic Science—that is, the critical improvement of our fundamental ideas about nature—is an activity that has flourished vigorously just twice in human history, each time for some 400 or 500 years. The first great burst of critical speculation about Nature, in classical antiquity, lost its self-confidence and momentum at the turn of our era, in Alexandria. It did so in familiar-sounding circumstances. It did so at a stage when, firstly, the central philosophical and intellectual ambitions of science were abandoned as unattainable, and its inquiries fragmented into a dozen or more unrelated debates and bodies of technique; and when, secondly, the concern of the larger educated public with scientific ideas flagged and fell away, to be replaced by a preoccupation with technology on the one hand, and with Oriental religions on the other. Science as we have come to think of it in the modern period revived only after a further 1200 years had passed, when Nicolaus Copernicus—as we saw at the outset—at last challenged the skepticism of the Alexandrians, and reinstated the older intellectual claims of scientific inquiry, as originally asserted in Ionia and Athens.

Once before in human history the enterprise of natural science proved a fragile one, in a situation that was in some respects not unlike our own. Fragmented into a diversity of subsciences, confused in the public mind with technology and craft know-how, divorced from the broader questions of natural philosophy, which had been the source of its original impulse and interest to educated men at large, the clear-spring water of the scientific debate inaugurated in Ionia and Athens disappeared into the Egyptian sands.

The problem that I have raised here about the health of late-20th-century science accordingly requires us to ask ourselves a string of questions about the relation between contemporary science and its broader human context. How far, then, has the health of science depended, historically and culturally, upon the preservation of a favorable environment outside science itself? Can we discover reasons why the work of natural science should have flourished in the way it did, only during those two periods of 500 years that we associate with classical Greece and with the modern period since Copernicus? Are we now in danger of losing that robust intellectual confidence that Copernicus re-introduced into the Western intellectual tradition in the early 16th century? To the extent that this may be so, does that mean that we are in serious danger of seeing the vigor of science as we have known it over the last 400 years, dying away, with the rise of other less rational attitudes towards Nature? To speak personally: When I look at the fragmentation of the scientific enterprise today, at its divorce from any broader concern with natural philosophy, and at the wide-spread confusions of attitude that one finds both inside and outside the profession, as a result of which science has become so firmly linked in men's minds to technology—rather than to its traditional allies, such as cosmology and theology—I find the threat of Alexandrianism never far from my mind. While Copernicus was, undoubtedly, the man who initiated the *rise* of modern science, may we in our own day be in danger of witnessing the beginning of its *fall*?

This may sound a gloomy way of posing a question that we should be considering in a cool and detached manner, but I do not think that we can afford to take the conditions for the survival of science as we have known it entirely for granted. In this respect, little could be more important to the future of the scientific enterprise, than to understand much better than we do the nature of those external factors on which the health of the scientific tradition depends. For only then shall we be in a position to understand what we can do to help preserve that scientific morale and that faith in man's rational potential that Nicolaus Copernicus gave back to the Western world some 430 years ago.

MAINSPRINGS OF SCIENTIFIC DISCOVERY

Gerald Holton

Science has been under a siege from some parts of the general public. A recent issue of a widely read news weekly proclaimed that it has found a "deepening disillusionment with science and technology," and diagnosed that on a "philosophical level, the reversal is the result of a new mood of scepticism about the quantifying, objective methods of science." New books that declare the supposed philosophical evil of science (above all, its "rationality") in florid counterculture language, are sure to be spread about by the book clubs and the paperback houses, and to give courage to those who look for popular reasons for decreasing the support for scientific research. Nor are the responses to such barrages always inspiring. Thus a new report by physicists on the state of their field refers briefly to what it calls "the al-

I wish to express my indebtedness to Miss Helen Dukas and the Estate of Albert Einstein for permission to cite Einstein's writings; to Professors Owen Gingerich and A. I. Miller for useful conversations; and to Professor Edward Rosen for prepublication use of several new Copernican translations.

leged public disenchantment with science," and warns darkly that the very discussion of the matter "serves as a self-fulfilling prophecy."

Caught in this crossfire, one's mind travels back to that staunch disciple of Copernicus, Johannes Kepler. Writing from Linz in October 1626 to Guldin, he tells what life was like when the peasants laid siege to the town. As luck would have it, Kepler (who was the local official mathematician) had been given an apartment in the County Hall, located in the city wall, with a fine view of the outskirts. Consequently when the war reached the town "a whole company [of soldiers] descended on our house. Constantly the ear was attacked by gunfire, the nose by foul smoke, the eye by the flash of firing. All doors had to be kept open for the soldiers, who through their going back and forth ruined the sleep by night and my studies by day." Precisely in these terrible circumstances, Kepler concludes, he raised his eyes above the battle to the astronomical heavens, and settled down to work. How well this image describes our temptation today!

Could we perhaps take comfort from the fact that so far, despite the attacks, the quantity and quality of scientific work today seem better than ever before? Consider the most recent meeting of the American Physical Society in Washington, D.C., the biggest meeting in its history. A good fraction of the nearly 50,000 physicists in the United States was at hand. Over 1200 papers were presented. In one session alone I noted that each of the 10-minute papers had 13, 18, 19, 22, or 26 listed authors; and one paper is announced simply as a collaboration of three universities and two national laboratories. This is no Silent Spring! Some of the new research results are exhilarating, as tomorrow's session of our symposium will show.

We know that the cycles of antiscientific agitation have come and gone away before. The phrase "the bankruptcy of science" characterized a widely fashionable tendency of attack around the turn of the last century—and in the very middle of it came the discoveries of x-rays, radioactivity, the quantum theory of radiation, and relativity. Similarly, the publication of the essay of Babbage on the *Decline of Science in England* (1830) [1] was followed within a year by Faraday's discovery of electromagnetic induction and, with it, by the start of a whole new science. Examples of this sort can be cited all the way back to Plato, who was determined to eradicate science in reaction against the Democritan tradition, but who himself (as Sosigenes tells us) launched the quest for astronomical explanations that led to the work of Ptolemy and hence to Copernicus.

Yet we cannot smugly turn our backs on current controversy concerning the place of science in national and intellectual life. At the very least, we have to ask ourselves what the conditions may be that conduce to the strengthening or weakening of science. I shall steer away from questions of funding, or manpower, or the politics of science, and other worthy issues that have been debated often enough. I want to concentrate instead on the precious center of scientific work, the nature of the process of scientific discovery. To begin with, I want to sketch some features of this process which, it seems to me, must be preserved and strengthened at this time of external and internal stress, if science is to continue to flourish.

It is not underrating the work of the few sociologists and psychologists who have interested themselves in this question to say that the state of knowledge about what happens in the minds of people as they do "new science" is far from satisfactory. Most of the studies made so far are chiefly applicable to the oldest of the three overlapping phases through which modern science has passed, i.e., to the heroic achievement of individuals such as Copernicus—rather isolated, working over decades on one great problem, at most weakly related to other contemporary scientists, or, for that matter, to the social conditions within which they do their individual work. Stage Two, characterized by collegial relationships, came with the rise of the *Accademia dei Lincei* and its cousins, bringing greater interaction with collaborators and with social institutions, greater use of instruments, eventually a greater flow of fairly direct applications issuing from basic speculations.

In Stage Three we have reached "strong coupling." I am thinking of the intense collaboration in industrial laboratories, and equally of the exhausting and exciting interactions between such men as Niels Bohr, Werner Heisenberg, Wolfgang Pauli, Max Born, Erwin Schrödinger, or within Rutherford's and Fermi's laboratories. I am thinking of the multiple and interdisciplinary interactions that were at work during the unfolding of the Watson-Crick model of DNA; of the 26 authors whose collaborative paper was read at the American Physical Society meeting; and, to some degree, of most of the rest of us doing science today. Not only each of the individuals, but also an organism composed of individuals, seems to be doing research. While happily there are at this time also Stage One and Stage Two populations at work, the major new trend is to interactive groups, strongly coupled to the rest of the profession and to society at large, to a research style that in some sense is as far from the Copernican model as modern technological agriculture is from paleolithic food-gathering.

As the spectrum of processes of "science-making" changes—and we have by no means reached the end of these changes—one necessarily encounters new questions. How adequately have the mental processes of the individual's share in discovery been adapting in the face of intensified coupling? How fragile are the ecological links within the science community, and between science and its macroenvironment? And, to specify the question which I shall be chiefly concerned with in what follows: What particular characteristics of the process of scientific discovery (though first developed long ago in science at Stage One) may be essential today both for the understanding and for the protection of the creative potential of contemporary science?

A MODEL FOR SCIENTIFIC REASONING

The study of the ways scientists think has not been marked by a cheerful collaboration on the part of its subjects. Until fairly recently, scientists were reluctant, almost embarrassed, to be introspective or observant about what Albert Einstein called their own "personal struggle" during the process of discovery. Recently, Sir Peter Medawar correctly observed:

> Scientists are usually too proud or too shy to speak about creativity and creative imagination; they feel it to be incompatible with their conception of themselves as "men of facts" and rigorous inductive judgments. The role of creativity has always been acknowledged by inventors, because inventors are often simple unpretentious people who do not give themselves airs, whose education has not been dignified by courses on scientific method.[2]

Until recently, philosophers of science have also preferred not to be involved. For most of them, Hans Reichenbach's dictum was typical: "The philosopher of science is not much interested in the thought processes which lead to scientific discoveries . . . that is, he is interested not in the context of discovery, but in the context of justification." Even the few scientists who *are* sympathetic to an analysis of the context of discovery, often use imagery that shows how skeptical they are of finding ways of understanding it. Thus Max Born wrote, "I believe that there is no philosophical high-road in science, with epistemological signposts. No, we are in a jungle and find our way by trial and error, building our roads behind us as we proceed."

The chief result of this attitude has been to discourage the study of this "jungle." Happily, the situation has recently been changing for the better. It has become more common for working scientists—and indeed for the best of them—to allow themselves to be interrogated by well-prepared and sensitive historians, psychologists, and sociologists of science, and to save and make available their drafts, notes, research apparatus, personal scientific correspondence, and other documents. They have begun to realize that they have a serious duty, both to current scholarship that depends on their cooperation and to the writing of the contribution of science to contemporary culture. What has been found so far? Virtually every one of the few modern studies of scientific thinking has shown that it still relies on some variant of the centuries-old hypothetico-deductive scheme, whether we look at children or established scientists of genius, or the majority of all scientific work that is of the run-of-the-mill, "Research & Development" type. To put into bold relief some of the essential features of the process at its best, we can use the searching, personal testimony of a man in the class of Copernicus as a starting point.

Einstein discussed his view on the nature of scientific discovery, and of theory construction in particular, on several occasions, notably in his essays "Motives of Research" (1918), "Physics and Reality" (1936), and "Autobiographical Notes" (1946). He gave what was perhaps his clearest and most succinct presentation of his thoughts on scientific reasoning in a letter (written on 7 May 1952) to his old friend Maurice Solovine. Einstein began this portion of the letter by explaining that Solovine had misunderstood certain of Einstein's previous statements concerning epistemology. Einstein apologized, and asserted: "I probably expressed myself badly. I view such matters schematically thus. . . ." [3]

There followed a diagram—not entirely surprising. As we know from Einstein's autobiography and many other evidences, he preferred to think visually. Einstein went on to explain:

(1) The *E* (experiences) are given to us [represented by the horizontal line along the bottom of the figure]. (2) *A* are the axioms, from which we draw consequences. Psychologically the *A* rest upon the *E*. There exists, however, no logical path from the *E* to the *A*, but only an intuitive (psychological) connection, which is always 'subject to revocation' [disavowal].

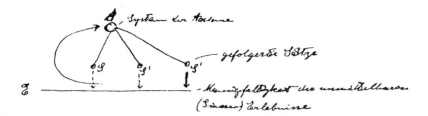

From a letter of A. Einstein to M. Solovine, 7 May 1952. (Courtesy of the Estate of Albert Einstein.)

We note that this point is one of the most persistent methodological remarks of Einstein. Even in 1918, when he was still doing his best consciously to toe the positivistic line, he wrote, "There is no logical path to these laws; only intuition, resting on sympathetic understanding of experience, can reach them . . . there is no logical bridge between phenomena and the theoretical principles." Later, going beyond even this, he wrote that the axioms are "free inventions of the human intellect." (Spencer Lecture, 1933,[4] and similarly in many of his letters). To return to the letter to Solovine:

(3) From the *A*, by a logical route, are deduced the particular assertions *S*, which deductions may lay claim to being correct. [As he had said in Spencer lecture: "The *structure* of the system is the work of reason."] (4) The *S* are referred [or related] to the *E* (test against experience). This procedure, to be exact, also belongs to the extralogical (intuitive) sphere, because the relation between concepts that appear in *S* and the experiences *E* are not of a logical nature. [In his "Reply to Criticisms" (1949)[5] Einstein elaborated on this point: The distinction between sense impressions or experience on one hand and ideas or concepts on the other is a necessary distinction, regardless of the reproach that using it makes one "guilty of the metaphysical 'original sin.'"]

These relations of the *S* to the *E*, however, are (pragmatically) much less uncertain than the relations of the *A* to the *E*. (For example, the notion "dog" and the corresponding experiences.) If such correspondence were not obtainable with great certainty (even if not logically graspable), the logical machinery would be without any value for the comprehension of reality (example, theology).

The quintessence is the eternally problematic connection between the world of ideas and that of experience. . . .

Now let us suppose that a connection *can* be made between the prediction S and the experiences E that are at hand. Does this constitute an adequate test of the theory being examined? Einstein discussed this question in his "Autobiographical Notes." At least in the case of the grand theories of greatest interest to him—those whose "object is the totality of all physical appearances"—he asserted that comparing the predictions of a theory with experiment is but one of *two* criteria according to which one can "criticize physical theories at all." [6]

The first criterion is that of "external confirmation." This is the easier one to meet, since one can often ("perhaps even always") make an adequate connection by suitable ad hoc "artificial additional assumptions." Moreover, Einstein phrased this criterion in a remarkably generous way: "The theory must not contradict empirical facts." This is, of course, very different from the much stronger injunction we usually associate with scientific "confirmation" by empirical test. Just how faithfully and effectively he followed this first criterion was shown repeatedly, e.g., in his steadfast and unswerving adherence to his ideas when, from time to time, evidence came which purported to show that his predictions, while not in unambiguous flat *contradiction* to the "facts of experience," at the very least were not being *supported* by experimental test.

Going on to the second criterion, Einstein explains that it "is concerned not with the relation to the material of observation, but with the premises of the theory itself, with what may briefly but vaguely be characterized as the 'naturalness' or 'logical simplicity' of the premises (of the basic concepts and of the relations between these which are taken as a basis)." Here is a clear place for individual aesthetic or other preferences—although as soon as this is raised, Einstein feels he, too, has to apologize a little for it. "The meager precision of the assertions contained in the last two paragraphs I shall not attempt to excuse by lack of sufficient printing space at my disposal, but confess herewith that I am not, without more ado, and perhaps not at all, capable to replace these hints by more precise definitions."

THEMATIC PROPOSITIONS AS GUIDES TO SCIENTIFIC THOUGHT

We now need an important clarification. At the heart of the method of scientific discovery shown schematically in Figure 1, there was the leap up from the plane of experience E to the premises A. That leap, it was stressed, is logically discontinuous. But it cannot be entirely "free" after all, if the premises later are to pass the tests of naturalness and simplicity (and the like) in order to meet the second criterion for a good theory. In fact, the leap *is* channelled and guided.

One such guide, at least for Einstein himself, was given by the fact that he attained the concepts for use at the A level by a form of mental play with visual materials "to a considerable degree unconsciously"—by a powerful iconographic rationality which he added to the more conventional semantic and quantitative ones. Another guide in the leap from E to S is one shared by all scientists engaged in a major work on novel ground: the guidance provided by explicit or, more usually, implicit preferences, preconceptions, presuppositions.

Einstein himself saw this and commented on it repeatedly. A good statement occurs in his unpublished essay "Induction and Deduction in Physics" (probably written in the early 1920s):

The simplest conception [model] one might make oneself of the origin of a natural science is that according to the inductive method. Separate facts are so chosen and grouped that the lawful connection between them asserts itself clearly. . . . But a quick look at the actual development teaches us that the great steps forward in scientific knowledge originated only to a small degree in this manner. For if the researcher went about his work without any preconceived opinion, how should he be able at all to select out those facts from the immense abundance of the most complex experience, and just those which are simple enough to permit lawful connections to become evident?[7]

It has always been this way. Otherwise we could not understand how Kepler could pursue astronomy so successfully even though his pre-Newtonian physics was so ineffective. While his instinct for physical problems was sound, his tools were not, and the actual success depended on Kepler's ability to shift to, and trust himself to, frankly metaphysical presuppositions when his physical ones gave out. Kepler's effort would have been doomed if he had not supplemented the mechanistic image of the universe as a physical machine with two other, very different ones: the universe as a mathematical harmony, and the universe as a central theological order. These three guiding ideas or grand themata continued to echo in the work of the 17th-century scientists who followed Kepler, and indeed up to the delayed triumph of the purely mechanistic view in the completion of Newton's work by Laplace.

In this, as in every other case of major scientific discovery, we are led to recognize the existence of (and even the necessity at certain stages in scientific thinking of postulating and using) precisely such unverifiable, unfalsifiable, and yet not arbitrary conceptions or hypotheses, a class to which I have referred as thematic presuppositions. They lie, as it were, along a dimension orthogonal to the other two that are equally necessary for scientific work, those referring to the empirical and analytical content, respectively. In Einstein's own scientific papers we can watch him stating his presuppositions boldly, as for example when he first announces his basic two postulates of relativity, almost brusquely declaring them to be hunches that he wishes to elevate to the status of postulates—without even bothering to connect them plausibly with the experimental material on the E level.

There is, of course, another side to this thematic origin of scientific thought. Dedicating oneself to some presuppositions or themata means one is likely to exclude others, as Einstein indeed did when he refused to accept the themata that were so basic in the work of the Copenhagen school on quantum mechanics. Just because they are not contingent on empirical ground, one can expect contrary themata to be vigorously held by opposing sides (as in the case, for example, of the theme-antitheme couple of atomism and the continuum). In the thematic conflict between scientists during the rise of quantum mechanics in the 1920s, some looked to Erwin Schrödinger's introduction of wave mechanics as "a fulfilment of a long baffled and insuppressible desire" (as one physicist expressed it in 1927). Others abhorred this continuum-based approach and found satisfaction only in fundamental explanations rooted in the thema of discreteness. Both groups faced, on the whole, the same experimental data. The passionate pursuits of their antithetical quests show the strength that the thematic attachment often has.

When one lists the general themes that have guided the process of scientific discovery of individual scientists and of the profession as a whole, one is struck by the antiquity and relative paucity of themata—by the remarkable fact that while the range and scale of recent theory, experience, and experimental means have multiplied vastly over the centuries, the number and kind of chief thematic elements have changed little. Since Parmenides and Heraclitus, the members of the thematic dyad of constancy and change have view for loyalty, and so have, ever since Pythagoras and Thales, the efficacy of mathematical forms versus the efficacy of materialistic or mechanistic models. The (usually unacknowledged) presuppositions pervading the work of scientists have long included such thematic preconceptions as these: simplicity, order, and symmetry; the primacy of experience versus that of symbolic formalism; reductionism versus holism; discontinuity versus the continuum; hierarchical structure versus unity; the animate versus the inanimate; the use of mechanisms versus teleological or anthropomorphic modes of approach.

These, and not many more—a total of a few dozen singlets, dyads, or triads, none of them right or wrong in any logical sense—seem historically to have sufficed to direct the energy of the creative leap in discovery. Our pool of these imaginative tools is characterized by a remarkable parsimony at the fundamental level, joined by fruitfulness and flexibility in actual practice. Only occasionally (as in the case of Niels Bohr's complementarity concept) does a qualitatively new theme enter into science, or is an old one discarded.

I do not have the space here to discuss the origins of these thematic presuppositions;[8] but we may well be bringing many of them to the science laboratory from our kindergarten. From a study by the school of Jean Piaget we know that in the psychological development of young children such ideas as conservation, or invariance persisting through apparent change, are an essential element of thought. So ingrained is that conviction that Henri Poincaré once said, when the law of conservation of energy was challenged: "We can be sure we shall find something that stays constant, and that we shall be able to call it energy."

We can now see that much of the fight of the priests of the counterculture against what they attack as overly rationalistic science is a sham: it is largely a fight against straw-men of their own making. They conceive of scientific rationality as limited to strictly quantitative and semantic-logic processes, but that applies at most, and only to a degree, to Public Science, i.e., to science as a pedagogical or as a consensus-seeking activity. What they attack is, however, only a poor caricature of the aspect of science I am here discussing—Private Science, the process by which reasoning men and women make discoveries. For there the discontinuous and thematic characteristics cannot be overlooked.

To allow those to be called "irrational" is to play with words. On the other hand, to let oneself be frightened into doubting the validity of thematic choices during discovery would indeed endanger the very process of discovery itself. So, also, would any limitation of individual thematic choice in order to accommodate differences in a collaborating group. In science at Stage Three, the chances of a team doing truly seminal work would be much decreased if the team had to limit itself to problems in which thematic presuppositions are not important, or where an adherence to a kind of thematic consensus, or lowest common denominator, is required.

SCIENCE AS A CHARISMATIC ACTIVITY

I now turn to another characteristic of scientific discovery that assures its innovative vigor. That trait, to put it briefly, is the scientist's confidence stemming from his perception of an overall goal that transcends the narrow task at hand—and, in the limit, that serves science in its function as a charismatic activity.

To recapture this conception, we can do no better than look at the work of Copernicus himself. To be sure, Copernicus, like any other good astronomer, relied on observation and calculation. He greatly advanced mathematical astronomy in the technical sense. But that is hardly the whole reason why he came to write the work for which he has been honored so long; nor does that explain its full power. In Copernicus' Dedication to Pope Paul III in the *De revolutionibus* (1543), he spells out his purposes and themes which reflect that other side. Copernicus writes that he wanted to understand "the movements of the world machine, created for our sake by the best and most systematic Artisan of all." In concluding Chapter 10 of Book I, Copernicus exults: "So vast, without any question, is this divine handiwork of the most excellent Almighty." Nature is God's temple, and human beings (it is implied) can, through the study of nature, discern *directly* both the reality and the design of the creator. This was a daring and dangerous idea; and it is significant that, when Copernicus' book was put on the *Index* of "Books to be Corrected," this last-quoted sentence was one of the relatively few deletions which were insisted upon as necessary; for it was equally clear to Copernicus and to his opponents that when the purpose of science is perceived large enough, it can become charismatic.

I use this now much debased word "charisma" (from the Greek word for gift) in its meaning of a special endowment conferred upon a believer as an evidence of the experience of grace, fitting him for the life and work to which he feels called. To be sure, it has not been fashionable during the past few decades to describe a motivation for scientific discovery in these terms. But without it we would miss the sense of exaltation in the beauty of the system as Copernicus might have seen it. In his forsaken corner of a miserable planet, Copernicus was persuaded he had found proof that the world is designed as a "beautiful temple," one which (he thought) could only have been made by the Divine Architect himself. Again and again he declares his delight, "the unbelievable pleasure of mind." His very first sentence in the first book of the *De revolutionibus* is: "Among the many and varied literary and artistic studies upon which the natural talents of man are nourished, I think that those above all should be embraced and pursued with the most loving care which have to do with things that are very beautiful." In such passages we sense the source of energy of a major scientific idea. It is not some pedestrian piecing together of a corner of the puzzle. Nor does the work give us just better astrometry and applications such as calendar correction, valuable though these are. Rather, the discovery is on a scale that produces an expansion of human consciousness—and it was so perceived by those who were converted to Copernicus' idea, above all by Kepler, who really became quite intoxicated by it. It is through that kind of energy that a scientific idea can change cultural evolution; and we have seen it again in the effect of a Darwin, Freud, or Einstein.

The themata that predominate in Copernicus' work are those of *simplicity* and *necessity,* and they appear in a stern manner that seems to have been new at the time but has become basic to all science since. As Copernicus proudly writes in the Dedication of the work, his heliocentric scheme for the system of planets has the property that

not only their phenomena follow therefrom, but also the order and size of all the planets and spheres, and heaven itself are so linked together that in no portion of it can anything be shifted without disrupting the remaining parts and the universe as a whole.

Thus, if the orbits are measured with respect to the sun, Mercury has the smallest relative radius and the smallest period of revolution; and as one goes to more distant planets one finds their periods increase, until one arrives at the "fixed" stars, having an infinitely long period.

The power of this solution was precisely its restrictiveness. There is nothing arbitrary, no room for the smallest ad hoc rearrangement of any orbit, as had been quite possible in pre-Copernican work. Copernicus' system as a whole revealed a sparse rationale, a *necessity* that binds each detail to the whole design. Hence it carries the conviction that we understand why the planets are disposed as they are, and not otherwise.

One is reminded here of Einstein's remarks to his assistant, Ernst Straus: "What I'm really interested in is whether God could have made the world in a different way; that is, whether the necessity of logical simplicity leaves any freedom at all." It was not idly that Einstein once called himself a deeply religious unbeliever. He defined religiosity as confidence in the rational order, and confidence that the human mind, at least to some degree, is open to the construction or discovery of reality.

In his essay on "Religion and Science" (1930) Einstein took pains to describe his idea of scientific religiosity, and it was one close to Spinoza's conception of religion. Einstein wrote that beyond the two more primitive stages of religion—that based on fear and that based on social and moral imperative—

. . . there is a third stage of religious experience which belongs to all of them, even though it is rarely found in a pure form. I shall call it cosmic religious feeling. It is very difficult to elucidate this feeling to any one who is entirely without it, especially as there is no anthropomorphic conception of God corresponding to it [nor any dogma or church] . . . . I maintain that the cosmic religious feeling is the strongest and noblest motive for scientific research . . . . What a deep conviction of the rationality of the universe and what a yearning to understand, were it but a feeble reflection of the mind revealed in this world, Kepler and Newton must have had . . . .[9]

This kind of terminology, now rarely heard, is somewhat-embarrassing to most scientists today. There are good sociological, psychological, and even political reasons why this should be so, why our usual list of motivations stresses the Baconian side of the heritage of modern science, the relief of man's estate, the discovery of cures, and the perfection of appliances, or simply the provision of a decent way to spend one's days on this earth. But the Baconian ethos, while necessary, is not sufficient; and by itself it does not help us understand fully the nature of high discovery. No one would argue that the personal language of Einstein should be introduced into our current scientific papers, not to speak of the far more theistic language of Copernicus, Kepler, and Newton. In my opinion, it is quite enough that a quiet underground current exists along the lines described by Einstein. It would be far more ominous if this cosmological or charismatic tradition were to dry up altogether, for that would indeed signal the decline of science.

Happily, it is not too difficult to glimpse the continued existence of this motivating tradition in scientific discovery of the fundamental level today, even if it may come to us in somewhat disguised form. Thus Steven Weinberg, on receiving the Robert Oppenheimer Memorial Prize, said:

Different physicists have different motivations, and I can only speak with certainty about my own. To me, the reason for spending so much effort and money on elementary particle research is not that particles are so interesting in themselves—if I wanted a perfect image of tedium, one million bubble chamber photographs would do very well—but rather that as far as we can tell, it is in the area of elementary particles and fields (and perhaps also of cosmology) that we will

Albert Einstein, c. 1905, while with the Patent Office in Bern. (Courtesy of the Einstein Archive.)

find the ultimate laws of nature, the few simple general principles which determine why all of nature is the way it is . . . .

The reason I take such an optimistic view of where we are now is that relativity and quantum mechanics, taken together but without any additional assumptions, are extraordinarily restrictive principles. Quantum mechanics without relativity would allow us to conceive of a great many possible physical systems. Open any textbook on non-relativistic quantum mechanics and you will find a rich variety of made-up examples—particles in rigid boxes, particles on springs, and so on—which do not exist in the real world but are perfectly consistent with the principles of quantum mechanics. However, when you put quantum mechanics together with relativity, you find that it is nearly impossible to conceive of any possible physical systems at all. Nature somehow manages to be both relativistic and quantum mechanical; but those two requirements restrict it so much that it has only a limited choice of how to be—hopefully a very limited choice.[10]

Copernicus would have understood this. And while any individual attempt of this sort is by no means guaranteed to be right, no other, less cosmological, approach is likely to lead to the truth either—and particularly not to a truth having the kind of exalting sweep that historically has given the scientific enterprise its real intellectual mandate.

The progress of science is threatened today not only by loss of financial support and of good people—which is bad enough; not only by diversion of too much of its energy to applied or engineering work that may not yet be bolstered by enough basic knowledge; and not only by the confusion and disenchantment of the wider public—and that, too, demands our concern, because some of it is surely due to a lack of proper attention on the part of scientists. No, what seems to me to be the most sensitive, the most fragile part of the total intellectual ecology of science is the understanding, on the part of scientists themselves, of the nature of the scientific enterprise, and in particular the hardly begun study of the nature of scientific discovery. In this pursuit, our own day-by-day experience as scientists will help us if we set it into the historic framework provided by those who went before us. The celebration meetings for Copernicus' 500th anniversary remind us that these forerunners of modern science have set for us the scale on which both scientific discovery in its most serious sense and science as a vocation are to be measured, and that they also gave us many of the themata that guide our work. Each time we make a speculative leap, channelled by such essential themes as harmony, ordered simplicity, or necessity—none of which are taught explicitly or appear in the index of our textbooks—Copernicus (and all the others who painfully tested the adequacy and limits of the themata we now use so naturally) stands unseen at our side, and guides us.

Notes

[1] Charles Babbage, *Reflections on the Decline of Science in England, and on Some of Its Causes* (London, 1830).

[2] Sir Peter Medawar, *Induction and Intuition in Scientific Thought (Memoirs of the American Philosophical Society*, vol. 75), (Philadelphia, 1969), p. 55.

[3] Albert Einstein, "Motives of Research" (1918), reprinted in translation in Albert Einstein, *Ideas and Opinions* (New York, 1954); "Physics and Reality," *Journal of the Franklin Institute*, vol. 221 (1936), pp. 313–347; "Autobiographical Notes" (1946), in Paul Schlipp, editor, *Albert Einstein, Philosopher Scientist* (La Salle, Illinois, 1949); Letter from Einstein to Maurice Solovine, 7 may 1952, in A. Einstein, *Lettres à Maurice Solovine* (Paris, 1956), p. 120.

[4] Albert Einstein, *On the Method of Theoretical Physics* (Oxford, 1933), [The Herbert Spencer Lecture delivered at Oxford, 10 June 1933].

[5] Albert Einstein, "Reply to Criticisms" (1949), in Paul Schlipp, editor, op. cit. [note 3].

[6] Einstein, Letter to M. Solovine [note 3].

[7] Albert Einstein, "Induction and Deduction in Physics," unpublished manuscript, ca. 1920.

[8] Gerald Holton, *Thematic Origins of Scientific Thought: Kepler to Einstein* (Cambridge, Mass., 1973).

[9] Albert Einstein, "Religion and Science," (1930) in *Ideas and Opinions* [note 3], pp. 36-40.

[10] Steven Weinberg, "Where We Are Now," *Science*, vol. 180 (1973), pp. 276-278.

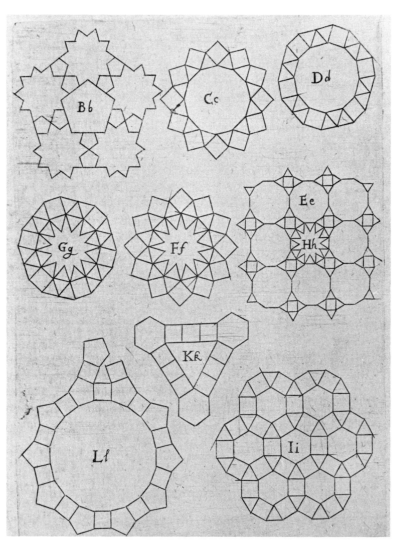

Construction of geometric solids, from Johannes Kepler, *Harmonice mundi* (Linz, 1619). (Permission of Harvard College Library.)

TRADITION IN SCIENCE

Werner Heisenberg

When we celebrate the 500th birthday of Copernicus, we do it because we believe that our present science is connected with his work; the direction that he had chosen for his research in astronomy still determines to some extent the scientific work of our time. We are convinced that our present problems, our methods, our scientific concepts are, at least partly, the result of a scientific tradition that accompanies or leads the way of science through the centuries. It is therefore natural to ask to what extent our present work is determined or influenced by tradition. Are the *problems* in which we are engaged freely chosen according to our interest or inclination, or are they given to us by a historical process? To what extent can we select our scientific *methods* according to the purpose? To what extent do we again follow a given tradition? Finally, how free are we in choosing the *concepts* for formulating our questions?

Any scientific work can only be defined by formulating the

Discussion of this paper begins on page 556.

questions that we want to answer. But in order to formulate the questions we need concepts by which we hope to get hold of the phenomena. These concepts usually are taken from the past history of science; they suggest a possible picture of the phenomena. But if we are going to enter into a new realm of phenomena, these concepts may act as a collection of prejudices, which hamper progress rather than foster it. Even then we have to use concepts, and we can't help falling back on those given to us by tradition. Therefore, I will try to discuss the influence of tradition first in the selection of problems, then in the scientific methods and, finally, in the use of concepts as tools for our work.

TRADITION IN THE SELECTION OF PROBLEMS

To what extent are we bound by tradition in the selection of our problems? When we look back into the history of science, we see that periods with intense activity alternate with long periods of inactivity. In ancient Greece the philosophers started asking questions of principle with respect to the phenomena in nature. There had been a considerable practical knowledge long before; great skill had been developed in building houses, cutting and moving big stones, constructing ships and so on; but it was first in the period after Pythagoras that this skill was supplemented by scientific inquiry. The relevance of mathematical relations in natural phenomena was discovered by Pythagoras and his pupils, and a great development in mathematics, in astronomy, and in natural philosophy followed. The decline of Greek science after the Hellenistic period marked the beginning of a long period of inactivity which lasted until the Renaissance in Italy.

During this period of stagnation, an admirable development of practical knowledge led to a high civilization in the Arab countries, but it was not accompanied by a corresponding development in science, by a deeper understanding of nature. More than a thousand years later, when humanism and the Renaissance had shown the way to a more liberal trend of thought, when the explorers had demonstrated the possibility of expansion on our earth, then a new activity in science was inaugurated by the discoveries of Copernicus, Galileo, and Kepler. This activity has lasted until our present time, and we do not know whether it will continue for long or will give way to a new period in which the interest will go into very different directions.

Looking back upon history in this way we see that we apparently have only little freedom in the selection of our problems. We are bound up with historical process. Our lives are part of this process and our choices seem to be restricted to the decision whether or not we want to participate in a development that takes place in our time, with or without our contribution. Without such a favorable development, our activity would probably be lost.

If Einstein had lived in the 12th century, he would have had very little chance to become a good scientist. Even within a fruitful period a scientist does not have much choice in selecting his problems. On the contrary, one may say that a fruitful period is characterized by the fact that the problems are given, that we need not invent them.

This seems to be true in science as well as in art. In the 15th century when painters in the Netherlands discovered the possibility of portraying men as active members of society, many gifted people were attracted by this possibility and competed in solving the problem. In the 18th century, Haydn tried in his string quartets to express emotions that had appeared in the literature of his time, in the work of Rousseau and in Goethe's *Werther*; and, then, the musicians of the younger generation— Mozart, Beethoven, Schubert—gathered in Vienna to compete in the solution of this problem.

In our century the development of physics led Niels Bohr to the idea that Lord Rutherford's experiments on alpha rays, Max Planck's theory of radiation, and the facts of chemistry could be combined in a theory of the atom. And in the following years many young physicists went to Copenhagen in order to participate in the solution of this given problem. One cannot doubt that in the selection of problems the tradition, the historical development, plays an essential role.

This may also sometimes be true in a negative sense. It can happen that traditional themes have been exhausted and that the gifted people turn away from a field in which they see no more objects for their activity. After Thomas Aquinas, the philosophers got tired of the theological and philosophical problems of scholastics and turned to humanism. In our time the traditional themes of art seem to be exhausted. In 1972, one of the most popular yearly exhibitions of modern art in Germany, which is held in Kassel and called "Documenta," was a center of political propaganda rather than of art. On the outside of the building of the exhibition, the young artists had fixed a huge poster with the text: "Art is superfluous."

In a similar way we cannot exclude the possibility that after some time the themes of science and technology will be exhausted, that a younger generation will be tired of our rationalistic and pragmatic attitudes and will turn their interest to an entirely different activity. In the present situation, however, many problems still exist in pure and in applied science. No effort is needed to invent them, and they will be passed on from the teachers to their pupils.

In this connection it is important to emphasize the very great role of personal relations in the development of science or art. It need not only be the relationship between teacher and pupil, it may simply be personal friendship or respect between people working for the same goal. This is probably the most efficient instrument of tradition. Among the many examples which could be mentioned for this kind of tradition I will only recall some of the personal relationships which have shaped the history of physics in the first half of our present century.

Einstein was well acquainted with Planck; he corresponded with A. J. Sommerfeld about the theory of relativity and about quantum theory; he was a dear friend of Max Born, although he could never agree with him on the statistical interpretation of quantum theory; and he discussed with Niels Bohr the philosophical implications of the relations of uncertainty. A large part of the scientific analysis of those extremely difficult problems, arising out of relativity and quantum theory, was actually carried out in conversations between those who took an active part in the research.

Sommerfeld's school in Munich was a center of research in the early 1920s. Wolfgang Pauli, Gregor Wentzel, Otto Laporte, W. Lenz and many others belonged to this group, and we discussed almost daily the difficulties and paradoxes in the interpretation of recent experiments. When Sommerfeld received a letter from Einstein or Bohr, he read the important parts in a seminar and started at once a discussion on the critical problems. Niels Bohr held a close association with Lord Rutherford, Otto Hahn, and Lise Meitner, and he considered the continuous exchange of information between experiment and theory as a central task in the progress of physics. The enormous influence of Niels Bohr on the development of physics in his time was not primarily due to his papers, but to his way of discussing again and again with his partners the fundamental difficulties of quantum theory, which, as he knew, did not allow for any cheap solution.

When wave mechanics was introduced by Erwin Schrödinger, Bohr saw at once that this was a very important new aspect of quantum theory; but that a simple replacement of the electronic orbits in the atom by three-dimensional matter waves could not solve the real difficulties. Again, the only way of analyzing the problem seemed to be personal discussion with the author. Schrödinger was invited to Copenhagen. In two weeks of most intense discussions, the way was prepared for the later development in the interpretation of quantum theory, for Bohr's concept of complementarity, and for the relations of uncertainty. I need not enlarge upon these examples. It is obvious that personal relations play a decisive role in the progress of science and in the selection of problems.

There are, of course, other motives for the selection of problems; motives that have played their role in the history of science. The best known of these is the practical applicability of science. In ancient times the interest in astronomy and mathematics was stimulated by the fact that knowledge in these fields was helpful for navigation and for the surveying of land. Navigation played a very important role in the 15th century, when the explorers left Europe and the Mediterranean and sailed westward.

When Galileo defended the ideas of Copernicus he made use of a newly invented instrument, the telescope, thereby demonstrating that a practical tool may be helpful in the progress of science, and science may be helpful in leading to the invention of practical tools. Galileo and his followers were strongly interested in the practical side of science. They studied mechanical devices, for example, the mechanical clock; they invented optical instruments, and so on.

It has always been a tradition in science, guiding the activity of many generations, that science should be applied to practical purposes and that the practical application should be a check on the validity of the results and a justification for the efforts. The atomic physicists of the first half of our present century followed this old tradition of science when they looked for practical applications of atomic physics. It was, of course, extremely disappointing for them that the first practical application was for warfare. Still the fact that one now could transmute chemical elements into others in large quantities was justly considered as a real triumph of science.

Interest in the practical application of science is frequently misunderstood as the trivial attempt of the scientist to acquire economic wealth. It is true that this trivial motive does play a role, depending of course on the individuals. But this motive should not be overestimated. There is another much stronger motive that fascinates the good scientist in connection with the practical application, namely, to see that one has correctly understood nature.

I remember a conversation with Enrico Fermi after the war, a short time before the first hydrogen bomb was to be tested in the Pacific. We discussed this plan, and I suggested that one should perhaps abstain from such a test considering the biological and political consequences. Fermi replied, "But it is such a beautiful experiment." This is probably the strongest motive behind the applications of science; the scientist needs the confirmation from an impartial judge, from nature herself, that he has understood her structure. And he wants to see the effect of his effort.

From this attitude one can also easily understand the motives that determine the line of research for the individual scientist. Such a line of research is usually based on some theoretical ideas, on conjectures concerning the interpretation of the known phenomena, or on hopes for finding new ones. But which ideas are accepted? Experience teaches that it is usually not the consistency, the clarity of ideas, which makes them acceptable, but the hope that one can participate in their elaboration and verification. It is the wish for our own activity, the hope for results from our own efforts, that leads us on our way through science. This wish is stronger than our rational judgment about the merits of various theoretical ideas. In the early 1920s we knew that Bohr's theory of the atom could not be quite correct; but we guessed that it pointed in the right direction, and we hoped that we would be able some day to avoid the inconsistencies and to replace Bohr's theory by a more satisfactory picture.

TRADITION IN THE SCIENTIFIC METHOD

The role of tradition in science is not restricted to the selection of problems. Tradition exerts its full influence in deeper layers of the scientific process, where it is not so easily visible. Here we should first of all mention the scientific method. In the scientific work of our present century we still follow essentially the method that had been discovered and developed by Copernicus, Galileo, and their successors in the 16th and 17th centuries. This method is sometimes misunderstood by terming it empirical science, as contrasted to the speculative science of former centuries. Actually Galileo turned away from the traditional science of his time, which was based on Aristotle, and took up the philosophical ideas of Plato. He replaced the descriptive science of Aristotle by the structural science of Plato. When he argued for experience, he meant experience illuminated by mathematical constructs. Galileo, as well as Copernicus, understood that by going away from immediate experience, by idealizing experience, we may discover mathematical structures in the phenomena, and thereby gain a new simplicity as a basis for a new understanding.

Aristotle had correctly stated that light bodies fall more slowly than heavy bodies. Galileo claimed that all bodies fall with the same speed in empty space, and that their fall could be described by simple mathematical laws. Fall in empty space could not be observed accurately in his time; but Galileo's claim suggested new experiments. The new method did not aim at the description of what is visible, but rather at the design of experiments and the production of phenomena that one does not normally see and at their calculation on the basis of mathematical theory.

Two features are, therefore, essential for the new method: (1) the attempt to design new and very accurate experiments, which idealize and isolate experience, and thereby actually create new phenomena; and (2) the comparison of these phenomena with mathematical constructs, called natural laws. Before we discuss the validity of this method, even in our present science, we should perhaps briefly ask for the basis of confidence that led Copernicus, Galileo, and Kepler on this new way. Following a paper of C. F. von Weizsäcker,[1] I think we have to state that this basis was mainly theological. Galileo argued that nature—God's second book (the first one being the Bible)—is written in mathematical letters, and that we have to learn this alphabet if we want to read it. Kepler is even more explicit in his work about world harmony. He says, God created the world in accordance with his ideas of creation. These ideas are the pure archetypal forms that

Plato termed ideas, and they can be understood by man as mathematical constructs. They can be understood by man, because man was created as the spiritual image of God. Physics is reflection on the divine ideas of creation; therefore physics is divine service.

We are in our time very far from this theological foundation or justification of physics. We still follow this method, however, because it has been so successful. The essential basis for this success is the possibility of repeating the experiments. We can finally agree about the results because we have learned that experiments carried out under precisely the same conditions do actually lead to the same results. This is not at all obvious. It can only be true if events exactly follow a causal chain, a sequence of cause and effect. But on account of its success, in the course of years this kind of causality has been accepted as one of the fundamental principles of science. The philosopher Kant has stressed the point that causality in this sense is not an empirical law, but it belongs to our method of science. It is the condition for the kind of science that was inaugurated in the 16th century and which has been elaborated ever since.

A consequence of this attitude in science is the assumption that we study nature as it "really is." We imagine a world that exists in space and time and follows its natural laws, independent of any observing subject. Therefore, in observing the phenomena we take great care to eliminate any influence from the observer. When we produce new phenomena by means of our experimental equipment, we are convinced that we do not really produce new phenomena; that is, we believe that actually these phenomena occur frequently in nature without our interference, and our equipment is just made to isolate and to study them. In all these points, we still follow confidently the tradition from the time of Copernicus and Galileo. But are we really entitled to do so, considering the well-known epistemological difficulties of quantum theory? In the big accelerators, for example, we study the collisions between elementary particles; and we imagine that even if we had not built the accelerators, such phenomena would occur in our atmosphere on account of cosmic radiation. But would there be waves or particles coming from the outside? Would they produce interference patterns or tracks? What does actually happen when we do not observe? And, do we know what the word "actually" means in this context? These are hard questions, and we see that tradition can lead us into difficulties.

It is generally believed that our science is empirical, and that we draw our concepts and our mathematical constructs from empirical data. If this were the whole truth, when entering into a new field, we should introduce only such quantities that can directly be observed and formulate natural laws only by means of these quantities. When I was a young man I believed that this was just the philosophy that Einstein followed in his theory of relativity. Therefore, I tried to take a corresponding step in quantum theory by introducing the matrices. But when I later asked Einstein about it, he answered: "This may have been my philosophy, but it is nonsense all the same. It is never possible to introduce only observable quantities in a theory. It is the theory which decides what can be observed." What he meant by this remark was that when we go from the immediate observation— a black line on a photographic plate or a discharge in a counter— to the phenomena we are interested in, we must make use of theory and of theoretical concepts. We cannot separate the empirical process of observation from the mathematical construct and its concepts. The most conspicuous later demonstrations of Einstein's thesis were the uncertainty relations.

This new situation in quantum theory does not necessarily question the traditional method in science; it only questions the assumption that concepts and mathematical constructs can simply be taken from experience. It is true that in quantum theory we cannot rely on strict causality; but by repeating the experiments many times we can finally derive from the observations statistical distributions, and by repeating such series of experiments we can arrive at objective statements concerning these distributions. This is a standard method in particle physics, which may be considered as a natural extension of the traditional method.

With regard to the scientific method, it seems that we follow strictly the tradition inaugurated in the time of Galileo. In spite of the many different fields that have been developed—physics, chemistry, biology, atomic and nuclear science, etc.—the fundamental method has always been the same. One has the impression that in this period most scientists believed that this was the only acceptable method which could lead to correct statements concerning the behavior of nature.

There has been one attempt to work on an entirely different line, which I should mention. The German poet, Goethe, tried to return to a descriptive science, a science that is interested only in the visible natural phenomena, not in experiments that produce artificial new effects. He objected to the separation of the phenomena into their objective and their subjective side, and he was filled with fear of the destruction of nature by an overflowing of technical science. In our time, when we know of the contamination of air and water, the poisoning of the soil by chemical fertilizers and atomic weapons, we understand Goethe's fear better than his contemporaries could. But Goethe's attempt did not really influence the course of science. The success of the traditional method was too overwhelming.

TRADITION IN SHAPING CONCEPTS

Besides the effect of tradition in the selection of problems and in the scientific method, the influence of tradition is perhaps strongest in shaping or passing on the concepts by which we try to understand phenomena. The history of science is not only a history of discoveries and observations, it is also a history of concepts. Therefore, I will try to discuss briefly the history of concepts during the period following Copernicus and Galileo, and the role of tradition in this history.

The new science started with astronomy, and therefore the positions and the velocities of bodies were natural first concepts for describing the phenomena. Newton, in his "mathematical principles of natural philosophy," added the concepts of mass and of force. He introduced the "quantity of motion," which is essentially what we call momentum and, later, such concepts as kinetic and potential energy completed the conceptual basis of mechanics.

These concepts remained for more than a century the basis of exact science as a whole, and their success was so convincing that whenever the phenomena suggested new concepts the scientists tried to follow the tradition and to reduce them to the old ones. The motion of fluids was pictured as motion of the infinitely many smallest parts of the fluid. Their dynamic behavior was successfully treated according to Newton's laws. When in the second half of the 18th century the interest was concentrated on electricity and magnetism, the concept of force was used for describing those phenomena. In this context, force was meant in the sense of mechanics, a force acting instantaneously and depending only on the positions and the velocities of the bodies concerned. To understand the different states and the chemical behavior of matter, Gassendi had revived the idea of its atomic constitution, and his followers used Newtonian mechanics to describe the motion of the atoms and the resulting properties of matter. A beam of light could be considered as consisting either of small, quickly moving particles or of waves; but even the waves would be the waves in some kind of material, and one could hope that finally the smallest parts of this material could be treated according to Newton's laws.

As in the case of the scientific methods, nobody doubted that reduction to the mechanical concepts could finally be effected. But here history decided otherwise. In the 19th century it gradually became clear that electromagnetic phenomena are of a different nature. Faraday introduced the concept of the electromagnetic field, and, after the completion of the theory by

Maxwell, this concept gained more and more reality. The physicists began to understand that a field of force in space and time could be just as real as a position or a velocity of a mass, and that there was no point in considering it as a property of some unseen material called "ether." Here tradition was more a hindrance than a help. Actually it was not before the discovery of relativity that the idea of the ether was really given up, and thereby the hope of reducing electromagnetism to mechanics.

A similar development can be recognized in the theory of heat; but here the alienation from the mechanical concepts could be seen only in rather subtle points. To begin with, everything looked very simple. A piece of matter consists of many atoms or molecules; statistical considerations about the mechanical motions of these many particles should be sufficient to describe the behavior of matter under the influence of heat or chemical changes. The concepts of temperature and of entropy seemed just adequate to get hold of this statistical behavior.

I think it was the American physicist Willard Gibbs who first understood what an abyss had been opened up in physics by these concepts. His idea of the canonical ensemble demonstrates that the word "temperature" characterized our degree of knowledge of the mechanical behavior of the atoms, but not the objective mechanical behavior. The word refers to a certain kind of observation, namely, it requires an exchange of heat between the system and the measuring equipment, the thermometer. It requires a thermodynamic equilibrium. Therefore, if we know the temperature of a system, we cannot know its energy accurately, the inaccuracy depending on the number of degrees of freedom in the system. Of course, tradition worked very strongly against this kind of interpretation, and I believe that the majority of physicists did not accept it until in our century quantum theory was completed. But I would like to mention that when I entered Niels Bohr's institute in Copenhagen in 1924, the first thing Bohr demanded was that I should read the book of Gibbs on thermodynamics. He added that Gibbs had been the only physicist who had really understood statistical thermodynamics.

In the theory of relativity and in quantum theory we had to learn that some of the oldest traditional concepts did not work satisfactorily and had to be replaced by better ones. Space and time are not as independent of each other as Newton had believed; they are related by the Lorentz transformation. The state of a system in quantum mechanics can be characterized mathematically by a vector in a space of many dimensions, and this vector implies statements concerning the statistical behavior of the system under given conditions of observation. An objective description of the system in the traditional sense is impossible.

I have to ask whether tradition has really been only a hindrance in these developments, whether it has just filled the minds of the scientists with prejudices or preconceptions, the removal of which was the most important condition for progress. At this point the problem comes from the word "prejudice." When we speak about our investigations, about the phenomena we are going to study, we need a language, we need words, and the words are the verbal expression of concepts. In the beginning of the investigations, it cannot be avoided that the words are connected with the old concepts; the new ones do not exist yet. Therefore, these so-called prejudices are a necessary part of our language and cannot simply be eliminated.

We learn language by tradition. Traditional concepts form our way of thinking about the problems and determine our questions. When the experiments of Lord Rutherford suggested that the atom consisted of a nucleus surrounded by electrons, one could not help asking: What is the location or the motion of the electrons in these outer parts of the atom? What are the electronic orbits? Or, when one observed events on very distant stars, it was only sensible to ask: Are these events simultaneous or not? To realize that such questions have no meaning is a very difficult and painful process. It should not be belittled by the word "prejudice." Therefore, one may say that in a state of

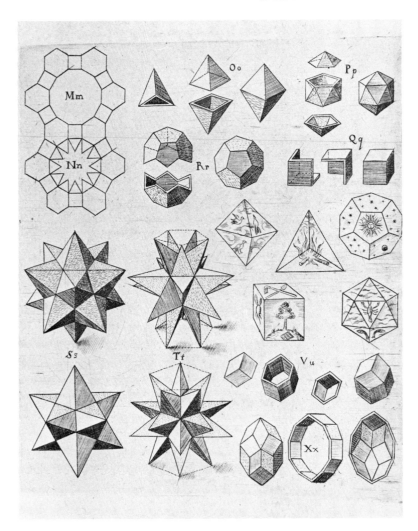

Kepler invented the stellated dodecahedron (*Ss*) and icosahedron (*Tt*). From his *Harmonice mundi* (Linz, 1619). (Permission of Harvard College Library.)

science where fundamental concepts are to be changed, tradition is both the condition for progress and a hindrance. For this reason it usually takes a long time before the new concepts are generally accepted.

Let me finally apply these ideas to the present state of physics. In our time the fundamental structure of matter is one of the central problems, and the concept of the elementary particle has dominated this problem since the time of Democritos. This can be clearly recognized in our pictures and in our questions. A lump of matter consists of molecules; a molecule consists of atoms; an atom consists of a nucleus and electrons; a nucleus consists of protons and neutrons. A proton—well, that could be an elementary particle. But we would term it "elementary" only if it could not be divided again; we would then wish that it would be a point of mass and of charge. A proton has a finite size, however, and can be divided. From a collision between two energetic protons many pieces may emerge. These pieces are not smaller than the proton; they are just particles like the proton. For example, the charge of any object out of the whole spectrum of particles is (if it is not zero) not smaller than that of the proton. So, what we see in such a collision should perhaps not be called a division of the proton; it is the creation of new particles out of the kinetic energy of the colliding protons.

If the proton is not elementary, what does it consist of? Of matter? But matter consists of particles, and so on. We see that we do not get a sensible answer to those questions, which we have asked and do ask, according to the tradition—a tradition going back 2500 years to the time of Democritos. We cannot help asking these questions, however, since our language is bound up with this tradition. We must use words like "divide," or "consist of," or "number of particles," and at the same time we learn from observations that these words have only a very limited applicability.

It is extremely difficult to get away from the tradition. In one of the most recent papers on elementary particles I saw the statement: "From the results of J. D. Bjorkén we can conclude that the proton in its electric properties has a granular structure." It did not occur to the author that such words as "granular structure" have perhaps no other meaning here than just the scaling law of Bjorkén, that they do not carry any further information. Or another example. Many experimental physicists nowadays look for "quark" particles, particles with a charge of one-third or two-thirds of the charge of the proton. I am convinced that the intense search for quarks is caused by the conscious or unconscious hope to find the really elementary particles, the ultimate units of matter. But even if quarks could be found, for all we know they could again be divided into two quarks and one antiquark, etc., and thus they would not be more elementary than a proton. You see how extremely difficult it is to get away from an old tradition.

What is really needed is a change in fundamental concepts. We will have to abandon the philosophy of Democritos and the concept of fundamental elementary particles. We should accept instead the concept of fundamental symmetries, which is a concept out of the philosophy of Plato. Just as Copernicus and Galileo in their method abandoned the descriptive science of Aristotle and turned to the structural science of Plato, so we are probably forced in our concepts to abandon the atomic materialism of Democritos and to turn to the ideas of symmetry in the philosophy of Plato. Again we would return to a very old tradition. As I said before, such changes are extremely difficult. Even with the change many complicated details will have to be worked out, both experimentally and theoretically, in elementary particle physics; but I do not believe that there will be any spectacular breakthrough, except for this change in concepts.

After going through the three most important influences of tradition in science—those in the selection of problems, in the method, and in the concepts—I should perhaps, in conclusion, say a few words about the future development of science. Of course, I am not interested in futurology; but since we can scarcely work on other problems than those that are given to us by the historical process, we may ask where this process has led to new and interesting questions. In physics I would like to mention astrophysics. In this field, the strange properties of the pulsars and the quasars, and perhaps also the gravitational waves, can be considered as a challenge. Then there is the new and wide field of molecular biology, where concepts of very different origin, namely, physical, chemical and biological concepts meet and produce a great wealth of interesting new problems. Finally, on the practical side, we have to solve the very urgent problems put by the deterioration of our environment. I have mentioned these points not in order to make predictions about the future, but in order to emphasize that we need not invent our problems. The scientific tradition, that is, the historical process, gives us many problems and encourages our efforts. And that is a sign for a very healthy state of affairs in science.

Note

[1]C. F. von Weizsäcker, *Die Einheit der Natur* (Munich, 1971), especially Chapter 1.

INTRODUCTION:
DOES SCIENCE HAVE A FUTURE?

Owen Gingerich

If we look back to the science curriculum of the late 1400s, when Copernicus was a student in Cracow, we find astronomy in the forefront—but its picture of the universe strikes us as quaint and outmoded. A weightless firmament, relatively nearby, whirled daily around the solid, central earth; birds and clouds would surely fly off into space if the earth moved, the 15th-century professors declared. Even more curious was the attitude toward celestial observations—systematic records were rarely made, and the often glaring discrepancies with the tables went almost universally unnoticed.

In 1543, the year of Copernicus' death, his great work *The Revolutions of the Heavenly Spheres* was printed. That was a vintage year in scientific publishing: Besides the *Revolutions* and Vesalius' masterfully illustrated volume on human anatomy, Archimedes appeared in print for the first time and Tartaglia produced his brilliant commentary to the first vernacular Euclid. The year marked a watershed in science, for in the century that followed we find emerging the far more familiar ideas that form the foundation of today's science. A great change—indeed, a revolution—took place in that age, with Copernicus at its forefront. Far more than political struggles, with their attendant fleeting changes, the work of Copernicus and his intellectual successors has shaped our contemporary world. Thus, in observing this quinquecentennial, we celebrate far more than the genius of a Renaissance astronomer; indeed, we celebrate the origins of modern science itself.

Consequently, in planning this symposium, we elected to include relatively little about Copernicus himself, but rather, to examine more broadly the nature of scientific discovery and the intellectual conditions that have encouraged scientific progress. Under this rubric, our Copernican cornucopia has poured forth a brilliant variety of insights illuminating different facets of the conditions that promote creative scientific inquiry. On Monday we had our 16th-century day, where we gained rich insights into the Renaissance milieu of which Copernicus was a part. Yesterday Professors Toulmin, Holton, and Heisenberg gave us perceptive reports on the conditions surrounding our present scientific enterprise. And now, on this concluding day of our symposium, we must in a sense look to the future.

In fact, it has become my assignment to ask if science *has* a future. This is a very serious question indeed. Will we have science and scientific discovery as we know it at the 1000th anniversary of Copernicus? Or even a century hence? I am sure that many of you view the growth of science and the understanding of our natural world as an inevitable, unsuppressible force. This idea of progress is deeply rooted in our Judeo-Christian heritage; but it is an idea that has come into ascendancy only in our own culture and within the past few centuries. I must remind you of the narrowness of our data base; for in looking back over the course of human history, we see that scientific discovery has flourished magnificently only in two widely separated epochs—in Greek and Hellenistic times, and again in our own age with Copernicus standing near the beginning of the era—facts that

Professors Toulmin and Heisenberg reminded us of yesterday. Our symposium has provided powerful hints that creative science is a delicate blossom that unfolds only with the proper conditions—open communication, comparative freedom from prejudice, and the support from society that makes inquiry possible.

There are at least two frighteningly pessimistic views concerning the future—or perhaps I should say the nonfuture—of science. It is not necessary to have a cyclic view of history to imagine that man will descend again into the Dark Ages. The means of mass destruction and of overkill hang over us like the Sword of Damocles. The problem of adequate food for a burgeoning world population seems to have retreated in urgency before the spectre of an energy crisis that may curtail the use of our automobiles. We are, in a very real sense, faced with the possible collapse not merely of science, but of society itself.

There are, on the other hand, those who anticipate the demise of science, not out of fear, but out of desire. Repelled by the technological mayhem and exploitation created by man's greed and rapacity, they mistakenly blame science itself for the failure to curb man's selfishness and irrationality. Disillusioned by a technology that all too often fails to meet ever more idealistic standards, alienated by a seemingly narrow and heartless mechanistic philosophy spawned by the Newtonian revolution, and barred from an appreciation of the intrinsic beauty of scientific discovery—and here we scientists have been regrettably negligent—many young people have turned to the occult or to an antiintellectual mysticism in their quest for a meaningful world view. Astrology, called in the 1771 *Encyclopaedia Britannica* "a just subject of contempt and ridicule," now brings the comforting belief to many persons that the stars seem to care in what otherwise seems to be a vast, meaningless, and even capricious universe. Like a small drop of ink that can discolor a whole bucket of water, these minority views, though greatly diluted, can permeate our society and can effectively push science toward the Dark Ages by cutting off both its moral and its economic support.

It seems clear to me that we would not be celebrating this quinquecentennial today if Copernicus had not been economically supported for the many years by the Catholic Church. His position as canon with the Frombork Cathedral chapter gave him the independence and security for his long-range astronomical researches. And the Copernican Revolution would surely have been greatly retarded if King Frederick of Denmark had not offered such generous aid to Tycho Brahe. The Danish astronomer once boasted that his Uraniborg Observatory had cost the king more than a ton of gold. Coming to more modern times, a little calculation shows that the great California telescopes entrepreneured by George Ellery Hale cost Andrew Carnegie about the same amount.

I remember an instance a few years ago when I organized a panel discussion for my natural sciences class at Harvard on "the imperfections of science." One of the participants listed among his bill of imperfections that science was too expensive, to which Philip Morrison, another of the panelists, replied that science was too cheap. I cannot estimate, and I doubt that our next speaker could easily estimate, how much it has cost to discover the nature of quasars. But in a nation that annually spends a billion dollars on pet food and five billion on cosmetics, the price of quasars must be cheap enough.

You will have to excuse the digression to Tycho Brahe and to the quasars, but frankly, I found myself very uncomfortable playing the role of the prophet of gloom. Let me therefore turn to a different group of pundits who also believe that scientific discovery will surely come to an end—I call them the optimists because they believe that everything will be discovered. I have heard one of my esteemed colleagues discuss in all seriousness the Golden Age when the natural world will be fully understood and scholars will be reduced to studying the plots of Japanese novels. It is true that the astonishing progress of scientific discovery in the last five centuries opens the door for considerable optimism. Nevertheless, I worry that our baseline is too small for

The great astronomical entrepreneur George Ellery Hale with his leading benefactor, Andrew Carnegie, in front of the Mt. Wilson 60-inch reflector, then the world's largest telescope. (Courtesy of the Niels Bohr Library, American Institute of Physics.)

an adequate projection. If we were to make a time-lapse motion picture of the history of the world in which we took one photograph per century, the film would run for more than three weeks before *Homo sapiens* would show up on the screen; the entire time that man has existed on earth would pass in about seven minutes; and the age of modern science initiated by Copernicus would be a brilliant spit-second flash at the end of the show.

By even the most modest extrapolation, the amount of scientific knowledge potentially available a hundred years from now boggles the imagination, not to mention the possibilities at the Copernican millenary celebrations. It is fairly clear that we cannot go on doubling the rate of production of scientific papers every 15 years nor can scientists expect to receive an ever larger proportion of the world's total productivity. But to argue from this that we will run out of scientific discoveries appears to me to be unwarranted. Critics of my view, and there will no doubt be many, will probably be annoyed if I quote the opinion of the U.S. Patent Commissioner Henry L. Ellsworth, who, much impressed in 1844 at the rate with which the Patent Office was running out of capacity, stated, "The advancement of the arts taxes our credulity and seems to presage the arrival of that period when human improvement must end."

Likewise, Professor A. A. Michelson's statement in the University of Chicago's catalogue for 1898-1899: "While it is never safe to affirm that the future of Physical Science has no marvels in store even more astonishing than those of the past, it seems probable that most of the grand underlying principles have been firmly established and that further advances are to be sought chiefly in the rigorous application of these principles to all the phenomena which come under our notice. An eminent physicist has remarked that the future truths of Physical Science are to be looked for in the 6th place of decimals."

As an astronomer working during the 1960s, I can attest to the marvelous and generally unexpected discoveries that filled the past decade with such excitement. The quasars, the pulsars, the three-degree blackbody background radiation that seems to be the vestige of the initial explosion of the universe, and the infrared "hot spots" with their attendant clouds of organic molecules —these were hardly glimpsed in 1960. Happily, some of these discoveries have fallen into place with marvelous swiftness, confirming or elaborating views of the universe hitherto only hazily defined. But with the answers came new questions and new puzzles, and I think that it is safe to say that astronomy had more unanswered questions in 1970 than it did in 1960.

I can only suppose that the 1970s will hold as many surprises for astronomers as the previous decade did. X-ray astronomy, now only in its infancy, has provided us with nearly a dozen new classes of objects. Thus, it is my short-range prediction that in 1980 we will have still more unanswered astronomical questions than in 1970.

No doubt those optimists who believe that all scientific discoveries will eventually be made will scorn the timidity of my seven-year projection and say that I have missed the point concerning the inevitable consequences of the accumulation of scientific knowledge centuries hence. Permit me, then, to mention Copernicus and the daring philosophical step he took in moving man's home from the center of the universe. The Copernican Revolution has put puny man in his place, we are told, and has taught him humility. It was, indeed, the first step that has led to the infinite universe, and to man's status as a mere speck on a small planet among billions of stars, billions of galaxies, and uncounted quasars. And yet, some say, this humble creature with his finite brain capacity will know all that is to be discovered about the physical universe. Personally, I cannot imagine a projection more geocentric, anthropocentric, egocentric, or alien to Copernicus' contributions to man's place in the universe.

Consequently, I opt for science with a future, for unlimited scientific discovery, indeed, for the challenge of endless horizons. Nevertheless—here I doff my hat as astronomer and take on the perspective of the historian of science—it seems to me that what constitutes an adequate scientific explanation may be far different in the future than we know it today. Greek science was mostly tied to the ideal of geometry. Arithmetic has gained ascendancy only in comparatively recent times, and to think of physical explanation in terms of equations is a relatively novel phenomenon. I should perhaps remind you that the equals sign was invented only in the generation immediately preceding Galileo. Group theory, a seemingly esoteric and incredibly abstract branch of mathematics, found rapid applications in the explanation of symmetries in the atom and its nucleus, and as Professor Heisenberg told us, this may provide the basis for a new, anti-Democritan concept for understanding the basic nature of matter. Today the advent of high-speed computers has already altered the requirements for an acceptable scientific explanation. Tomorrow, the extension of holography may give a more visual formulation of basic concepts of the universe. Where are the scientific applications of the mathematical logic worked out by such scholars at Kurt Gödel? Can we claim with any confidence that all the branches of mathematics have been discovered? If not, think of the untold new forms that our conceptualization of the universe can take.

The history of science teaches us that what passes for an acceptable explanation at one time, or even what are construed as the pressing questions, can change radically in the next era. As an explanation for the multiplicity of languages, the Tower of Babel will now hardly satisfy us. Aristotle's explanation of gravity in terms of natural rectilinear strivings of matter was no longer suitable to Newton, but his own mathematization of the phenomenon can hardly be considered any closer to an ultimate explanation. Thus, I see not merely an extension of man's knowledge of the universe, but an evolution of his entire mode of assimilating his discoveries. This I feel is inevitable as we take on the increasingly complicated subjects of biology, psychology, and sociology.

I have now perhaps trod far enough on the treacherous

ground of prediction. I was about to say that I was ready to pack up my crystal ball, but the metaphor itself gives me some pause. We can scarcely imagine anything less scientific than gazing for a flickering image in a magic sphere. Yet, as Professor Temkin told us on Monday, the magus played a significant role in the intellectual furnishings of the Renaissance. Today science is far more exclusive. Magic and love, music, poetry, and religion all lie outside its bounds; but that is not to say that science encompasses all that there is to know, for such intimations can only alienate some of the most creative spirits among us. Perhaps by the year 2473 the rather novel word "scientist" will once again have been replaced by "philosopher," and perhaps scientific philosophy will differ from ours in ways more profound than we can now imagine, encompassing some of these other important dimensions of the human experience. With that remark it is indeed time for me to pack up my crystal ball to make way for two of the most distinguished scientific practitioners of our age, men in whose work we can more truly glimpse the outline of the science of the future.

QUASARS AND THE UNIVERSE

Maarten Schmidt

It is my privilege to review our present knowledge of the universe at large as derived from quasars. To supply some continuity with the preceding talks, which dealt mostly with Copernicus' reconstruction of the solar system, I will first briefly discuss the location of the solar system in the galaxy, the interpretation of the "nebulae" as galaxies, and the evidence for the expansion of the universe from the redshifts of galaxies. Then we will turn to the discovery of quasars, their distances, their nature and their remarkable distribution in the universe.

LOCATION OF THE SOLAR SYSTEM

After Copernicus had placed the sun at the center of our planetary system, it replaced the earth as the center of the universe in man's mind. Even when the Milky Way galaxy was subjected to thorough study early in the 20th century this privileged position of the sun was not challenged for several decades. Inspection of the Milky Way on a clear, dark night shows no strong concentration of light in any direction that could be interpreted as its center. Instead there are maxima of brightness in various directions. Robert Trumpler discovered in the late twenties that there is interstellar dust in the plane of the Milky Way. Our view is limited to only a few thousand light years in the plane of the Milky Way, but it is hardly affected at all when we look perpendicular to the Milky Way. The situation may be compared to ground fog or smog, which badly limits horizontal viewing yet allows us to see the stars. Harlow Shapley studied in the twenties the so-called globular star clusters, which are visible well away from the dusty Milky Way band. He found that these clusters were concentrated around a point some 50,000 light years distant in the direction of the constellation Sagittarius.

Bertil Lindblad and Jan Oort showed in 1926 that our local part of the galaxy is rotating about a point in the same direction at a similar distance. It became clear that this point represented the center of the star system which we locally see as the Milky Way. This displaced the sun from its central location in the universe. Instead, the sun is one average star out of some 100 billion that constitute our galaxy.

THE NATURE OF THE NEBULAE

Photographs of the sky outside the Milky Way band show, besides stars, extended nebulous spots that were called "nebulae." Just 53 years ago a debate took place before the National Academy of Sciences between Shapley and Heber Curtis. The subject of the debate was the size of our Galaxy and the nature of the nebulae. Shapley at that time believed that the galaxy had a diameter of about 300,000 light years and that no star systems of similar size existed. Curtis advocated a size of 30,000 light years for our Galaxy and identified the nebulae as similar star systems far outside our Galaxy. Shapley was influenced by the observations of Adriaan van Maanen, subsequently found to be erroneous, that showed motions along the arms of several bright spiral nebulae.

Four years later, Edwin Hubble showed that the nebulae were indeed outside our Galaxy. He observed the Andromeda nebula at Mount Wilson and detected in that nebula stars of variable light. He recognized these stars as Cepheid variables of which the properties were known from studies of nearby examples. In particular, by comparing the apparent brightness of the Cepheids in Andromeda with the absolute luminosity of Cepheid variables, he derived a distance of about one million light years for the Andromeda Cepheids—thus placing the Andromeda nebula far outside our Galaxy. Any remaining doubts that the Andromeda nebula was a galaxy like our own were removed by Walter Baade at Mount Wilson in 1944, when he resolved the central regions of the nebula into individual stars. Modern determinations place the size of our Galaxy at around 100,000 light years, at the geometric mean of those advocated by Shapley and by Curtis, and the distance of the Andromeda galaxy at two million light years.

EXPANSION OF THE UNIVERSE

Spectra of galaxies show that their absorption lines are usually somewhat shifted toward longer wavelengths. This redshift of the lines is attributed to the Doppler effect and interpreted as a velocity of recession. Milton Humason observed many bright galaxies at Mount Wilson and found recession velocities of several hundred, up to a thousand, kilometers per second.

Hubble determined the distances for these galaxies by studies of their Cepheid variables or other bright stars. He announced in 1929 that there was a proportional relationship between the distances and the recession velocities, $v = Hr$, where v is the velocity and r the distance of the galaxy. The proportionality constant H is now called Hubble's constant. Its value has been revised frequently. The most recent determination by Allan Sandage makes it 17 kilometers per second per 1,000,000 light

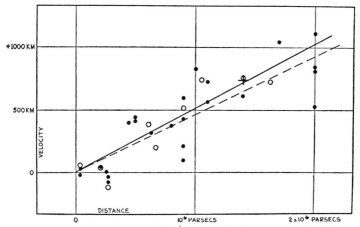

The velocity-distance relation for galaxies as published by Edwin Hubble in *The Realm of the Nebulae* in 1936. The scale of distances of the galaxies has been revised substantially since that time. (Courtesy of Yale University Press.)

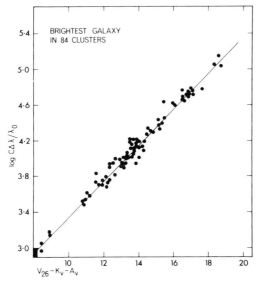

The redshift-magnitude relation for the brightest member of clusters of galaxies. Plotted is the redshift (on a logarithmic scale) versus the apparent magnitude. The tight relation confirms that Hubble's velocity-distance relation is valid on a scale of billions of light years. (From A. Sandage in the *Astrophysical Journal*, vol. 178 (1972), p. 12.)

years. Hubble's law shows that at some time in the past all galaxies must have been very close together. This "age" of the universe is around 12 billion years.

The galaxies used by Hubble in his studies were rather nearby, so that he could study individual objects in them. As a consequence, Hubble's law was derived from a minute fraction of the universe, and one might still argue that the expansion found by Hubble was a local anomaly. Humason was able to determine redshifts for progressively fainter galaxies and managed to obtain a redshift of 20 percent for a faint cluster of galaxies on the basis of observations carried out with the 200-inch Hale telescope at Palomar in 1951. A diagram in which are plotted the redshift of a cluster of galaxies versus the magnitude (observed brightness) of its brightest member galaxy shows a remarkably well-defined relationship with little scatter. This diagram is easily understood if there exists a velocity-distance relationship and if the absolute luminosity of each of these galaxies is essentially the same.

There exists, however, an alternative interpretation of the diagram. The tight correlation could be due to a relation between redshift and absolute luminosity, as might perhaps be caused by a gravitational redshift. In that case the distances of all the clusters of galaxies in the diagram would have to be the same (within 15%) in order to understand the small scatter. Having all observed clusters of galaxies in a thin spherical shell around us with ourselves in the center is clearly pre-Copernican and anthropocentric. In modern astronomy this is immediately rejected and we conclude instead that the clusters of galaxies provide independent evidence that the universe is uniformly expanding on a large scale.

For an individual, faint galaxy no direct distance determinations are possible. Hubble's law provides the only way to derive the distance; hence the redshift is used extensively as a distance indicator in extragalactic astronomy. Its application to quasars will be discussed presently.

THE QUASARS

The Discovery of Quasars. Quasars were initially discovered as radio sources. Hundreds of radio sources were detected by radio astronomers in the early fifties. Accurate positions of these sources became gradually available, allowing in some cases the identification of a radio source with a usually faint galaxy. Many of the radio sources are probably very distant galaxies that are beyond easy reach of the optical telescopes.

The early sixties provided a few cases where the accurate position of the radio source agreed with that of a star rather than a galaxy. These were not accidental misidentifications since the star in each case was unusually blue in color and showed a spectrum with unidentified emission lines, different in each case. The mystery was resolved in 1963 when studies of the spectrum of the bright "star" identified with the radio source 3C 273 showed that it could be interpreted with a redshift of 16 percent, i.e., the emission lines were shifted toward the red by 16 percent of their laboratory wavelengths.

This discovery was remarkable for two reasons. First, finding a large redshift for a "star" was entirely new: The largest redshift or blueshift observed for an ordinary star is 0.2 percent corresponding to a velocity of 600 kilometers per second, close to the escape velocity from our galaxy. The second remarkable feature was that the quasar (as these objects were soon to be called) was so bright: Galaxies with a redshift of 16 percent are not brighter than 17th magnitude while the 3C 273 quasar is 13th magnitude, or 40 times brighter. This showed immediately that the absolute luminosity of the quasar would be around 40 times brighter than even the most luminous galaxies if the quasar redshift was cosmological.

Properties of Quasars. Quasars are defined as star-like objects of which the spectrum shows a substantial redshift. Redshifts are known for some 250 quasars at present. The largest redshift, recently announced by R. F. Carswell and Peter A. Strittmatter at Arizona, is 340 percent. In this object all lines are shifted to wavelengths that are 4.4 times their laboratory wavelengths: Lyman-α, normally at 1216 angstroms, is observed at around 5350 angstroms, in the green part of the visual spectrum. The shift is as large as would be observed for an object receding at 90 percent of the velocity of light.

The quasar 3C 273. The image of 3C 273 is enlarged by overexposure of the photograph; it, in fact, corresponds to the image of a point. The faint linear jet reaches out to 20 seconds of arc from the quasar. (Hale Observatories photograph by H. C. Arp.)

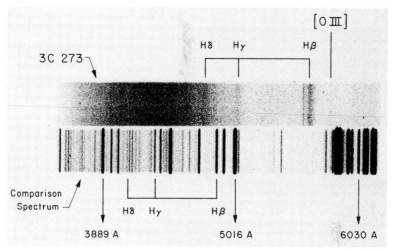

Spectrum of the quasar 3C 273. The lower spectrum consists of hydrogen and helium lines and serves to establish the scale of wavelengths. The upper part is the spectrum of the quasar, a star-like object of magnitude 13. The Balmer lines Hβ, Hγ and Hδ in the quasar spectrum are at longer wavelengths than in the comparison spectrum. The redshift of 16 corresponds to a distance of two billion light years in the expanding universe. (Hale Observatories photograph by M. Schmidt.)

Many of the high-redshift quasars show absorption lines in their spectra in addition to the emission lines. The absorption lines have usually smaller redshifts than the emission lines and sometimes show multiple redshifts. They are probably due to material ejected by the quasar in our direction, although some of the absorption lines might be due to intergalactic gas at a smaller cosmological redshift than that of the quasar.

Almost all quasars show light variations, usually by 20-50 percent (but sometimes much larger), over time intervals as short as a week. As a consequence the variable part of the quasar cannot be more than one light-week in diameter.

Relatively few quasars are observable as radio sources. Their radio properties are almost indistinguishable from those of the radio galaxies. In particular, they often show the remarkable double-component structure, with typical separations of 3-30 seconds of arc, that still remains a puzzling feature of radio galaxies. Most of the quasars have no detectable radio emission and they are found on the basis of their blue color. There are, however, galactic stars of blue color such as the white dwarfs and the final confirmation of a quasar is only obtained from the spectrum. Studies of some sample areas on the sky show that the total number of quasars is between one million and ten million.

Distances of Quasars. If the large redshifts of quasars are a consequence of the expansion of the universe, as are galaxy redshifts, then their distances are very large and hence their absolute luminosity must be very large—in some cases at least a factor of 100 larger than that of the giant galaxies. The observed variations in light suggested that the high luminosity could be produced by active parts in the quasar as small as a light-week. It was this combination of high luminosity and small size that seemed at first baffling, and as a consequence other explanations of the redshift were considered.

An alternative explanation was suggested by James Terrell of Los Alamos in 1964. He proposed that the redshifts were indeed caused by large recession velocities, but that the quasars had very much smaller distances than given by Hubble's law. Quasars would have been ejected in an explosion from the center of our Galaxy relatively recently, some 10 million years ago. All quasar distances would be reduced by a factor of 1000 and the luminosities by a factor of a million. Analysis of this proposal showed that it was impossible for our Galaxy to eject more than one million quasars at speeds that are a large fraction of the velocity of light, because the total energy required for this explosion is more than can be available in the nuclear part of our Galaxy.

Alternatively, quasar redshifts might be of gravitational origin: In strong gravitational fields a photon will lose energy on its way out and this will be observed as a redshift of the spectral lines. A large mass must be concentrated in a very small volume so as to yield a substantial redshift. The problem with this explanation is that some of the emission lines observed in quasar spectra can only be produced in a very tenuous gas, with a density of not more than 10,000 particles per cubic centimeter. At such low density the gas is an inefficient radiator of emission lines and a large volume of gas is needed to produce the observed lines. This is in conflict with the requirement that the quasar be very small so as to produce a large gravitational redshift. This can only be resolved by making the mass of the quasars very large, as much as 100 times the mass of our Galaxy. In that case, however, the Schwarzschild radius of the object is six light years and it is unlikely that variations of light over times as short as a week or a month would be observable.

Both alternative redshift hypotheses lead to very severe problems and the cosmological redshift hypothesis emerged as the most likely one. Support for the cosmological redshift is still scarce. The most impressive evidence is provided by two cases where a quasar, observed in the direction of a cluster of galaxies, has the same redshift as the cluster of galaxies. Also, Jerome Kristian of the Hale Observatories has found that it is possible that quasars are in galaxies by showing that nebulosity near the quasar is present in those cases where it would be expected.

Arguments against the cosmological interpretation of quasar redshifts have accumulated over the past years, particularly through the efforts of Halton Arp and Geoffrey Burbidge. They argue that objects of different redshift are seen associated in the sky and hence must be at the same distance from us, in violation of Hubble's law. It is the reality of the claimed associations in the sky that is at issue here. As an example consider the apparent chain of galaxies VV 172 studied by Wallace Sargent at the Hale Observatories. Four of the galaxies have a radial velocity of 16,000 kilometers per second, but the fifth galaxy has 37,000 kilometers per second. Either the fifth galaxy is a background galaxy at twice the distance of the other four galaxies, or it is at the same distance and violates Hubble's law. Another interesting case is that of NGC 7603 studied by Arp at the Hale Observa-

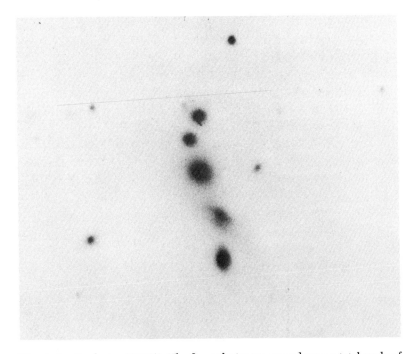

The chain of galaxies, VV 172. The five galaxies are spaced over a total angle of about one minute of arc. Four of the galaxies have a radial velocity of 16,000 km/sec. One galaxy, the second one from the top, has a radial velocity of 37,000 km/sec. If it could be proved that all five galaxies were at the same distance from us, then the second galaxy would violate Hubble's law. No such proof exists at present. (From W. L. W. Sargent in the *Astrophysical Journal (Letters)*, vol. 153, no. 2 (1968), pl. L5.)

tories. The small galaxy appears to be connected to the large one by two spiral arms. The large galaxy has a radial velocity of 8800 kilometers per second, the small one of 16,900 kilometers per second. Hubble's law is violated unless the smaller galaxy is much more distant than the larger galaxy. The apparent connection between the two galaxies argues against different distances. Of course, the small galaxy could be accidentally projected behind the end of the spiral arm.

Clearly, in individual cases such as we have described, no clear decision can be made as to whether the "associated" objects are at the same or at different distances. Only by a statistical evaluation of a substantial number of associations can we ascertain whether their number is larger than would be expected from accidential supposition of foreground and background objects. Such studies can only be undertaken if the "association" between objects is fully described and defined (as to types of objects involved, limits on their brightness and on their angular distance, etc.). It is this required precise description of an association that is missing in most cases. Instead, the statistical studies have often been done on the same objects that led to the discovery of a particular type of association. The resulting high significance of the associations is illusionary.

The debate is continuing. The one side keeps pointing out peculiarities in positions, or distributions of redshifts, etc., that would tend to dismiss the redshift as a distance indicator, and the other side insists that all of these phenomena could be coincidences and that only careful statistical studies can tell. It should be pointed out that if the former side is correct then there must exist an unknown cause of redshift that does not fit in with present-day physics. I find the astronomical arguments for this point of view very weak at present and I see no good reason to doubt the cosmological nature of galaxy and quasar redshifts.

The observations that I would consider most critical for the cosmological redshift hypothesis of quasars are those of variability at radio wavelengths. Radio astronomers are now carrying out interferometric observations with two or more radio telescopes that are separated by thousands of miles, from the United States to Sweden, the U.S.S.R., or Australia. Exceedingly fine structure, on a scale of one-thousandth of a second of arc, has been detected in the radio image of some quasars and galaxies. This fine structure is observed to be different from month to month. At first it was believed that angular motions were observed that, at the large cosmological distances for quasars, would correspond to tangential velocities of up to about ten times the velocity of light. Subsequent observations have shown that the situation is very complicated and it may well be that the radio image of the quasars is flickering, with different parts dominating at different times, without any real angular motions.

The Structure of Quasars. Assuming that the redshifts of quasars are cosmological, what is their structure and their energy mechanism? The diminishingly small angular diameter of the optical quasar allows no direct information to be gained about its structure. The weak nebulosity observed near some quasars is hardly detectable in the glare of the dominating quasar. At radio wavelengths, quasars show either the puzzling two-component structure, or rapid variability on a very small angular scale. None of these structural phenomena has increased our understanding of quasars yet.

The emission-line spectra of the quasars show that there is an atmosphere of gas that must be very tenuous. This gas may well be distributed in filaments, as in the Crab nebula, over a volume of perhaps a few hundred light years diameter. The total mass of the gas is probably around a million solar masses.

Inside this shell of gas is the small mysterious object X. It may have a mass of a few billion sun's masses and is responsible for the copious amounts of energy radiated. Speculative theories have been based on object X being a supermassive star or a compact star cluster. Suggested mechanisms for the production of the energy include nuclear energy, gravitational energy, annihilation of matter and antimatter, and pulsar-like mechanisms.

Space Distribution of Quasars. The distribution of quasars in the universe can be derived from well-studied samples of quasars for which redshifts and hence distances are known. Nearby quasars will appear brighter and hence are more likely to be studied than distant quasars. This selection effect must be accurately removed in the derivation of the space distribution of quasars from the available observations. The results are remarkable: The space density of quasars increases with redshift and hence with distance. The density increase is very large, rising to a factor of between 1000 and 10,000 at a redshift of 2.5.

Since this density increase is seen in all directions, we seem to be located at a position of minimum density of quasars in the universe. Have we again landed in a preferred central position as in pre-Copernican times? The solution of this dilemma is contained in the finite velocity of light. As we are looking out to very distant objects we receive light emitted a long time ago. Let us assume that the expansion of the universe started 12 billion years ago. Then an object of redshift 2.5 is so distant that the light we observe departed 10 billion years ago when the universe was only 2 billion years old. The very large quasar density at redshift 2.5 refers to this early cosmic epoch. Our "preferred" position in the universe is caused by the fact that as we look out to larger distances we are in fact seeing the universe at earlier epochs.

Our picture of quasar evolution is then that there existed some 1000 to 10,000 times as many quasars two billion years after the universal expansion started than at the present time. What about cosmic times even earlier than two billion years? At the corresponding large redshifts of more than 2.5 only the intrinsically brightest quasars would be visible, at 19th or 20th magnitude. Only one or two out of some 25 such faint quasars have observed redshifts larger than 2.5. The number we would have expected ranges from 3 to 10 depending on the exact way in which the number of quasars in the universe decreases with cosmic time (or increases with redshift). Hence, there is only a marginal indication that quasars are rarer at redshifts exceeding 2.5 than we would have expected. Much further observational work on quasars is required until we can establish accurately their behavior at very large redshift.

The similarity between quasars and the active nuclei of Seyfert galaxies was noted by I. S. Shlovskii as early as 1964. It is attractive to think that quasars are, in fact, the birth stage of nuclei of galaxies. The strong concentration of quasars to early cosmic epochs agrees well with the general impression that galaxies were formed soon after the universal expansion started.

Kristian's finding that all quasars may be in galaxies fits in well with this point of view.

If this speculation is correct then we would now be able to study by direct observation the birth process of galaxies up to ten billion years ago in the form of quasars. Quasar astronomy will remain fascinating for many years as it gradually leads us to an increased understanding of the early stages of the universe and of the formation of galaxies.

THE UNIVERSE AS HOME FOR MAN

John Archibald Wheeler

"The last of life, for which the first was made, is yet to come." Can one change these words of Robert Browning from a statement about life to a question about discovery? Are the discoveries from Copernicus to today only the prelude to greater discoveries? Are we ever to clear up the greatest mystery of all, the origin of this universe that is our home?

Hang up a balance with its two pans (figure on p. 263). Into one load all the captured insights of the last five centuries, from the sun-circling earth of Copernicus to the double helix that transmits life. Add to it the choice gold of geology from the time of Hutton and Lyell to the today of continental drift and moon rocks. Pile on what Freud gave us, and psychology and physiology and sociology and all the other sciences. Cap the pile with the most glittering prize of all—the insight that Darwin won for us into the origin of man and life.

Look into the other scalepan, now quite empty. Ask what will

Discussion of this paper begins on page 575.

be laid upon it in the next 500 years. Are the riches that we see and celebrate in the one pan only the prelude to still greater prizes to pile up in the other in the next five centuries? Or are the discoveries yet to come destined to be secondary? Copernicus and Darwin brought us revolutionary insights into man's place in the universe. The quantum principle and relativity, the two overarching principles of 20th-century physics, brought a revolution to our thinking about the nature of matter and energy. Have we seen the last of the great revolutions in our view of man and the universe? If more than three-fourths of all the investigators who ever lived are now alive, are they and their successors to achieve nothing more than extensions and applications of what we already know?

A few weeks ago I put before my sophomore class this question of whether in time to come the now empty scale will outweigh the one filled yesterday. Yes, 31 of them voted; and 4 voted no. Many of these young people are going to commit themselves to one or another branch of science. They will be giving not only university years but the highest hopes of life itself. They do this at a time when a wide choice of career is open to them, and when influential voices are disenchanted with the world of science. What do they see that others do not? For he who believes in discoveries is surely more likely to make a discovery, and he who believes in applications and extensions is more likely to make applications and extensions. What does the student see, or what does anyone see, to encourage him to believe that the greatest discoveries *are* yet to come?

FACTORS MAKING FOR MORE DISCOVERY

Many are impressed with the sheer ascending curve of discovery today. Many are inspired by the new tools we have to see farther than we ever saw before—farther in space, farther into molecular biology, farther in time, farther in the spectrum, farther in the scale of energy. But over and above these indicators of today

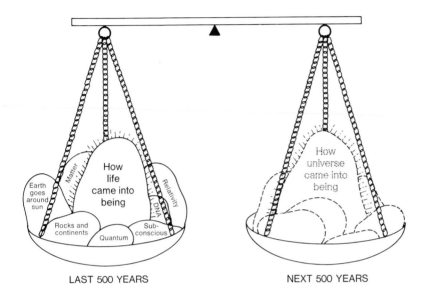

LAST 500 YEARS   NEXT 500 YEARS

Will the discoveries of the next 500 years outweigh those of the past 500 years? Nothing so much argues "yes" as these two facts: (1) Man has proven his ability to answer questions that once seemed unanswerable. "Preposterous" would have been the reaction of anyone before Darwin to the statement that by 1973 one would understand in broad outline how life came into being. (2) How the universe came into being is the greatest question of all, and today, more than ever, calls insistently for an answer.

there stand as lighthouses for the reflective mind the writings of the greatest thinkers of the past, who foreshadowed so much that has been brought to light, and anticipated so much more that has not yet been uncovered.

To assess the prospects of all the branches of knowledge, from biology to psychology, from chemistry to medicine, and from geology to sociology, would be too much to expect of anyone in this age of specialization. It may not be out of place, however, for an investigator to say something about the signs that he sees in the fields closest to his heart that the greatest discoveries are yet to come. Thus limiting myself, let me say a little about the pace of discovery, a little about the tools, a little about the human forces I see propelling us, and then focus on some of the mysteries of the universe in which we make our home.

The Pace of Discovery

Forty years ago we knew three elementary particles. Ten years ago we knew more than a score. Today each new increment of energy gives us new particles. We see no limit. We have even given up using the word "elementary" for particles.

Ten years ago came the first distance determination for a

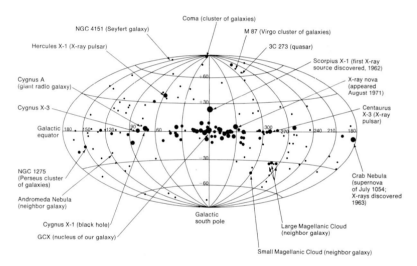

Location of x-ray sources on the celestial sphere. The equator of the diagram follows the Milky Way through the sky. For the most part, objects located near that equator are local to this galaxy; and those well removed from it, located outside the Milky Way. Especially interesting are (1) the Crab Nebula located at the extreme right, most of the x-rays from which originate in the cloud of gas, now three light years in radius, which was ejected at the time of the supernova explosion of July 1054 A.D., and at the center of which is a residual neutron star spinning on its axis 30 times a second ("pulsar"); (2) Hercules X-1 and Centaurus X-3, the only two neutron stars so far discovered that are located in double star systems (see Table 1); and (3) Cygnus X-1, a member of a double star system and generally believed to be a black hole. (Diagram from R. Giacconi, derived from the 3U Catalog of x-ray sources.)

quasi-stellar object by Maarten Schmidt.[1] Today we have scores of these hundred-galactic candlepower objects as beacons to light up the far away and long ago. The year 1954 brought the realization and proof that certain sources of radioemission are as distant as faraway galaxies.[2] In 1956 understanding dawned that such an object has circulating within it high energy particles with the unbelievable energy of 10^{60} ergs—the energy of annihilation of half a million suns.[3]

A decade ago we saw the first x-rays coming from a source in the sky not the sun.[4] Today we know many emitters of x-rays and emitters of many kinds.[5] Some sources are so powerful as to show up even at distances as great as the radio galaxy Virgo A[6] and the even greater distance of the quasi-stellar source 3C 273.[7] Most detectable x-ray sources, however, lie close to the plane of the Milky Way and, like the Crab Nebula, are local to this galaxy (figure on p. 264).

New Tools for the Observer

Above and beyond these sources of radiation are the tools that discovered them and promise new discoveries. The image intensifier multiples the power of the optical telescope.[8] The radio-telescope (figure on p. 266) outdoes the optical telescope in resolving power.[9] It also picks up sources that never are seen and never can be seen by optical means. The x-ray telescope, brought into being by the pioneer work of Herbert Friedman and his collaborators, and impressively developed by them, by the group of Riccardo Giacconi and Herbert Gursky and their collaborators, and by others, increased in power more in one decade than did the optical telescope in two centuries. The infrared telescope has opened up a new domain for astrophysics.[10] Surprisingly, many sources turn out to radiate more powerfully in the infrared than in any other part of the spectrum.

Detectors are already in operation today looking for the two most elusive emissions on the books of physics, neutrinos and gravitational radiation. In 1953 and 1954 Reines and Cowan and their collaborators detected neutrinos for the first time.[11] They came from a nuclear reactor; but sources of neutrinos in space, even explosive sources of gigantic energy, are too far away to trigger a detector of any reasonable size here on earth.[12] The largest neutrino detector ever constructed (figure on p. 267), however (a tank containing 390 cubic meters of perchlorethylene[13]), is believed to be well within an order of magnitude of penetrat-

The world's greatest array of radiotelescopes, located at Westerbork, Netherlands, provides an interferometer to analyze the structure of the radio sources associated with distant galaxies.

The 390 m³ tank of perchorethylene set up in the Homestake Gold Mine in South Dakota by Raymond E. Davis in his continuing search to detect solar neutrinos. (Courtesy of Brookhaven National Laboratory.)

ing to what goes on at the heart of that massive object, our sun. This feat is beyond the power of any other detector of any other radiation.

Gravitational Radiation and Second-Generation Detectors. Gravitational radiation is of no use for exploring the core of the sun or any static star. According to all calculations, however, it will signal to the outside the asymmetric and dynamic collapse of the center of a star, even a distant star. When such a supernova-associated collapse occurs almost anywhere in the Milky Way, it should be detected by a first-generation gravitational wave detector, such as has been constructed by Joseph Weber, the pioneer, and his collaborators and nine other groups in six countries. Whatever may be the geophysical or other explanation of the many events that Weber et al. have reported in the last year,[14] two points stand out: (1) In this time no optical telescope has seen any supernova anywhere in the Milky Way. (2) No other group has found evidence on their detectors for anything clearly above noise.

Fortunately second-generation devices[15] of far greater detecting power are now under construction at Stanford, Baton Rouge, and Rome. Ultimately such equipment should be able to detect of the order of one event a month from star collapses in our Milky Way and in nearby galaxies. The future of gravitational wave astronomy appears bright. It looks at the dynamic interior of a collapsing object. Little does it care that surrounding clouds of obscuring matter block all traditional telescopes.

New technology in other fields of knowledge gives hope there too, our colleagues tell us, of dramatic new advances. It helps in thinking of the future to think of the past, and of Joseph Henry, who in his pioneer experiments in electromagnetism had to hand wrap his wires with cloth. Insulated wire was not for sale on any market. Half a century later there was nowhere one could turn to buy a vacuum pump. These circumstances of the past remind us afresh how new science leads to new tools, and new tools lead to new science.

The Force Structure of Science

In addition, science more insistently today than ever demands answers to its questions. It pushes the investigator through a highly effective "force structure" (figure on p. 369). The central force-transmitting element is the association of like-minded investigators, the "collegium." This association is as much a source of guidance, pressure to create, and moral support for the individual worker as it ever was; but today it has become more, an instrument of society,[16] both responding to needs and pointing out possibilities.

MYSTERIES FORESHADOWING GREATER DISCOVERIES

We can believe that the discoveries of the future will outnumber those of the past, because of the pace in today's science, the tools, and the pressure; but will they be greater? Nothing so much encourages the answer "yes" as the mysteries encountered wher-

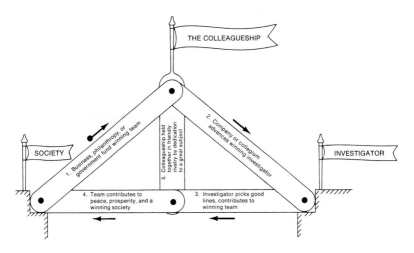

The force structure of science. Society pressures the investigator to find out. This "force for forwarding," like the bridge above, collapses when any one of its five struts weakens. Fortunately, never has this bridge stood stronger than it does in advanced countries today.

ever we turn. Each one of us has his own catalogue of the great unknowns. Let me beg your indulgence to mention three at the top of my own list: the mind, the universe, and the quantum. I know no area where the mystery is greater than it is in these three fields, or where the linkage between observer and system observed is stranger, or any more suggestive of things hidden beyond present imagination.

The Mystery of the Mechanism of the Mind

Wilder Penfield has told us in earlier years[17] of the patient lying on the operating table with brain case open, following past and present in an amazing stream of double consciousness, a consciousness moreover strangely observing consciousness. Today no mystery more attracts the minds of many distinguished pioneers from the field of molecular biology than the mechanism of brain action. Participating in the exploration are workers from fields as far removed from one another as neurophysiology, anatomy, chemistry, circuit theory, and mathematical logic. Many feel that the decisive step forward is waiting for an idea, an as-yet-undiscovered concept, a central theme and thesis. Whatever it will prove to be, we can believe that it will somehow touch the tie between mind and matter, between observer and observed.

Mind and Universe

The brain is small. The universe is large. In what way, if any, is it, the observed, affected by man, the observer? Is the universe deprived of all meaningful existence in the absence of mind? Is it governed in its structure by the requirement that it give birth to life and consciousness? Or is man merely an unimportant speck of dust in a remote corner of space? In brief, are life and mind irrelevant to the structure of the universe—or are they central to it? Lack of conclusive evidence on so cosmic an issue suggests that something is still to be learned about how the universe came into being.

On the subsequent dynamics of the universe we know more. Table I summarizes some of the large scale features of the universe, "the home of man," as we envisage them today. In arriving at this picture one combines the astrophysical observations with Einstein's standard 1915 geometric theory of gravity,[18] otherwise known as general relativity. That theory includes not only his dynamic equation for the change of geometry with time, but also a point he viewed as too central even to require statement in equation form: the universe must be closed. A spherical model universe starts with a big bang, expands, reaches a maximum radius, recontracts, and undergoes a final catastrophic collapse (Table 1). This sphericity is argued by two considerations: (1) As seen from the earth, the distribution of galaxies (averaged over the scale of many clusters of galaxies) is nearly isotropic in direction. (2) According to the standard "Copernican assumption," we occupy no special location. Therefore, the same isotropy should obtain in all other locations, but no system can be everywhere isotropic without being homogeneous. Departures from

TABLE 1.

A typical cosmological model.

Time from start to now	10×10^9 yr
Hubble time now	20×10^9 yr
Hubble expansion rate now	$49.0 \dfrac{\text{km/sec}}{\text{megaparsec}}$
Rate of increase of radius now	0.66 lyr/yr
Radius now	13.19×10^9 lyr
Radius at maximum	18.94×10^9 lyr
Time, start to end	59.52×10^9 yr
Density now	14.8×10^{-30} g/cm^3
Amount of matter	5.68×10^{56} g
Equivalent number of baryons	3.39×10^{80}

These illustrative values are all derived via Einstein's theory from two key astrophysical data, each believed to be uncertain by an amount of 20 percent: (1) the actual time, $\sim 10 \times 10^9$ yr, back to the start of the expansion, as determined from the evolution of the stars and the elements; and (2) the "Hubble time" or time linearly extrapolated back to the start of the expansion, $\sim 20 \times 10^9$ yr; that is, the time that would have been taken by galaxies to get to their present distances if they had always been receding from us at their present velocities.

uniformity complicate the mathematics but do not save the system from singularity.[19]

The First Cycle of Doubt and Test of Geometrodynamics. The fantastic prediction that the universe is dynamic forced itself on Einstein in 1915 as a by-product of his geometric account of gravitation. No consequence could have seemed more unwelcome. To him the belief was as strong as to others at his time that the universe endures from everlasting to everlasting. Something had to be wrong with his great geometrodynamic equation. He reviewed the derivation. The original arguments of correspondence with Newtonian physics and of simplicity were so compelling that there was no natural way to change the equation. Therefore, he changed it in the least unnatural way that he could find: he introduced into the equation a so-called cosmological term. Its sole purpose was to prevent the expansion. Then in 1929, 14 years later, came Edwin Hubble at the Mt. Wilson telescope and the conclusion that the universe actually is expanding (figure below). Thereupon, Einstein rejected the cosmological term, calling it "the biggest blunder of my life."

Nothing gives one more faith that we will someday understand the mystery of creation than the ability of the human mind to predict, and predict correctly, and predict against all expectation, so fantastic a phenomenon as the expansion of the universe. The astrophysical evidence makes it hardly possible to doubt that this expansion has slowed, as predicted. It has been only ten billion years or so since all physics began, as we know from the abundance of the elements and the evolution of star clusters and of stars themselves.[20] Starting with the big bang, the galaxies had to get to their present distances in this short time. Yet moving away at their presently observed and slowed-down rates of recession, they would have taken 20 billion years, twice as long as the available time, to get to where we find them. So they must have had higher recession velocities in the past.

The Second Cycle: The Expansion is Slowing. The cosmological term marked one cycle of doubt and test of relativity. A second cycle of doubt and test came when the Hubble time (the

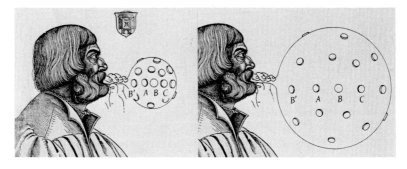

Expansion of the universe symbolized by expansion of a rubber balloon with pennies affixed to its surface to represent the galaxies. In the expansion no atom expands, no meter stick expands, no solar system expands, and no galaxy expands. Rather the distance from galaxy to galaxy increases. Each galaxy *A* may well consider itself to be the center of the expansion because the distance from *A* to *C* increases twice as fast as the distance from *A* to *B*, when *C* is twice as remote from *A* as *B* (Hubble's relation between distance and recession velocity).

time today determined to be 20 billion years plus or minus 20 percent or so), by reason of mistaken distance determinations, came out to be only two or three billion years—an apparent sign that the expansion had been speeding up, rather than slowing down, as predicted. This was the era of theories of a "steady state universe," and of a "continuous creation of matter," theories contradictory in spirit to general relativity and now ruled out by the newer determinations of galactic distances and by direct microwave radio evidence for the big bang.[21]

The Third Cycle: The Missing Matter. Underway today is a third cycle of doubt and test of Einstein's theory. From the two figures of ~10 billion years (from start to now) and ~20 billion years (to expand to present separations at present rates), and from these two figures alone, one can predict (Table 1) what should be the average density of matter in space. Determinations of the masses of galaxies revealed only of the order of one-thirtieth of this amount.[22] This discrepancy gives rise to "the mystery of the missing mass" and a continuing search for unrecognized sources of mass. Investigators today ask, "How much hydrogen floats in the space between the galaxies?" and, "How many dark stars are there?" This Copernican year has seen a decisive step forward. Ostriker and Peebles conclude[23] that the typical galaxy contains 3 to 20 times as much matter as previously believed, most of it in dark matter, outside the distance where the galaxy had previously been thought to end. The mystery of the missing mass and the resulting third cycle of doubt and test, far from being a discomfiture to relativity, would seem to be an inspiration to astrophysics.

The Paradox of Collapse and the Black Hole. Expansion, and later attainment of a maximum volume, according to Einstein's theory, are prelude to recontraction and complete gravitational collapse. Collapse confronts us with what in my view is the greatest crisis in the history of physics, because in it we hear physics apparently saying that physics must come to an end.

Physics stops, but physics must go on; that is the paradox. We do not know the way out. We get directly opposed signals. Well established and never contradicted theory tells us that a time must arrive when physical law ends: There is no law of physics that does not demand space and time for its statement. With the collapse of space and the end of time, the very framework for every law of physics therefore also collapses.[24] On the other hand physics has always meant that which goes on its eternal way despite all the surface changes in the appearance of reality. Physics cannot stop—but physics stops! How are we ever to work our way through this paradox? No cheap way out has ever been found. One can believe that a revolution lies in wait, rich with new understanding.

"There is no hope of advance in science without a paradox." No one who agrees with Bohr on this point can fail to reflect again and again on the paradox of collapse. Three thoughts come to the fore as one reflects. First, every black hole that we can observe will provide us with a kind of "observational model" for the collapse of the universe. Second, in collapse long-established conservation laws of physics are one by one transcended. Third, one can imagine other cycles of the universe, endowed with other values of the physical constants; but even minor changes in some of these constants would rule out life as we know it, and quite possibly any life at all.

No one believes more the possibility of a big disaster than the man who has seen a little disaster. It is reasonable, therefore, to test for gravitational collapse of the universe, later, by looking for a gravitationally collapsed star, now.

A star is predicted to undergo gravitational collapse when in the course of its normal astrophysical evolution it arrives at an interior density so great that the normal "elastic" forces of the medium are no longer able to sustain it against the pull of gravity.[25] It collapses, at first slowly, then more and more rapidly. Two fates we know for this infall. One is a neutron star, a kind of way station on the road to complete collapse. A neutron star is as dense as the nuclear heart of an atom. It measures 10 or 30 or 100 kilometers across, the dimensions of a town, but it has the mass of a sun. The other fate is complete gravitational collapse, disappearance from view, formation of a back hole.

The "Horizon" of a Black Hole. The cat in *Alice's Adventures in Wonderland* disappeared from view, all except its grin. The infalling furnace of gas leaves the world of the touchable, all except its gravitational pull. What happened to the completely collapsed matter? That mystery is hidden from sight by an intangible boundary, a horizon. From within, no photon has the power to climb out and into view against the pull of gravitation. The distance around this horizon is only a few miles, roughly ten times smaller than the distance around the equatorial belt of a neutron star.

It "would not, in consequence of its attraction, allow any of its rays to arrive at us; it is therefore possible that the largest luminous bodies in the universe may, through this cause, be invisible." These are the words of Pierre Simon Laplace in 1795. His was the first prediction of a black hole.[26] The idea was clear to him. A rocket of low velocity can escape from the moon. It takes a higher velocity to break away from the earth, and a still higher velocity to get free of the pull of the sun. If the sun were much denser and more compact, however, or (with its present density) much more massive, a rocket trying to escape and moving even with the speed of light would never make it.

The first treatment of a black hole within the framework of Einstein's geometric theory of gravity came within our time.[27] The first impressive observational evidence for one came still more recently, in 1972–1973, from x-ray astronomy. Leading to this evidence is a fascinating trail.

The year 1932 saw the discovery of the neutron. The next year, on 15 December, trying to understand how a star can flare up into a supernova, Walter Baade and Fritz Zwicky of Pasadena proposed[28] "that supernovae represent the transitions from ordinary stars into *neutron stars*, which in their final stages consist of extremely closely packed neutrons." Baade and Zwicky thought of the center of the Crab Nebula, the center of the debris from the supernova of A.D. July 1054, as being one of the best places to look for such a neutron star. From that point 34 years later, Comella, Craft, Lovelace, Sutton, and Tyler[29] found that periodic radio pulses were emitted 30 times a second. The way for this finding was paved by the 1968 discovery of the first pulsars by Hewish, Bell, Pilkington, Scott, and Collins.[30] In 1969, Cocke, Disney, and Taylor[31] and Lynds, Maran, and Trumbo[32] found flashes of visible light 30 times a second. Today, no one sees any reason to doubt the identification of the source as a neutron star that, like the earth, has an obliquely oriented magnetic field and whirls about its axis today, only 900 years after the time of formation, at 30 turns a second.

Today we know of more than 100 of these "pulsars" or neutron stars; however, they have presented a continuing mystery. Why do they always occur singly? We know that roughly half of all the stars in the skies are in the married state. Why then do we not see any neutron star that is companion to a normal star? We now do! The last two years have brought two cases (Table 2) of a neutron star in a double star system. The flashing, excited in the gas streaming in from the normal star by the rapid rotation of the highly magnetized neutron star, comes (in these cases) not in the radio spectrum (where one had been looking with radiotelescopes), but in the x-ray spectrum, brought within our reach by x-ray telescopes of new power in orbit above the earth's atmosphere.

Cygnus X-1 as First Black Hole. The greatest event in x-ray astronomy in 1973 was the discovery by R. Giaconni and his collaborators of evidence that the object Cygnus X-1 (one member of a double star system) is a compact x-ray source and may be a black hole.[33] The optical component of the pair moves back and forth a distance of 5.2×10^6 km in the line of sight with a 5.6-day period. Its mass, from two lines of evidence (spectral character and absolute luminosity), is concluded to be of the order of 25 solar masses. The invisible component, in order to swing by its

TABLE 2.

Two cases of a neutron star in a double star
system, observed as an x-ray pulsar.
(M⊙ stands for the mass of the sun, 1.989 × 10³³ g.)

Name of X-ray Source	Cen X-3	Her X-1
Optical partner	Not yet known	HZ Her
Distance	Not yet known	8000 to 15000 lyr
Radio identification?	Not yet	Not yet
X-ray source periodically eclipsed?	Yes	Yes
Period of eclipses	2.1 days	1.7 days
Time totally eclipsed	0.488 day	0.24 day
Time from full to zero intensity	0.035 day	$< \sim$10 min
Average period of pulses	4.8 sec	1.24 sec
Orbital velocity of x-ray source in line of sight as deduced from Doppler shift of pulse period from its average velocity	415 km/sec	169 km/sec
Motion of x-ray source in line of sight from center of mass of double star system	$\pm 1.191 \times 10^{12}$ cm	$\pm 3.95 \times 10^{11}$ cm
Mass of optical component	$\geqq$15 M⊙	<2.5 M⊙
Mass of x-ray component (neutron star)	~0.2 to 0.3 M⊙	<1.3 M⊙

Table data from summary by R. Giacconi, *Physics Today*, vol. 26, no. 5 (May 1973), pp. 38-47; mass data from R. W. Leach and R. Ruffini, *Astrophysical Journal (Letters)*, vol. 180 (1973), pp. L15-L18.

gravitational pull so big a visible mass back and forth so great a distance and so quickly, has to be of the order of ten solar masses, and certainly greater than five solar masses, one reasons. An ordinary star of this mass would be quite visible in the optical, quite invisible in the x-ray spectrum. This is not an ordinary star. Moreover, it is too heavy to be a white dwarf or a neutron star. No one sees any natural and reasonable interpretation for it except as a black hole.

X-radiation does not come out of a black hole. It comes out of gas on its way towards the black hole from the normal star. Gas is drawn in towards the compact component by its powerful gravitational attraction. In the ensuing "traffic jam" it is compressed and heated. It rises to temperatures calculated by Ya. B. Zel'dovich and I. D. Novikov to be of the order of 10^{10} to 10^{11} degrees Kelvin.[34] The temperature is so high that the gas cannot avoid emitting x-rays before it reaches the horizon of the black hole.

We can well believe that Cygnus X-1 is not the only black hole to be made into an x-ray source by infall of matter from a normal star companion. We already know five more x-ray sources that are possible candidates for black holes in double star systems. The black hole seems to have become an inescapable component of present day astrophysics.

Evidence that a "Black Hole Has No Hair."—Make a black hole by dropping in objects and particles and radiations of the greatest variety (figure on p. 279). All details of what goes in get washed out in the resultant object. It is characterized by mass and electric charge and angular momentum, but by no other adjustable parameter, no other lines of force, no hills and valleys, no other particularities, so far as we can judge.

This conclusion that in this sense "a black hole has no hair" is quite strange from the point of view of elementary particle physics. There, many a test, many a calculation, and many an experiment have left us no choice except to say that the net number of heavy particles or "baryons" (neutrons, protons, etc.) emerging from a collision or a radioactive decay or any other process is equal to the net number entering the process. We have also a similar law for the conservation of the number of light particles or "leptons" (neutrinos and electrons) in any interaction, however violent. Now make two black holes, one by dropping in baryons and leptons, the other by pouring in electromagnetic radiation; tailor them to have identical mass, electric charge, and angular momentum. Then there is no way to distinguish the one black hole in any way whatsoever from the other; no experiment has ever been proposed; no method or principle has ever been conceived. In this sense all memory is lost of anything that might be called the "particle content" of a black hole. The law of con-

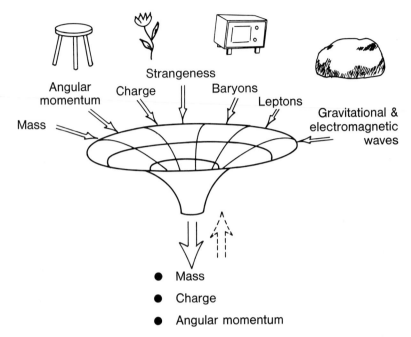

The black hole as "eraser." Whatever the object or particle or radiation dropped into the black hole, it leaves not a single trace (at the *classical* level of observation) behind of any particularity it carried in, except through the addition it makes to the mass, charge, and angular momentum of the black hole, according to the best presently available reasoning.

servation of particles is not "violated"; rather, in this context, it loses all usefulness and meaning. It is *transcended*.[35]

All of these conclusions depend upon an analysis of black holes in the context of classical physics. In the context of quantum physics, according to Jacob Bekenstein and Stephen W. Hawking, a black hole of solar mass has an effective temperature of the order of 10^{-7} degree Kelvin. Therefore it is not absolutely black, but emits a fantastically weak thermal radiation. It could be conceived that this thermal radiation, which includes neutrinos and gravitons as well as photons, might bear some imprint of the baryon and lepton number before collapse, but at the *classical* level of observation one must still conclude that baryon number and lepton number are transcended; that "a black hole has no hair."

"Mutability" as Lesson of the Black Hole. "Number of particles" loses all meaning in the collapse at the center of a black hole. Therefore, no one sees how "particle number" can have any meaning for the universe itself either in the collapse at the end of time or in the big bang at the beginning.

Put this startling "lesson of the black hole" in perspective. Before general relativity, one thought of the universe as eternal. Einstein's theory tells us that it is mortal. Before gravitational collapse and black hole physics, one thought of a baryon or a lepton as immutable. The universe might be dynamic but at least, one thought, such a central property as net particle number must be preserved. Today mutability looks more and more like being the universal feature of nature, showing up at level after level of structure.[36]

The wood in my desk is a "fossil" from a photochemical reaction in some tree 20 years ago. Heat it, not very much, and its entire molecular constitution will change. The nuclei of iron in my watchband were created in a thermonuclear reaction in a star some billions of years ago at temperatures of 10 to 20 million degrees. It too is a fossil. Put it into another star and cook it more to change it into a new element. Today we find ourselves challenged to think of the "elementary particles" that make up these molecules and nuclei as also fossils, fossils likewise subject to reprocessing, fossils from the most extreme conditions of all, at the big bang itself.

What strange picture of physics are we coming to? In time past, when asked for an example of a law, a law that is immutable, and always obeyed, we have been accustomed to cite some familiar law of physics. Or when desiring to give a woman a token of eternal fidelity, we present her with a diamond, with its crystal planes at angles that are forever fixed; but heat that diamond up to conditions hot enough and the crystal dissolves into gas. The crystal planes disappear. Every law of physics, we think today goes back in one way or another to some symmetry of nature; but these symmetries derive from forces that in their turn can be overcome by still greater forces.

Step by step workers of the past have climbed the staircase of law, and law transcended (figure on p. 282). Each new level of law has an enlarged domain of supremacy and demands conditions more extreme than ever to overpower it; but each time a way is found to supply these still more formidable circumstances. At the top of the stairs lie the most extreme conditions of all, gravitational collapse, and collapse not only of the star, but of the universe itself.

Collapse is dissolution. There is not one law of physics that does not require spacetime for its statement. With the collapse of the universe the framework falls down for everything one has ever called a law. At the end, so far as we can see today, there is no law left—except the "law of mutability": there is nothing that does not change. And as at the end, so it must be at the big bang beginning. There is dissolution at one end of time, "coming into being" at the other.

Life and Mind and the Structure of the Universe. If I said that mutability is the greatest lesson emerging from collapse, and the mechanism of "coming into being" and dissolution is the greatest outstanding problem, I would be misrepresenting the situation, because I have not spelled out the greatest issue of all: What role, if any, does any future requirement for life and mind play in the structure of the universe?

When one gives a number for an answer to any question, it is agreed in science today, one is not doing enough. One must specify in addition the limits of error of that answer. On no question do these limits of error encompass a wider range of uncertainty than on the importance of life and mind for the constitution of the world: Zero? Or everything?

Philosophy debated this issue before it began to come within the purview of science. Parmenides, great forerunner of Socrates,

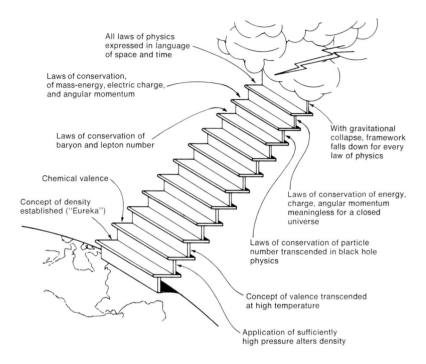

The staircase of law, and law transcended. Archimedes, rushing out of his bath, dramatized the concept of density; but later work showed that any law of conservation of density is transcended by the application of sufficiently great pressure. The concept of valence won successes from the moment of its introduction in 1868; but in time temperatures could be raised so high that the very idea of valence lost its applicability. The principle of the fixity of nuclear charge and mass, cornerstone of solid state physics and chemistry under ordinary conditions, is transcended in thermonuclear reactions. The principle of the conservation of lepton number and baryon number has been discovered to be central to elementary particle physics, but it is transcended in black hole physics. The black hole, like other systems, is guided in its dynamics by the laws of conservation of mass-energy, electric charge, and angular momentum. But none of these three would-be conserved quantities has any meaning when one turns from a finite system to a closed universe. There is no platform on which to stand to make the measurement. There is no place "outside" in which to put a planet in orbit about a closed universe in the fruitless effort to give a meaning to "total mass." With the collapse of the universe, space and time themselves are transcended as categories, and the framework collapses for everything one has ever called a law.

Plato, and Aristotle, taught—to put it in the words of today—that "What is . . . is identical with the thought that recognizes it";[37] or, more briefly, "No one to think, no one to know? Then no world!" In the time of Newton, George Berkeley spoke to the effect[38] that, "No object exists apart from mind." The issue about the role of the knower has been taken up with new life in the past dozen years, and already in part translated into quantitative terms, by workers in cosmological observation, theory, and experiment. The concept of mutability furnishes a key to these endeavors.

Mutability implies adjustability. Ask then this question: Is the "initial adjustment of the structure of the universe" made in such a way as to render possible the existence of the knower? If so, this is a testable prediction, in the sense that it exposes itself in principle to destruction in a dozen ways. To make the test, examine the effect of a change in the original adjustment and see if it would rule out life.

Table 3 examines three conceivable readjustments in the physics of the universe. Each turns out, according to present reasoning, to exclude life as we know it, and even to exclude varieties of life reasonably far from anything we know. There will be more tests in days to come. No longer is it possible to throw the question out as meaningless, though it is stranger than science has ever met before: Has the universe had to adapt itself from its earliest days to the future requirements for life and mind?

Until we understand which way truth lies in this domain, we can very well agree that we do not know the *first* thing about the universe.

One view has it that as we keep on investigating matter, we will work down from crystals to molecules, from molecules to atoms, from atoms to particles, from particles to quarks, and so on, and mine to forever greater depths. A very different concept might be called the "Leibniz logic loop." On this view the analysis of the physical world, pursued to sufficient depth, will lead

TABLE 3.

Comparison between this universe and a hypothetical
universe that has been readjusted in one or another way,
in size or structure, to learn what kind of a universe it
takes to allow life and mind to develop.

Readjustment	Consequence

INITIAL CONDITIONS OF EXISTENCE

Give up the large scale homogeneity of the universe.

> Large portions of the universe so dense that they would already have undergone gravitational collapse; other portions so thin that they would not be able to give birth to galaxies and stars. (For more detail, see C. B. Collins and S. W. Hawking.[1])

Make a substantial reduction in the number of particles in the universe ($\sim 10^{80}$) and the general scale of the universe at maximum expansion ($\sim 10^{40} \times$ "classical electron radius").

> No life (R. H. Dicke[2]). [Dicke's reasoning in brief: No mechanism for life has ever been conceived (G. Wald[3]) that does not require elements heavier than hydrogen. The production of heavier elements from hydrogen demands thermonuclear combustion. Thermonuclear combustion requires several billion years of cooking time in the interior of a star, but no universe can provide several billion years of time, according to general relativity, unless it is several billion light years in extent. So why, on the "no-knower-no-world" view, is the universe as big as it is? Because we are here!]

Make a huge reduction in the mass of the universe. [Argue the ridiculous disproportion that would appear between the agency employed, a universe with of the order of 10^{22} stars, and the "goal" achieved, consciousness of that universe on one planet of one star. Cut back the amount of matter invested in the universe. Try to get better "cost effectiveness," but be moderate. Do not cut by a factor as extreme as 10^{22}. Reduce the number of particles by the lesser trial factor of 10^{11}. The resulting universe still contains enough matter to make one galaxy, or 10^{11} stars. Why is that not more than enough?]

The universe runs through its entire cycle of expansion and recontraction in a time only of the order of one year. There is no opportunity to form stars, let alone heavy elements, planets, and life. [The cutdown investment in original matter, far from giving a better return, gives no return at all. From this point of view any purported extravagance in our universe is far from obvious.]

Other changes in initial conditions.

No analysis of consequences available.

DIMENSIONLESS CONSTANTS OF PHYSICS

Readjustments of the dimensionless ratios of the volumes of the various atoms and their chemical compounds.

Impossible. All of these volumes are derived from one and the same standard basic dimension, the Bohr radius, $\hbar^2/me^2 = 0.529 \times 10^{-8}$ cm.

Readjust the dimensionless ratio known as the reciprocal fine-structure constant,

$$\frac{\left(\begin{array}{c} quantum\ of\ angu\text{-} \\ lar\ momentum \end{array}\right)\left(\begin{array}{c} speed \\ of\ light \end{array}\right)}{\left(\begin{array}{c} elementary\ unit \\ of\ electric\ charge \end{array}\right)^2} = \hbar c/e^2 = 137.039.$$

All of the stars would be red stars if the reciprocal fine-structure constant were changed a few percent in one direction; blue stars, if it were changed by a few percent in the other direction, Brandon Carter[4] reasons. In neither case would a star like the sun be possible.

Readjustment of other dimensionless constants of physics.

No analysis of consequences available. The prime difficulty is ignorance of which constants are the independently adjustable "handles" and which constants are secondary to and derived from these primary constants.

LAWS OF PHYSICS

Readjust the laws of physics themselves, in such a way for example that there are no particles; or so that there are particles but these particles are not identical.

> Unknown. [The idea of mutability is new to us. We may speak of the laws of physics as one by one transcended in collapse, or as one by one coming into force in the big bang. We do not really know, however, what it would mean to readjust one or another law of physics. Still less do we know the consequences of any change for the possibilities of life and mind. For example, how could life exist if there were no particles?]

---

[1] C. B. Collins and S. W. Hawking, *Astrophysical Journal*, vol. 180 (1973), pp. 317-334.

[2] R. H. Dicke, *Nature*, vol. 192 (1961), pp. 440-441.

[3] G. Wald, "The Origin of Life," *Scientific American*, vol. 191, no. 2 (August 1954), pp. 44-53; also in *Proceedings of the National Academy of Sciences*, vol. 52 (1964), pp. 595-611; and pp. 185-192 in *Frontiers of Modern Biology* (Boston, 1962). See also G. Wald, pp. 127-142 in M. Kasha and B. Pullman, editors, *Horizons in Biochemistry* (New York, 1962), and pp. 12-51 in A. I. Oparin, editor, *Evolutionary Biochemistry, Proceedings of the Vth International Congress on Biochemistry, Moscow, 1961* (London, 1963). See also L. J. Henderson, *The Fitness of the Environment* (New York, 1913).

[4] B. Carter, "Large Numbers in Astrophysics and Cosmology," unpublished preprint (1968); an updated version was reported at the International Astronomical Union Symposium, No. 64, Cracow, Poland, September 1973. Carter's detailed and thoughtful analysis has not yet been probed and reprobed to sufficient depth by others to establish his result about "137" as a battle-tested conclusion.

back in some now hidden way to man himself, to conscious mind, tied unexpectedly through the very acts of observation and participation to partnership in the foundation of the universe. To write off the power of observation and reason to make headway with this question would seem to fly against experience.

The World of the Quantum

No more hopeful sign do I see that we can and will make our way into this unknown land than the immense progress we have already made into the world of the quantum, where the observer and the observed turned out to have a tight and totally unexpected linkage.

The quantum principle has demolished the view we once had that the universe sits safely "out there," that we can observe what goes on in it from behind a foot thick slab of plate glass without ourselves being involved in what goes on. We have learned that to observe even so miniscule an object as an electron we have to shatter that slab of glass. We have to reach out and insert a measuring device. We can put in a device to measure position or we can insert a device to measure momentum, but the installation of the one prevents the insertion of the other. We ourselves have to decide which it is that we will do. Whichever it is, it has an unpredictable effect on the future of that electron. To that degree the future of the universe is changed. We changed it. We have to cross out that old word "observer" and replace it by the new word "participator." In some strange sense the quantum principle tells us that we are dealing with a participatory universe.

Quantum Principle as "Merlin Principle." Quantum principle? What an inadequate name for an overarching feature, or *the* overarching feature, of nature. We understand any other principle of physics in enough completeness to summarize it, beginning with a good name, in a dozen words—but not this. It continually unfolds with fresh meaning. It might almost better be called the "Merlin principle." Merlin the magician, pursued, changed first to a fox, then to a rabbit, and finally to a bird fluttering on one's shoulder. The "Merlin principle" has put on a new face in each decade of its history, beginning with Planck's original discovery in 1900 (Table 4). This principle finds its most sophisticated expression in our times in terms of a "lattice of propositions," in the language of the calculus of logical propositions, central structure of mathematics. Is mathematical logic the last change of clothes before we find the quantum bird fluttering on our shoulder? Are those two words, "mathematical logic," a

TABLE 4.

Quantum principle as Merlin principle. (Each new discovery about the quantum principle, each additional way of expressing its contents, seems almost like one of the changes of clothes of Merlin the magician as he was being pursued.)

Descriptive Feature	*Key Figures*
Quantization of energy	Planck[1]
Probability law for the emission of a particle or a photon	Rutherford,[2] Soddy,[3] Einstein[4]
Quantization of angular momentum in an atom and quantum theory of atomic stability	Bohr[5]
Wave description of particle motion	de Broglie,[6] Schrödinger[7]
Particle diffraction and interference	Davisson and Germer,[8] G. P. Thompson[9]
Physical quantities described mathematically as noncommuting matrices or "q-numbers"	Heisenberg,[10] Dirac[11]
The "waves" of quantum mechanics interpreted as probability amplitudes	Born[12]
Uncertainty principle	Heisenberg[13]
Complementarity	Bohr[14]
Probability amplitude to go from A to B viewed as "sum over all conceivable histories" leading from A to B	Feynman[15]
The "relative state" or "branching history" or "many worlds" interpretation of quantum mechanics	Everett[16]
Lattice of yes-no propositions about the physical system	Birkhoff and von Neumann[17]

[1] M. Planck, *Annalen der Physik,* vol. 4 (1901), pp. 553-563.

[2] E. Rutherford and F. Soddy, *Philosophical Magazine,* series 3, vol. 5 (1903), pp. 441-445.

[3] Ibid., pp. 445-457; pp. 576-591.

[4] A. Einstein, *Zeitschrift für Physik,* vol. 18 (1917), pp. 121-128.

[5] N. Bohr, *Philosophical Magazine,* vol. 26 (1913), pp. 1-25.

[6] L. de Broglie, *Philosophical Magazine,* vol. 47, (1924), pp. 446-458; *Annales de Physique,* vol. 3 (1925), pp. 22-128.

[7] E. Schrödinger, *Annalen der Physik,* vol. 79 (1926), pp. 361-376.

[8] C. Davisson and L. H. Germer, *Physical Review,* vol. 30 (1927), pp. 705-740.

[9] G. P. Thomson and A. Reid, *Nature,* vol. 119 (1927), p. 890; G. P. Thomson, *Proceedings of the Royal Society of London,* vol. A117 (1928), pp. 601-609.

[10] W. Heisenberg, *Zeitschrift für Physik,* vol. 33 (1925), pp. 879-893.

[11] P. Dirac, *Proceedings of the Royal Society of London,* vol. A109 (1925), pp. 642-653.

[12] M. Born, *Zeitschrift für Physik,* vol. 37 (1926), pp. 863-867.

[13] W. Heisenberg, *Zeitschrift für Physik,* vol. 43 (1927), pp. 172-198.

[14] N. Bohr, especially as expressed in his book, *Atomic Theory and the Description of Nature* (Cambridge, 1934).

[15] R. P. Feynman, *The Principle of Least Action in Quantum Mechanics,* doctoral dissertation, Princeton University, 1942; R. P. Feynman and A. R. Hibbs, *Quantum Mechanics and Path Integrals* (New York, 1965).

[16] H. Everett III, *Reviews of Modern Physics,* vol. 29 (1957), pp. 454-462, and the more extensive version of his Princeton doctoral thesis that appears on pp. 3-149 in B. DeWitt and N. Graham, editors, *The Many-Worlds Interpretation of Quantum Mechanics* (Princeton, 1973).

[17] G. Birkhoff and J. von Neumann, *Annals of Mathematics,* vol. 37 (1936), p. 823. For a recent review, see the article by E. G. Beltrametti and G. Cassinelli in *Colloquio Internazionale sulle Teorie Combinatorie,* Accademia Nazional dei Lincei (Rome, September 1973).

special augury of new understanding ready to unfold?

The Revolution in Mathematics. Over a hundred years ago an earlier question of "mere logic" furnished the starting point for a revolutionary development in physics: "Parallel lines never meet." Is this statement a consequence of the other axioms of Euclidean geometry or is it an independent postulate of its own? Pursuing the implications of this question, Bernhard Riemann developed the geometry of curved space. On it Einstein based his general relativity theory of gravitation. On that foundation rose in turn the impressive structure of modern physical cosmology.

Mathematical logic did not end in the last century. In the past few decades it has experienced a revolution without parallel. Discoveries[39] about the consistency of axioms and the undecidability of certain propositions have shaken the old foundations, not only of logic, but of all of mathematics. If in the year 5000 there still exist universities, and if at that time the light of learning still shines, then, as one colleague puts it, professors will still be expounding these discoveries of Kurt Gödel and Paul Cohen as central to the center of all knowledge.

The heart of the Milky Way toward the constellation Sagittarius. A dark lane of obscuring dust conceals most of the brilliant star clouds in the direction of our Galaxy's nucleus. (Harvard College Observatory photograph.)

Mathematics and physics are not two worlds: they are two aspects of one world. It is difficult to believe that the revolution in the foundations of mathematics will be without consequence for our understanding of the quantum principle, and for our answers to two key questions about that principle.[40]

Is Physical Law Legislated by Cosmogony? We cannot believe that "Day One" of creation, to put it figuratively, brought space and fields and particles, and "Day Two" saw the quantum principle instituted and imposed on these dynamic entities. Did Day One then see the enactment of the quantum principle; and Day Two, the construction of these entities out of, and in consequence of, this principle? As if that were not question enough, there is a greater one. Of all imaginable ways to bring a universe into being, why cannot a single one succeed that does not embody the quantum principle from the start? Until we see how it comes about that the quantum principle is indispensable, it may be questioned whether we understand the first thing about it.

THE HEART OF DARKNESS

Three mysteries we have passed in review that call out for clarification: the quantum, the universe, and the mind. All three lie at that point where, in the phrase of Fred Hoyle, "mind and matter meld." All three threaten that clean separation between observer and observed, which for so long seemed the essence of science. Consciousness can analyze the world around; but when will consciousness understand consciousness? The universe runs its course from big bang to collapse; but what part do the future requirements for life and mind have in fixing the physics that comes into being at that big bang? The quantum principle says, "No physics without an observer"; but from what comes the necessity of this principle in the construction of the world? Of all challenges ever confronted by the adventurous fellowship of science I know none more deserving than these three to be called the heart of darkness.

Many another issue we and our colleagues could have listed, at many another frontier of knowledge; on them there may be faster and easier progress. Surely the ever increasing pace and tools and pressures of science augur more discoveries in the next 500 years than in the last 500 years. That the discoveries will be greater, nothing argues so strongly as the depth of the mystery in which we live. We can believe that we will first understand how simple the universe is when we recognize how strange it is.

In closing, turn to the man (figure opposite) to whom more than any other we give the credit for initiating the last 500 years of science:

> Copernicus, you were watchman of your day
> And told us of the course
> That this planet swings around the sun.
> Give penetration to the keen eyes
> Of the fellowship of science
> That now keep watch for all mankind
> On what there is ahead of peril and promise.
> Light up the hundred skies we search,
> From the genetic control of replication
> To the engineering of molecules,
> And from the geology of the planets
> To the sociology of achieving societies.
> You identified and solved a mystery of motion
> Where many could see no mystery.
> Remind us each day of the greatest mystery of all,
> Why there is something rather than nothing.
> You used numbers to test ideas.
> Tell us that some day magic mathematics
> Will unveil the central mystery of the universe
> And we will see all things, great and small,
> Glittering with new light and meaning.
> Remind us that there is no other universe
> Than the universe of mind and man,
> The universe that is our home.

Nicolaus Copernicus (Torun, 19 February 1473–Frombork, 24 May 1543). The flower that Copernicus holds, for him a symbol of medicine, may serve for us to symbolize a central mystery of today: What part do the future requirements for life and mind have in fixing the physics that comes into being at the big bang and fades away at collapse? Are the universe and life related as flower case and flower?

Notes

[1] M. Schmidt, "3C 273: A Star-like Object with Large Red-Shift," *Nature,* vol. 197 (1963), p. 1040.

[2] W. Baade and R. Minkowski, "Identification of the Radio Sources in Cassiopeia, Cygnus A, and Puppis A," *Astrophysical Journal,* vol. 119 (1954), pp. 206-214; "On the Identification of Radio Sources," *Astrophysical Journal,* vol. 119 (1954), pp. 215-231.

[3] G. R. Burbidge, "On Synchrotron Radiation from Messier 87," *Astrophysical Journal,* vol. 124 (1956), pp. 416-429.

[4] R. Giacconi, H. Gursky, F. Paolini, and B. Rossi, "Evidence for X Rays from Sources outside the Solar System," *Physical Review Letters,* vol. 9 (1962), pp. 439-443; "Further Evidence for the Existence of Galactic X Rays," *Physical Review Letters,* vol. 11 (1963), pp. 530-535.

[5] R. Giacconi, S. Murray, H. Gursky, E. Kellogg, E. Schreier, and H. Tananbaum, "The *UHURU* Catalog of X-Ray Sources," *Astrophysical Journal,* vol. 178 (1972), pp. 281-308.

[6] E. T. Byram, T. A. Chubb, and H. Friedman, "Cosmic X-Ray Sources, Galactic and Extragalactic," *Science,* vol. 152 (1966), pp. 66-71.

[7] C. S. Bowyer, M. Lampton, J. Mack, and F. De Mendonca, "Detection of X-Ray Emission from 3C 273 and NGC 5128," *Astrophysical Journal (Letters),* vol. 161 (1970), pp. L1-L9.

[8] W. C. Livingston, "Image-tube Systems," *Annual Review of Astronomy and Astrophysics,* vol. 11 (1973), pp. 95-114.

[9] M. H. Cohen, "High-resolution Observations of Radio Sources," *Annual Review of Astronomy and Astrophysics,* vol. 7 (1969), pp. 619-664.

[10] G. Neugebauer, E. Becklin, and A. R. Hyland, "Infrared Sources of Radiation," *Annual Review of Astronomy and Astrophysics,* vol. 9 (1971), pp. 67-102.

[11] F. Reines and C. L. Cowan, Jr., "Detection of the Free Neutrino," *Physical Review,* vol. 92 (1953), pp. 830-831; C. L. Cowan, Jr., F. Reines, F. B. Harrison, H. W. Kruse, and A. D. McGuire, "Detection of the Free Neutrino: A Confirmation," *Science,* vol. 124 (1956), pp. 103-104; and F. Reines, C. L. Cowan, Jr., F. B. Harrison, A. D. McGuire, and H. W. Kruse, "Detection of the Free Antineutrino," *Physical Review,* vol. 117 (1960), pp. 159-173.

[12] T. C. Weekes, *High-Energy Astrophysics* (London, 1969), pp. 171-172.

[13] J. N. Bahcall, "Neutrinos from the Sun," *Scientific American,* vol. 221, no. 1 (1969), pp. 28-37.

[14] J. Weber, M. Lee, D. J. Gretz, G. Rydbeck, V. L. Trimble, and S. Steppel, "New Gravitational Radiation Experiments," *Physical Review Letters,* vol. 31 (1973), pp. 779-783.

[15] See for example B. Bertotti, editor, *Experimental Gravitation,* Course 56, Italian Physical Society "Enrico Fermi" (New York, 1973).

[16] See for example R. K. Merton, *The Sociology of Science* (Chicago, 1973).

[17] W. Penfield, "Some Observations on the Functional Organization of the

Human Brain," *Proceedings of the American Philosophical Society,* vol. 98 (1954), pp. 293-297.

[18] For a systematic account, see C. W. Misner, K. S. Thorne, and J. A. Wheeler, *Gravitation* (San Francisco, 1973); see also S. W. Hawking and G. F. R. Ellis, *The Large Scale Structure of Spacetime* (Cambridge, 1973).

[19] R. C. Tolman, *Relativity, Thermodynamics and Cosmology* (Oxford, 1934); A. Avez, "Global Properties of Periodic Closed Space-Times," *Comptes Rendus de l'Academie des Sciences,* vol. 250 (1960), pp. 3585–3587; R. P. Geroch, *Singularities in the Spacetime of General Relativity: Their Definition, Existence, and Local Characterization,* doctoral dissertation, Princeton University (Princeton, 1967); S. W. Hawking and R. Penrose, "The Singularities of Gravitational Collapse and Cosmology," *Proceedings of the Royal Society of London,* vol. A314 (1969), pp. 529-548.

[20] For an up-to-date review see D. N. Schramm, "The Age of the Elements," *Scientific American,* vol. 230, no. 1 (1974), pp. 69-77.

[21] A. A. Penzias and R. W. Wilson, "A Measurement of Excess Antenna Temperature at 4080 Mc/s," *Astrophysical Journal,* vol. 142 (1965), pp. 419-421; R. H. Dicke, P. J. E. Peebles, P. G. Roll, and D. T. Wilkinson, "Cosmic Blackbody Radiation," *Astrophysical Journal,* vol. 142 (1965), 414-419.

[22] J. H. Oort, "Distribution of Galaxies and the Density in the Universe," *Onzieme Conseil de Physique: La structure et l'evolution de l'univers* (Brussels, 1958); see the analysis of this work in P. J. E. Peebles, *Physical Cosmology* (Princeton, 1971).

[23] J. P. Ostriker and P. J. E. Peebles, "A Numerical Study of the Stability of Flattened Galaxies: Or, Can Cold Galaxies Survive?," *Astrophysical Journal,* vol. 186 (1973), pp. 467-480.

[24] For a fuller statement, see J. A. Wheeler, "From Relativity to Mutability" in J. Mehra, editor, *The Physicist's Conception of Nature* (Dordrecht and Boston, 1973), pp. 202-247.

[25] See for example B. K. Harrison, K. S. Thorne, M. Wakano, and J. A. Wheeler, *Gravitation Theory and Gravitational Collapse* (Chicago, 1965).

[26] P. S. Laplace, *Exposition du système du monde,* vol. 2 (Paris, 1796). The relevant page is reproduced photographically in Misner, Thorne, and Wheeler, op. cit. [note 18], p. 623.

[27] J. R. Oppenheimer and H. Snyder, "On Continued Gravitational Contraction," *Physical Review,* vol. 56 (1939), pp. 455-459.

[28] W. Baade and F. Zwicky, "Supernovae and Cosmic Rays," *Physical Review,* vol. 45 (1934), p. 138 (abstract of paper presented at Stanford meeting of American Physical Society, 15-16 December, 1933); cf. also "On Super-novae," *Proceedings of the National Academy of Sciences,* vol. 20 (1934), pp. 254-259; "Cosmic Rays from Super-novae," *Proceedings of the National Academy of Sciences,* vol. 20 (1934), pp. 259-263.

[29] J. M. Comella, H. D. Craft, Jr., R. V. E. Lovelace, J. M. Sutton, and G. L. Tyler, "Crab Nebula Pulsar NP 0532," *Nature,* vol. 221 (1969), pp. 453-454.

[30] A. Hewish, S. J. Bell, J. D. H. Pilkington, P. F. Scott, and R. A. Collins, "Ob-

servation of a Rapidly Pulsating Radio Source," *Nature*, vol. 217 (1968), pp. 709-713.

[31] W. J. Cocke, M. J. Disney, and D. J. Taylor, "Discovery of Optical Signals from Pulsar *NP 0532*," *Nature*, vol. 221 (1969), pp. 525-527.

[32] R. Lynds, S. P. Maran, and D. E. Trumbo, "Photoelectric and Spectroscopic Observations Related to a Possible Optical Counterpart for Pulsar CP 1919+21," *Astrophysical Journal (Letters)*, vol. 155 (1969), p. L121.

[33] R. Giacconi, "Progress in X-ray Astronomy," *Physics Today*, vol. 26, no. 5 (1973), pp. 38-47.

[34] Ya. B. Zel'dovich and I. D. Novikov, *Relativiistic Astrophysics, Volume I: Stars and Relativity* (Chicago, 1971).

[35] J. A. Wheeler, "Transcending the Law of Conservation of Leptons," *Atti del Convegno Internazionale sul Tema: The Astrophysical Aspects of the Weak Interactions; Quaderno No. 157*, Accademia Nazionale dei Lincei, (Rome 1971), pp. 133-164; see also Misner, Thorne, and Wheeler, op. cit. [note 18], p. 640, for further citation of the literature, including papers by J. B. Hartle and J. Bekenstein.

[36] J. A. Wheeler, op. cit. [note 24].

[37] Parmenides of Elia in the Italy of the Greek world, poem called *Nature* (∼500 B.C.), part called "Truth," as summarized by A. C. Lloyd in article on Parmenides, *Encyclopaedia Britannica* (Chicago, 1959), vol. 17, p. 327.

[38] George Berkeley (1685-1753), as summarized in the article on him by Robert Adamson in *Encyclopaedia Britannica* (Chicago, 1959), vol. 3, p. 438.

[39] K. Gödel, "Über formal unentscheidbare Sätze der principia Mathematica und verwandter Systeme I," *Monatshefte für Mathematik und Physik*, vol. 38 (1931), pp. 173-198; translated by B. Meltzer, *Kurt Gödel: On Formally Undecidable Propositions of Principia Mathematica and Related System* (New York, 1962); with an introduction by R. B. Braithwaite; and P. J. Cohen, "The Independence of the Continuum Hypothesis," *Proceedings of the National Academy of Sciences*, vol. 50 (1963), pp. 1143-1148; "The Independence of the Continuum Hypothesis, II," *Proceedings of the National Academy of Sciences*, vol. 51 (1964), pp. 105-110; and *Set Theory and the Continuum Hypothesis* (New York, 1966).

[40] Wheeler, op. cit. [note 24].

III The Collegia

INTRODUCTION TO THE COLLEGIA

Owen Gingerich

Do scientific explanations reveal some ultimate reality of the universe? Or are they convenient constructions, fictitious models? These questions were recurring themes as six dozen scientists, historians, theologians, and philosophers gathered in three collegial groups during Copernicus Week to discuss issues raised by the nature of scientific discovery. The idea of relatively intimate working discussions in parallel with the public symposium was first suggested by Professor John Wheeler, Chairman of the Program Committee. Eventually a trio of collegia were formed:

- on science and society in the sixteenth century, organized by Owen Gingerich at Hotel Washington;
- on the interplay of literature, art, and science, convened by O. B. Hardison, Jr., at the Folger Shakespeare Library;
- on science, philosophy, and religion in historical perspective, organized by Walter Shropshire, Jr., at the Parish House of St. John's Church.

The collegia produced a stimulating intellectual fare; questions ranged from "forbidden knowledge" and the ethical limits of scientific discovery to the attitudes toward novelty at the time of Copernicus. Collegia scholars analyzed and criticized several of the symposium addresses; contradictory views clashed in the framework of reasoned discourse, and in more than one instance participants went back to do more homework before approving the text printed here.

The collegia provided such a fascinating interaction of various disciplines that we felt richly rewarded by our pre-symposium gamble to have each session recorded by a stenographer. I have edited these transcripts—eight in all—drastically reducing their length by more than half, severely pruning out redundancies, and sometimes even rearranging the statements into a more logical sequence. Each participant reviewed his own contributions, often honing, clarifying, or correcting his remarks; in some instances references have been added in the text or as footnotes so that the reader can more readily identify the material under discussion. Although the printed text accurately conveys the flavor of the original conversations, it does present a highly abbreviated and considerably more polished version of the proceedings.

In the preparation of this volume, particularly of the collegia transcripts, I was greatly aided by my secretary, Joan Jordan, and also by Carol Arnold and Penelope Gregory. At the same time I take this opportunity to thank Gordon Hubel, director of the Smithsonian Institution Press, Joan Horn, copy editor, and Natalie E. Bigelow, designer for this volume.

Because references appear so often in the symposium and in the collegia to Osiander's anonymous introduction to Copernicus' *De revolutionibus* it seemed reasonable to include it here in an English translation. It is found on the following pages together with an illustration of the original text, photographed from an unusual copy of the original 1543 Nuremberg edition in which Copernicus' student Rheticus has indignantly crossed out the unauthorized addition to the book. This copy has been in the collection of Harrison Horblit, and a similarly marked example is found in the Uppsala University Library.

OSIANDER'S INTRODUCTION TO
De Revolutionibus

To the Reader Concerning
the Hypotheses of This Work

Since the novelty of the hypotheses of this work—which sets the earth in motion and puts an immovable sun at the center of the universe—has already received a great deal of advance publicity, I have no doubt that some scholars will have taken grave offense and think it wrong to raise havoc among the liberal arts with their proper time-honored classical tradition.

If, however, they are willing to ponder the matter closely, they will come to the conclusion that the author of this work has done nothing that merits blame. For it is the specific task of the astronomer to chart the course of the stars and planets through patient and refined observations. On this basis he has to compute the causes or, rather, develop hypotheses, since he cannot possibly attain the final causes [*veras*] of these movements. Led by these assumptions, the past and future heavenly movements can be calculated with the help of geometry.

This author has done a first-rate job in both respects. After all, it is not necessary that his hypotheses should be true, nor even probable. This alone suffices, that they provide a computation that tallies with the observations. Perchance there is someone so ignorant of geometry and optics that he holds the epicycle of Venus as likely and believes this is the cause why Venus sometimes precedes and sometimes follows the sun by forty degrees or more. For who does not see that from this assumption it necessarily follows that the diameter of the planet in perigee should appear more than four times greater, and the body more than sixteen times greater, than in apogee? Yet the experience of every age contradicts this. And there are other things in this discipline not less absurd, which need not be treated here. For

AD LECTOREM DE HYPO.
THESIBVS HVIVS OPERIS.

ON dubito, quin eruditi quidam, uulgata iam de nobitate hypotheseon huius operis fama, quòd terram mobilem, Solem uero in medio uniuersi immobilè constituit, uehementer sint offensi, putétq; disciplinas liberales recte iam olim constitutas, turbari nō oportere. Verum si rem exacte perpendere uolent, inueniēt authorem huius operis, nihil quod reprehendi mereatur cōmisisse. Est enim Astronomi proprium, historiam motuum cœlestium diligenti & artificiosa obseruatione colligere. Deinde causas earundem, seu hypotheses, cum ueras assequi nulla ratione possit, qualescunq; excogitare & confingere, quibus suppositis, ijdem motus, ex Geometriæ principijs, tam in futurū, quàm in præteritū recte possint calculari. Horū autē utrunq; egregie præstitit hic artifex. Neq; enim necesse est, eas hypotheses esse ueras, imò ne uerisimiles quidem, sed sufficit hoc unum, si calculum obseruationibus congruentem exhibeant. ni si forte quis Geometriæ & Optices usq;adeo sit ignarus, ut epicyclium Veneris pro uerisimili habeat, seu in causa esse credat, quod ea quadraginta partibus, & eo amplius, Solē interdum præcedat, interdū sequatur. Quis enim nō uidet, hoc posito, necessario sequi, diametrum stellæ in περιγείω plusq; quadruplo, corpus autem ipsum plusq; sedecuplo, maiora, quàm in ἀπογείω apparere, cui tamen omnis æui experientia refragatur? Sunt & alia in hac disciplina non minus absurda, quæ in præsentiarum excutere, nihil est necesse. Satis enim patet, apparentiū inæqualium motuū causas, hanc arte penitus & simpliciter ignorare. Et si quas fingēdo excogitat, ut certe quàplurimas excogitat, nequaquã tamen in hoc excogitat, ut ita esse cuiquam persuadeat, sed tantum, ut calculum recte instituant. Cum autem unus & eiusdem motus, uarie interdum hypotheses sese offerant (ut in motu Solis, eccentricitas, & epicyclium) Astronomus eam potissimum arripiet, quæ compræhensu sit quàm facillima. Philosophus fortasse, ueri similitudinem magis re

gis requiret,neuter tamen quicquam certi compræhedet, aut
tradet,nifi diuinitus illi reuelatum fuerit. Sinamus igitur &
has nouas hypothefes,inter ueteres,nihilo uerifimiliores inno
tefcere,præfertim cum admirabiles fimul,& faciles fint,ingen
temç thefaurum, doctifsimarum obferuationum fecum ad-
uehant.Necp quifquam,quod ad hypothefes attinet,quicquá
certi ab Aftronomia expectet,cum ipfa nihil tale præftare que
at,ne fi in alium ufum conficta pro ueris arripiat, ftultior ab-
hac difciplina difcedat, quàm accefferit. Vale.

NICOLAVS SCHONBERGIVS CAR
dinalis Capuanus, Nicolao Copernico, S.

Vm mihi de uirtute tua,côftanti omniû fermone
ante annos aliquot allatû effet,cœpi tum maiorem
in modû te animo côplecti ,atcp gratulari etiâ no-
ftris hominibus,apud ĝs tâta gloria floreres.Intellexerâ enim
te nô modo ueterû Mathematicorû inuêta egregie callere,fed
etiâ nouâ Mûdi ratione côftituiffe.Qua doceas terrâ moueri:
Solem imû mûdi , adeoçp mediû locu obtinere:Cœlû octauû
immotû,atcp fixû ppetuo manere:Lunâ fe unâ cû inclufis fuæ
fphæræ elementis,inter Martis & Veneris cœlû fitam,anni-
uerfario curfu circû Solem côuertere.Atcp de hac tota Aftro-
nomiæ ratione cômentarios à te côfectos effe, ac erraticarum
ftellarû motus calculis fubductos in tabulas te côtuliffe,maxi
ma omniû cum admiratione.Quamobrem uir doctifsime,ni
fi tibi moleftus fum,te etiâ atcp etiâ oro uehementer , ut hoc
tuû inuentû ftudiofis cômunices,& tuas de mundi fphæra lu
cubrationes unà cû Tabulis,& fi quid habes præterea , ĝd ad
eandem rem pertineat , primo quoçp tempore ad me mittas.
Dedi autem negotiû Theodorico à Reden,ut iftic meis fum-
ptibus omnia defcribantur,atcp ad me transferantur.Quod fi
mihi morem in hac re gefferis,intelliges te cum homine no-
minis tui ftudiofo,& tantæ uirtuti fatisfacere cupiente rem ha
buiffe.Vale, Romæ, Calend.Nouembris,anno м. d.xxxvi,

ij

Osiander's anonymous introduction to *De revolutionibus* in the original 1543
Nuremberg edition. In this copy, in the Horblit Collection, Rheticus has struck
out the unauthorized addition. (Photographs by Owen Gingerich.)

it is quite clear that this field of study is thoroughly and frankly ignorant of the causes of the apparent irregular motions. And if scholars in the field construct and think up causes—and they have certainly thought up a good many—nevertheless they advance their models not to persuade anyone it is so, but to present a correct basis for calculation.

Since, then, for one and the same movement several hypotheses have been advanced in the course of time, such as an eccentricity or an epicycle for the movement of the sun, the astronomer will be inclined to accept the one that is easiest to grasp. The philosopher is more likely to insist on some semblance of truth; but neither of them will learn or teach anything certain, unless it has been divinely revealed to him.

Therefore, let us allow these new hypotheses to take their place among the old ones—which were by no means more probable—especially since these are both admirable and simple, and are based upon a vast amount of learned observations. But, as far as hypotheses go, let no one expect any certainty from astronomy, since this study cannot provide such. Otherwise, if one takes models—with their own purpose—to stand for reality, one leaves this discipline more ignorant than before he entered it. Farewell.

Based on a translation by Heiko A. Oberman. Professor Oberman points out the apparently unnoticed fact that the first paragraph almost directly reflects the statement from Martin Luther's *Tischrede* cited in note 9 of his paper, page 161 in this volume. Professor Oberman prefers the following translation for the fourth paragraph:

"Since, then, for one and the same movement several hypotheses, such as eccentricity or epicycle for the movement of the sun, have been advanced in the course of time, the astronomer will be inclined to accept that one which has the highest degree of *prima facie* consistency. The philosopher is more likely to insist on probability; but neither of them will be able to learn or teach anything that claims to be ultimate truth, unless it has been divinely revealed to him."

SCIENCE AND SOCIETY
IN THE SIXTEENTH CENTURY

Owen Gingerich, *Convener and Moderator*

LIST OF PARTICIPANTS

John Christianson
    Professor of History, Luther College, Decorah, Iowa
    Biographer of Tycho Brahe
Elizabeth Eisenstein
    Scholar, Washington, D. C.
    Printing in the history of ideas
Karin Figala
    Deutsches Museum, Munich
    History of alchemy
Owen Gingerich
    Professor of Astronomy and of the History of Science, Harvard
        University, Cambridge, Mass.
    Astronomer-historian of Kepler and Copernicus
Paul F. Grendler
    Associate Professor of History, University of Toronto
    Italian Renaissance

Marie Boas Hall
Lecturer in History of Science and Technology, Imperial
College, London
Renaissance science
Rupert Hall
Professor of the History of Science and Technology, Imperial
College, London
Renaissance science and technology
Willy Hartner
Professor of History of Science, Johann Wolfgang Goethe
University, Frankfurt
Early European and Islamic exact sciences
Janice Henderson
Doctoral candidate, Yale University, New Haven, Conn.
16th-century astronomy
Gerald Holton
Professor of Physics, Harvard University, Cambridge, Mass.
Physicist-historian of Einstein and Kepler
Melvin J. Lasky
Editor, *Encounter* magazine, London
Historian and essayist
Benjamin Nelson
Professor of Sociology and History, New School for Social
Research, New York
Civilizational structures
Edward Rosen
Professor of History, City College of New York
International dean of Copernican scholars
Noel Swerdlow
Assistant Professor of History, University of Chicago
Ancient to Renaissance astronomy
William Wallace
Professor of History and Philosophy of Science, The Catholic
University of America, Washington, D. C .
Philosophy of science of the Middle Ages

Deborah Warner
Associate Curator for Astronomy, National Museum of History
and Technology, Smithsonian Institution
History of astronomical charts and instruments
Robert S. Westman
Assistant Professor of History, University of California at
Los Angeles
The Copernican revolution and philosophy of science
Curtis Wilson
Professor of History, University of California at La Jolla
17th-century physical sciences

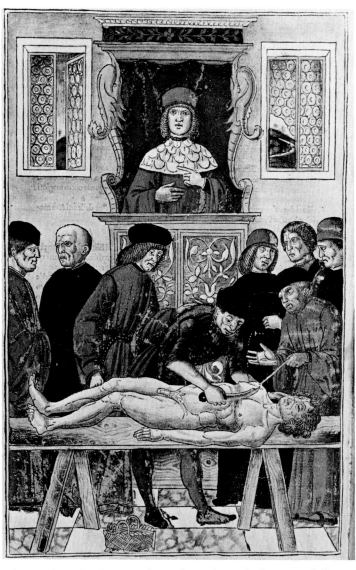

Johannes de Ketham's *Fasciculus medicine* shows the first printed illustration of a human dissection, attributed by some scholars to the artist Gentile
Bellini. Originally published in 1491 in Venice, shortly before Copernicus
became a medical student in nearby Padua. (Permission of Yale Historical
Medical Library.)

THE SPIRIT OF INNOVATION
IN THE SIXTEENTH CENTURY

Marie Boas Hall

It is not my intention here to discuss the content of 16th-century science, nor to assess how far scientists of the period made novel contributions to the progress of science. Rather, leaving aside entirely the question of the ways in which medieval science was transmuted into the "modern" science of the 17th century, I wish to examine the attitude towards innovation of Renaissance thinkers, both scientists and nonscientists, whether in intellectual life or in general activity, and, when possible, to consider the grounds upon which this opinion was based.

We are all, I imagine, familiar with examples of conservatism in 16th-century life and thought, and with the scandalized reaction to "dangerous novelties" in philosophy or, above all, in religion. Education, too, experienced an internal conflict between those who sought to continue in the tried and true ways of scholasticism, and those who, advocating the "new learning," tried to introduce humanism into established universities, or founded colleges and universities to promote humanism. In science one can find plenty of examples of this conservatism: for example, the case of Agricola who defended himself for writing in Latin upon such subjects as mining and assaying with the excuse that no one had previously attempted to describe what he was about to discuss. This conventional attitude was expressed in the well-known comment of Luther in 1539, "that is the way nowadays; whoever wants to be clever must needs produce something of his own, which is bound to be the best since he has produced it"— a classic instance of the pot calling the kettle black, but a comment often (wrongly, in my view) regarded as typical of ecclesiastical comment upon Copernican astronomy.

Examples more significantly indicative of academic conservatism are easily found. Medical historians frequently cite the case of Galen worship, the conviction expressed by men who had (it must be emphasized) recently had occasion to study his works profoundly, that everything Galen said must be true and that it was idle love of novelty that made a younger generation insist that they had discovered errors in the Master. A notable and similar reaction was that of Philipp Melanchthon (both one of Luther's closest disciples and an eminent academic), who, in his *Initia doctrina physicae* of 1549, challenged Copernican astronomy on both empirical and authoritarian grounds:

> The eyes are witnesses that the heavens revolve in the space of 24 hours. But certain men, either from the love of novelty, or to make a display of ingenuity, have concluded that the earth moves; and they maintain that neither the eighth sphere nor the sun revolves. Now, it is a want of honesty and decency to assert such notions publicly, and the example is pernicious. It is the part of a good mind to accept the truth as revealed by God and to acquiesce in it.

These examples are, of course, by no means unique. Yet it is relevant to note that probably a majority of these critics had accepted 15th-century humanism, which had itself been in conflict with traditional scholasticism. Notorious Galenists like the English Thomas Linacre and John Caius were famous translators of Greek texts, while Melanchthon would never have sported a Greek name had he not accepted the newest humanist fashion.

Although humanism was a backward looking movement, forever trying to revive a dead classical past and rejecting the living medieval present, it was not conservative in its origins, and was not so regarded. True, humanism was backward looking by definition, but at the same time it was inherently innovative. The early humanists deliberately rejected the values in which they had been educated and sought a new set of values in the rediscovery of a far-removed past. Moreover, a humanist felt himself to be doing something new when he "discovered" manuscripts neglected or forgotten by immediately previous generations, when he critically edited a text, or when he turned to sources of learning little regarded by his medieval ancestors. Certainly, in the 15th century the joyful study of classical Greek was novel even for Italians, and much more so for Northern Europeans. The search for Greek manuscripts in Constantinople before 1453, or for neglected Latin manuscripts in European monastic libraries was an excitingly novel sport, taking on something of the air of an academic competition. The techniques of classical erudition, and in particular of the establishment of good texts, the conjectural restoration of lost passages and the emendation of doubtful readings excite few people nowadays, but many more then, when as a discipline textual study was just being invented and all was yet to do. To try to revive the past glory of Greece and Rome in Florence, Lombardy, Germany, the Low Countries, Paris, or England produced curious consequences most unlike the culture of classical antiquity as we see it; but however quaint, it appeared highly novel, excitingly or dangerously so according to the point of view of the appraiser.

The Renaissance was semantically oriented to a high degree, so that it is not inappropriate here to point to a significant change in meaning of the very words we use to describe the kind of attitude I am here claiming for the 15th and 16th centuries. In classical Latin "*innovare*" meant "to renew"; in English and hence almost certainly in contemporary Latin and other vernaculars the words "innovation," "novel," and "novelty" all make their appearance in the late 15th century, definitely taking on their present significance in the course of the 16th century. Simultaneously their connotative force shifted from possible pejorative (for all novelties in authoritarian spheres like theology are potentially dangerous, by definition) to firmly praiseworthy and desirable. A generation that felt itself to be renewing the glories of the past began first tentatively, and then positively and even vaingloriously to assert its newness.

At first this claim for novelty was intermingled with renewal of the past. Consider Georg Peurbach, humanist professor of rhetoric at Vienna, who was dedicated to the belief that the errors and difficulties of astronomy could be easily and totally resolved by a return to Ptolemy's original writings and the establishment of a humanistically correct text and translation of the *Almagest*. While awaiting the opportunity to travel from Vienna to Rome, where he was sure he could find an accurate Greek manuscript, he wrote a review of Ptolemaic astronomy as best he could on the basis of the medieval version of the *Almagest* available to him—and called it *Theoricae novae planetarum* ("New Speculations about the Planets"). And indeed the word "new" occurs with ever-increasing frequency in the (often quite conventional) scientific treatises of the later 15th and 16th centuries.

Now it is true that the novelty that early 16th-century authors claimed was often merely a reinterpretation or reassessment of the past, a "renewing" as they saw it of ancient doctrine. This is very evident in the history of astronomy: First came reconsideration of Ptolemaic doctrines, as with Peurbach; next came rejection of Ptolemy in favor of a more remote past, as with Fracastoro's attempt to revive the homocentric spheres of Eudoxus. In this light, the work of Copernicus could be viewed as yet another renewal-rejection of Ptolemy in favor of the oldest of all Greek mathematical cosmologies, the Pythagorean. It is easy to dismiss Copernicus's own statements about the role played by humanism in the establishment of what we see as a totally novel system, but it must be remembered that the line of reasoning he outlines was one very natural in the 16th century, and explains the terms in which many of his contemporaries described his achievement. I do not in fact believe that Copernicus himself thought his achievement was limited to renewal; but I think that to him renewal and novelty were naturally associated. I further believe that the sense of novelty gained by taking a new look at the past was of considerable assistance in encouraging the creation of genuinely novel notions. Perhaps it is worth noting here that to many of his contemporaries a decidedly novel feature of the work of Copernicus was his use of new observations made by himself, few though these seem in modern eyes.

This point of view is perhaps easier to detect in the parallel case of 16th-century medicine. Historians of medicine were long accustomed to write of the authority of Galen as equally dominant from the first medieval foundation of medical schools until its vociferous climax among conservatives in the mid-16th or even 17th centuries. But we now have a clearer picture of the chronological development of Galenism, realizing that there is a world of difference between Galenic therapeutics—set texts of the medieval curriculum—and Galenic physiology and anatomy, hardly known before the early 16th century. Though it is true that *On the Use of the Parts* was translated into Latin for the first time (and that direct from Greek) in 1322, its main impact came nearly two centuries later—even Leonardo da Vinci seems to have met it only about 1506, and it was printed ten years later than that. The *Natural Faculties* was translated by Thomas Linacre and printed in 1523; the *Anatomical Procedures* (by Guinther of Andernach) only in 1531. These are deeply exciting and interesting works, and it is no wonder that the men who first studied them were so fired by their novelty that they lectured upon them to the exclusion of all else and devoted their efforts to editing the texts and inculcating their pupils with the glories of the "new" Galen. In this light even the extravagant claim of John Caius—that everything Galen said was true—is comprehensible, if not admirable.

But their teaching was to bear strange fruit, and what they inculcated was the spirit of novelty and rebellion. The first generation to be taught the new Galen in medical schools was the first generation to deny Galen's overriding authority and to begin anatomical research *de novo*. Of this generation the most notable was Vesalius: He listened to Guinther's lectures on Galen; he helped him to edit a new and authoritative edition of Galen (Venice, 1541); and then, having acquired a lecturing post at Padua, he insisted both upon seeing for himself and proclaiming a number of instances where he found Galen in error. His claims of originality run triumphantly throughout his great work, the *Fabrica* (Basel, 1543)—triumphantly, but not always accurately, for he was by no means always correct in his claims of originality or of deviation from Galen. A fellow-student of Vesalius at Paris, Michael Servetus, though a much less skilled and original anatomist, was yet more genuinely novel. A wild innovator in religion, Servetus called anatomy to his aid in a theological disquisition and hence, parenthetically, became the first European to publish an account of the pulmonary or lesser circulation— although he made no particular claim of novelty for it.

A wilder medical innovator than Servetus, though in therapeutics rather than in anatomy, was that strange figure Paracelsus. Paracelsus was nothing of a humanist, and not particularly learned, though it is possible that the barbarism of his Latin prose and his preference for his native German dialect were deliberate. Much about Paracelsus is confused by myth and legend, often of his own creation: It is not certain whether or not he had a medical degree; the name "Paracelsus" does not definitely refer to the Roman medical writer; it is not certain that he dramatically burned textbooks before his students at Basel, or whether if he did they were those of Galen or Avicenna; the obscurity of his teaching makes it difficult to decide how far he subscribed to the more mystic strains of early 16th-century Hermeticism; and the interpretation of his appeal to "experience" is fraught with complication—for experience was in contemporary usages mystical as well as sensual. Certainly his interest lay almost entirely in medical practice. Equally certain he had no interest in the revival of anatomy, or in anatomical research, discovery, or accuracy. He rightly claimed to be doing something totally new when he applied alchemical theory to medical physiology, and the iatrochemistry of his followers was an innovation indeed. Chemical or mineral drugs had been prescribed before Paracelsus, as a glance at Dioscorides or Pliny will show, but not so freely, nor so dangerously; and Paracelsus and, even more, his followers made free and generous use of both the new alchemy and the new industrial chemistry with the use of distilled products, antimonial preparations, mineral acids, and new salts.

Indeed, Paracelsan medical practice was in close touch with the developing technology of the Renaissance, and naturally so, for medical and surgical practice was more craft than science. No one I think would deny that 16th-century technology was producing much that was new, even if the roots of the tradition are to be found in the Middle Ages. Furthermore, when 16th-century writers referred to what one of them called "the triumphs of this New Age," it was almost entirely technological achievements that they named; whether genuinely contemporary or not, the chief exceptions were "the restoration of scholarship and the recovery of ancient manuscripts," that is to say the achievements of 15th-century humanism.

The most telling example of this fashion is the *Nova reperta* series of copper-plates published by the Flemish artist Jan van der Straet (Stradanus, 1523–1605) late in his career. These plates illustrate Renaissance discoveries, printing, longitude determination, exploration, engraving on copper, oil colors, guaiacum), medieval ones (the lodestone, gunpowder, distillation, stirrups, spectacles, windmills), and some ancient ones (silk, watermills, olive oil, sugar, the astrolabe)—and one, polishing armor, whose significance eludes me. What matters, however, is not his accuracy, but the fact that he thought that these prints

"Also invented were eyeglasses, which remove dark veils from the eyes," according to this scene from Stradanus' *Nova reperta* (Antwerp, c. 1580). (Courtesy of Burndy Library.)

would sell, that is, that there was a rather large market for pictures illustrating the *principle* of novelty.

This is only to dramatize what other contemporaries were already proclaiming in print. Fernel in 1548 instanced war machines, the printing press, paper, the magnetic compass, exploration, and navigation. Jean Bodin in 1566 listed the magnet, exploration, longitude determination, war machines, new metals, new fabrics, new medicines, printing. Blaise de Vigenère in 1571 was most impressed by the magnetic compass and printing. The repetition is perhaps tiresome, but no less earnest, as is the belief that these discoveries were truly novel. As Fernel said, "Our age today is doing things of which antiquity did not dream"; or again, Blaise de Vigenère declared, "Many infinitely beautiful things of which the ancients were ignorant have been discovered," while Bodin insisted, "The art of printing alone would easily be able to match all the inventions of the ancients."

Here indeed the learned men were on sure ground. Admittedly the credit for much of what they wrote about belonged to their despised medieval ancestors—notably the stirrup, the compass, paper, eye glasses—but by no means all. Agricola (a learned man) was advisedly cautious when he defended himself for writing on technical subjects by claiming that no one had treated of these things properly before, for he was not technologically experienced; but most writers on technological subjects saw no particular need to explain why they were moved to write upon obviously important and expanding crafts, novel though it is in our eyes. For the Middle Ages we have few documentary sources for technological processes. The *Diverse Arts* of Theophilus, the notebooks of Villard d'Honnecourt, the references in Roger Bacon, the work of Guido da Vigevano, the manuscript of the Hussite engineer—these assume immense importance because of their rarity. The vernacular writings on technology of all sorts —war machinery, architecture, mining and assaying, pyrotechnics, dyeing, pottery—and the spate of 16th-century machine books, all testify to the growth of technology and to the interest in it. The learned authors whom I have already cited testify to the impact of these books and the achievements they record. Above this there is the expanded insistence upon the growth of technology as the proper paradigm for the growth of science in the work of Francis Bacon and the belief common in the 17th century that science and technology must go hand in hand.

The area where the claim for novelty was most genuine was, indeed, in exploration and discovery, and here men had a right to feel that science had something useful to contribute. Whatever the ancients had done, they had not opened up a sea route to India and China from the Atlantic ports of Spain and Portugal; nor had they sailed boldly into the Atlantic and discovered a new world. The Western Hemisphere had clearly been reserved for later navigators to find, and here it was difficult to believe that "later" did not mean "better," that by virtue of successful discovery of new areas 15th- and 16th-century sailors were not better navigators than the ancients, and wiser seamen. There is some truth in this, of course, for Greek and Egyptian ships could not have sailed around the world as did Magellan's expedition, the improvements in ship building and sailing techniques being partly the result of medieval experience in the Mediterranean, partly carefully fostered development under Portuguese royal patronage.

There is a further point of interest here, for at the same time they were gathering experience of deep-water sailing, 15th- and 16th-century seamen were gathering experience of navigation—and here science, at least elementary mathematics and astronomy, could assist. It is difficult to realize that the navigational instruments adopted commonly during the 16th century—quadrant, mariner's astrolabe, cross-staff, log, back-staff; and the rules for finding latitude by their means—were unknown before 1450. There was now a need to determine one's position in the open sea, in response to which shore-based learned men devised tables and rules based on modifications of old instruments, while instrument makers and men with experience of the sea devised new instruments. None of these sighting instruments, simple though they were in principle, can have been easy to handle at sea, on the deck of a small and lively ship. It was far easier, as with Amerigo Vespucci, to land and take sightings on shore. The feat was possible, however, and was in fact frequently done, more frequently as increase in navigation meant that at last men were sailing where others had sailed before, and yet others would wish to follow. Nothing like this had happened before.

At the same time it was becoming increasingly feasible to represent the newly explored areas of the world on globes, maps and charts, in a way not possible before. The plane charts, the late-medieval *portolani* of the Mediterranean and European coastal waters, were inadequate to represent the larger areas of North and South Atlantic waters, the Western Hemisphere, or the Pacific. Here humanism and mathematical art contributed to theoretical cartography, based initially upon Ptolemy. Ptolemy's

42   The fyrſt Booke

Example.

Admit B *my ſtation where I place mine Inſtrument* A *the marke whoſe altitude I deſire a-*

Practical surveying from Thomas Digges' *Pantometria* of 1591. (Permission of Harvard College Library.)

Geography (though without maps) was discovered early in the 15th century in Constantinople, and brought to Italy in 1410 in Latin translation. It soon acquired maps, at first drawn following Ptolemy's directions, then redrawn based on new knowledge and even new methods of projection necessary to accommodate large areas of the curved earth upon a flat sheet of paper. There were many manuscript editions before 1475, and many printed editions after this. By the mid-16th century Ptolemy had been left behind and a new cartographic tradition—the atlas—had been founded. It is difficult to say how far the new cartography was useful to the seamen, but there is no doubt that it spread knowledge among landsmen who acquired printed maps for decoration, for pleasure, and for information, just as it is certain that some ship captains and their passengers did take globes with them on long sea voyages.

One totally new problem arose as a result of these long voyages—the problem of longitude determination. Ironically, this was ultimately to be solved primarily by technological means—the invention of the accurate chronometer—but this could not have been known to an age that had no accurate clocks (though the method was suggested by Gemma Frisius). The 16th century very naturally and logically sought the answer in astronomy. Columbus is said to have tried to determine longitude by observing a lunar eclipse and comparing the time of its occurrence with that predicted for Europe in the almanac of Regiomantanus (1474); but this was a fortunate opportunity upon which one could not rely. Peter Apian presciently argued for the method of lunar differences, a method which was highly successful in the 18th and 19th centuries, but which needed more accurate lunar tables than those available before the end of the next century. Still, the method was advocated; it clearly was practical in principle, and is probably that to which Bodin and others referred as something unknown to the ancients. (The other method suggested, the plotting of magnetic variation on the assumption that it was constant at any given point of the Earth's surface, was challenged almost as soon as proposed, though it must be said that it was equally novel.)

While cartography took its departure from the work of antiquity, the determination of latitude and longitude did so only very remotely. Ptolemy knew how to determine latitude by observation on land and in known parts of the world, but certainly not at sea when one's position was almost totally unknown. Equally, Ptolemy knew how to determine longitude differences by eclipse observation, but the Greek methods were those of scholars working under favorable conditions. What rightly impressed the later 15th century as novel was the devising of practical methods useful to and usable by barely literate seamen. A good part of the novelty lay in the fact that these methods were, for the most part, not invented by craftsmen and seafarers, but by mathematicians, learned in esoteric lore. No wonder that the 17th century had a prejudice in favor of the easy and rapid application of science to technology, to "the relief of man's estate" when it had so recently been shown to be practicable. Surely it is not fanciful to see in this, one of the genuinely novel aspects of the 16th-century scene. That it was a false dawn, in that science was not to be so useful again for nearly two centuries, does not affect the fact that it was rightly taken to be a sign of innovation and triumphant superiority, in knowledge, if not in intellectual powers, to the ancients.

Finally, I think one might consider an offshoot of the surprised discovery that all knowledge did not stop with the ancients, but that many important and interesting things remained to be discovered in "modern" times—the early stirrings of the idea of intellectual and material progress. This seems to me peculiarly important because of the obvious fact that it is so much easier to acquire or devise a new idea or make a new invention if one expects to be able to do so. Or put another way, novelty is an offspring of the idea of innovation, even more than it is the offspring of necessity. So perhaps I might, somewhat tautologically, ask you to consider the spirit of innovation the most novel aspect of the 16th century's cautious but genuine appreciation of novelty as both desirable and attainable.

DISCUSSION OF M. B. HALL'S PAPER

GINGERICH: Thank you for this very eloquent paper. I have asked Professor Curtis Wilson to introduce the discussion.

WILSON: I think that innovation often occurs where there isn't what I would call a spirit of innovation. Copernicus would be a case, I think; he is a man who certainly wants his science to be true to first principles. I think he is also moved, perhaps, by the fact that, in contrast to Ptolemy, he is a Christian who believes in creation, who believes the world is a *machina mundi* that should have a certain beauty in its construction. He is not, therefore, in servitude to the notion that the heavenly motions are finally incomprehensible and we should merely save appearances. Machiavelli insists he is being entirely novel, that he is opening a new route like the navigators, never before followed by anybody, whereas Copernicus insists that he does not depart from the ancients except for good cause and when the facts practically force him to do so.

One of the interesting things in the paper concerns the role of humanism. You can find people going back to the past and at the same time innovating at a very great rate. I think Francis Viète insists the ancients had the analytic art that he is talking about. He nevertheless claims to be doing it anew because the intervening period of barbarism has somehow closed over the pristine thing.

The 16th-century figure, Simon Stevin, is in some ways a more interesting case in that he is an engineer and financier. He is involved in all sorts of practical activities. He is not interested in returning to some original Greek wisdom because he feels the Greeks didn't quite have it, and he feels there must have been a *siècle sage*, a wise age long before. He can't prove it, but he has all sorts of signs of it like alchemy and algebra, and he thinks the Arabs were writing before Ptolemy and had algebraic notation and so forth. So he postulates this ancient wise age to which we should now return by all kinds of innovative investigation and experimentation. So I think that the humanism and the notion of return to the past does not necessarily interfere with innovation when people have other ideas.

Let me close on the general question of social roles. I feel that Needham has shown a huge amount of innovation in China, but I don't see that he has shown any signs of a spirit of innovation. I am sure that innovation occurred here and there and people who were doing it had a sense of the novelty of it. But it didn't get into a literary tradition. I wonder if maybe what went on in the Renaissance was finally that with printed books and with publishing, people got to talking about novelties, and that then people like Bacon did claim to present totally new procedures. Thus I wonder if we don't need to go toward the direction of social contexts to find when innovation comes to be aided and prompted by a spirit of innovation.

NELSON: I would just make this one observation concerning the last remarks. As far as China is concerned, the civilizational structures were so constituted as to offer little possibility that innovations in technology would be reinforced in a heightened spirit of innovation. This is a decisive element in the present discussion that has to be isolated, because the whole structure of Chinese civilization is based upon the idea of maintaining the mandate of Heaven and the continuous effort of preservation of the truth, the work of the Mandarin, the Confucian scholar, and so on. So we do not get the spirit of innovation there.

Part of Dr. Hall's thought seems to me to require reference to these comparative civilizational notions. We can have the spirit of innovation in a society that allows a place for the notion of a spirit of innovation. What I sense in Dr. Hall's remarks was that in the era under discussion the spirit of innovation becomes very much more vocal and general and even institutionalized as part of the cultural structures. A very great deal of innovation does go on without its becoming defined as an explicit sort of horizon of the society.

There was one other thought I had in this connection concerning the on-going controversy at this time between the ancients and moderns. What I would observe is that there are people who can very well be for the ancients and be for innovation. There is no absolute coincidence of being for the past and being against innovation. As Professor Wilson has observed, in the case of Copernicus, we have a person who does not continue to think of himself as being cut off from all previous ways of thinking and yet very definitely innovates. It seems to me that we have got to work at the problem of innovation against these other perspectives.

M. B. HALL: As I tried to say at the beginning, it wasn't my intention to talk about the content of science or how scientists do science. What I was hoping to air was the point that in the 16th

IMPRESSIO LIBRORVM.

Poteſt vt vna vox capi aure plurima: Linunt ita vna ſcripta mille paginas.

"Just as one voice can be heard by a multitude, so single writings cover a thousand sheets," according to this scene from Stradanus' *Nova reperta* (Antwerp, c. 1580). (Courtesy of Burndy Library.)

century, there is a very genuine belief that novelty, innovation, is both possible and desirable; and this has not really very much to do with whether they were genuinely novel.

EISENSTEIN: I have been wondering about this in connection with the question of the impact of printing. The question of novelty in the 16th century is extremely complex because ideas of what is new and what is innovative, what is old and what is traditional, are undergoing change at this very time.

In early printers' prefaces and blurbs you find the boast of being innovative asserted most strenuously. Before printing you do find some interesting concern with novelty expressed in Friars' sermons. There is a sermon collection of 1332, or perhaps before, stating that modern man admires novelty and disdains the wisdom of antiquity.

A list of inventions, compiled before printing, appears in a philological and humanist work by a Vatican librarian in the 1440s, which later ran through a great many incunabula editions. In connection with a dictionary he is compiling, the librarian inquires whether it is permissible to coin new words for the new things that have appeared in the postclassical period. Thus he lists all the postclassical inventions that cannot be defined by classical Latin. I think interest in innovation is largely rhetorical and philological before it gets implemented in a new way with the advent of printing. Thereafter one finds printers, engaged in competing with each other, saying: "My book is better than yours, and I have got an edition that contains something new in it." The places to look for a new spirit of innovation are in prefaces to arithmetic books and to other texts and guide-books. Map making and map publishing also encourage a new discrimination between the old and the new. The sense that the Ptolemaic "oecumene" is set apart from the new world comes from producing atlases containing old world maps and new ones.

In terms of actual innovation, it is often the reluctant innovator, such as Copernicus or Erasmus, who creates a revolution without necessarily intending or realizing what he is doing.

I would suggest that maybe the modern concept of innovation is the offspring of preservation after printing. You can see more clearly where you are innovating because previous steps are more permanently fixed and also made more visible.

GINGERICH: The concept of the reluctant innovator is an interesting one. I was about to ask Ed Rosen to comment on Copernicus as an innovator.

ROSEN: There is one trend of thought in this period that should be emphasized. It is that innovation is equal to renovation. In other words, new is old. This is connected with a periodization of history, namely, that the world has gone through three stages, the ancient wisdom, the recent stupidity, and the revival of the ancient wisdom. The search for novelty is, therefore, identical with the search for the ancient wisdom. This is Copernicus' view. Both he and his disciple Rheticus (even more emphatically) explicitly disclaimed that what they were doing was done for the sake of novelty. That would have been a great disgrace in their opinion. It was the search for truth that motivated them, and, of course, the search for truth was the key tendency also in ancient science. That was their point of view.

WALLACE: I am interested in some of the terms, for example, *innovare*. It would seem that in theology the term *novatores* had a considerable usage, being a common designation in Catholic circles for the reformers. Whether or not it was generally used in other contexts during the 16th century I do not know.

Another expression is *via moderna*, and here my interest arises from my studies of Ockham and the nominalist tradition generally. Nominalism, as proposed by Ockham, is a doctrine which maintains that only individual things exist and therefore denies any extramental existence to universals or natures that individuals share in common. The doctrine encourages a skeptical attitude towards intellectual knowledge and particularly its objective value, and so is somewhat akin to fictionalism, which views concepts and explanations arrived at by reasoning as mere fictions, and thus devoid of any real existence. Now the expression *via moderna* was generally used by writers in philosophy and theology to designate Ockham's school and so distinguish it from the *via antiqua* of the more realist Aristotelians. I wonder if the term *moderni* was also used to designate followers of the *via moderna*? If so, however, this was not the only usage, for I have noted in a manuscript of Dietrich von Freiberg, dating from about 1304, the statement that he and other "moderns" (the Latin is *moderni*) have used the astrolabe to fix the position of the Milky Way in the sphere of the fixed stars. Now Dietrich was a Dominican friar, in the tradition of Albert the Great, and too early to fall under nominalist influences.

Other terms, such as *recentiores* and *iuniores*, are used continually in the scholastic literature of the early 16th century, and these too have connotations of innovation. Yet there seems to be a change of attitude towards those designated by these terms in the late 15th and 16th centuries, as compared to the 14th century. In the 14th century they were generally combatted by the more orthodox philosopher and theologians, whereas in the later centuries there seems to have been a major effort to assimilate their thought. Thus it became a part of the scholastic Aristotelianism itself, which was characteristically eclectic in the early 16th century, though still differentiated from the Greek Aristotelianism of the classical humanists. Incidentally, it is not an easy matter to translate *recentiores* and *iuniores* into English, and perhaps it is noteworthy that both terms are usually translated simply as "moderns."

M. B. HALL: Curiously enough, in the 16th century, they don't so much talk about "us moderns." They talk about a new age and new discoveries and novelty. New, but there is less emphasis on the *moderni*, I think.

ROSEN: Those terms to which you refer, Father Wallace, are purely chronological in content, and they are ambivalent as regards approval or disapproval. All the words that you mentioned mean "postclassical." It depends on the content and on the author whether he is blaming or praising.

For example, in the astronomical treatises of the 16th century, you sometimes find *moderni* used as a term of contempt for people who don't even know what Ptolemy had explained in the *Syntaxis*. On the other hand, you find the word used to mean that these people had made observations that were unavailable to Ptolemy. In these instances the implication is that this is a term of praise.

WESTMAN: I want to make a remark that may seem to be a slight digression. I live in what is considered an old neighborhood in Los Angeles. Some of the buildings are as much as 35 years old. There is a restaurant that proudly proclaims that it was founded in 1956. Now, the sense of what is old depends upon where you live. Obviously, if you are living in a place where buildings are being torn down quite frequently, you cannot help but feel a sense of change going on all the time.

The rapidity of such change would not have been true of the Renaissance, but it seems to me these terms of novelty indicate some kind of changing consciousness, perhaps a vague sense that something is happening in the society.

Now the question I have is whether the *consciousness* of novelty precedes *actual* innovation or whether the two go hand-in-hand together. There must be some kind of interaction between the consciousness of novelty and the actual emergence of novelty itself. This is a hard matter to document.

NELSON: It seems to me this discussion tells us in many ways that Dr. Hall has opened a large area that has to be examined carefully in both wider temporal and spatial contexts.

There is a stratigraphy here that is intriguing. The return to the old or the insistence on preserving the old and, indeed, the unhappiness about innovation has very deep connections all through the medieval times, because it is the old way which is felt to guarantee certain kinds of orders, not just hierarchial and established ones, but also the orders of existence for communities. There are many rights that are undefined, and so we discover the following fundamental fact: Every time any new law is brought forward it has to be rediscovered and shown to have been an old law. If we study the attitudes among peasants and, for that matter, other members of the society at higher levels, we find that great numbers of all groups are continually insisting on the absolute necessity of preserving older ways.

What I am proposing is that there are many strata within the struggle over old and new, and they are not of one sort. One critical thing is that very many quite new things are done under old terms or old terms given new meaning.

WESTMAN: What Professor Nelson has just been saying suggests to me that there really is an ambivalence about novelty and change. You do something that is new, and as soon as you do it, you have to reassure yourself and others that it is really old. It is probably true in general—not only in the Renaissance period— that the attempt to assert an autonomous position is frequently accompanied by the statement that one is not really completely autonomous.

LASKY: I want to raise the question whether we are not discussing "new" and "old" too exclusively from the 20th-century secular point of view and, possibly, isolating it too much from the entire Christian polemical atmosphere of the time. I would say that orthodoxy and heresy has in itself the analogy between the Old and the New, and it has always been the explosive contradiction in Christianity. It has proved to be a very fruitful context for modern scientific—and social-scientific—developments simply because it always had that ambivalence in it.

When suddenly we get the very sharp 17th-century polemics about whether everything is "new"—after all, the Biblical text itself said we shall inherit a new Heaven and a new Earth and a new Man—and whether the very notion of newness is acceptable by tradition, isn't that the modern intellectual context in which all kinds of heretical notions can emerge?

It seems to me that the very sense of "newness" is one that was approved of in many of the great theological contexts. They should, properly, be read that way. I would think that all of these 15th-, 16th-, and 17th-century intellectuals—and that is what they were—lived fully in an intellectual atmosphere of polemic and dissent. Consequently, I would think you couldn't really divorce the sense of newness and of oldness, of orthodoxy and innovation of tradition and change and all of the intellectual dialectics that go with these ambiguities. A representative mind of the Copernican era may well have argued: What we are for is not "innovation," which is terrible; but we are surely for "the new."

There were also the theologians who said that they were neither for innovation, nor for the new; they were for "regeneration." Whatever the verbal strategies were, the actual historic change remains the same. But I have one final question: Whether the "revolutions" of science and the heavens aren't really, from a psychological point of view, fundamentally different from the revolutionary changes in society? Because, after all, if you dis-

discover Copernican truths about the distance in the skies, they were always *there*. Even if you discover something that hadn't been previously known about them, the facts were out there from all time, whether created by Nature or created by God.

However, if you are creating a new way of doing something, such as Gutenberg printing, you cannot really avoid its absolutely unprecedented character. Here you have an authentic, almost empirical, almost physical sense of some innovation, which is actually a breakthrough in European history.

R. HALL: I think this is surely very true. There must be a sense in which a man who assembles a mechanism for the first time feels that nobody has done this before. On the other hand, when the microscope is introduced, people have an idea that, after all, the ancients were aware of lenses and microscopes and so forth. In that sort of sense, you can get a justification, a historical precedent, for what you are doing, even though you know that nobody did it last week, so to speak.

ROSEN: Did you ever look at the meaning of the term "inventor primus?"

Galileo was called a liar because on the title page of the *Sidereal Message* he says he invented the telescope. This accusation is advanced by people who think "inventor" 20th century equals "inventor" 17th century, but it doesn't. The first English law protecting inventions is 1624, at the time of James I. If you look at that statute—and I was fascinated when I saw it to realize that it answers the Galileo problem—you will see that the claim of the first inventor is distinguished from the claim of the second inventor. That implies than an inventor is anybody who makes a thing for himself. If, however, he adds the claim of priority that nobody has done this before him, then he is not called "inventor" in the 17th century, but "first inventor." Having seen this in the English statute, I then examined the contemporary documents elsewhere. And I found "primus inventor" in Latin and equivalent expressions in Italian, Spanish, German, and so on.

HARTNER: I must say that Galileo evoked the impression when

SIDEREVS
NVNCIVS
MAGNA, LONGEQVE ADMIRABILIA
Spectacula pandens, suspiciendaque proponens
vnicuique, præsertim vero

PHILOSOPHIS, *atq̃* *ASTRONOMIS*, *quæ à*
GALILEO GALILEO
PATRITIO FLORENTINO
Patauini Gymnasij Publico Mathematico

PERSPICILLI
Nuper à se reperti beneficio sunt obseruata in LVNÆ FACIE, FIXIS IN-
NVMERIS, LACTEO CIRCVLO, STELLIS NEBVLOSIS,
Apprime verò in
QVATVOR PLANETIS
Circa IOVIS Stellam disparibus interuallis, atque periodis, celeri-
tate mirabili circumuolutis; quos, nemini in hanc vsque
diem cognitos, nouissimè Author depræ-
hendit primus; atque

MEDICEA SIDERA
NVNCVPANDOS DECREVIT.

VENETIIS, Apud Thomam Baglionum. M DC X.
Superiorum Permissu, & Priuilegio.

Title page of Galileo's *Sidereus nuncius* (1610). Galileo's claim to have
invented the telescope, discussed here by Professor Rosen, appears in the
words "PERSPICILLI Nuper à se reperti." (Permission of Harvard College
Library.)

he stood before the Senate of the City of Venice that he had found the telescope as the first man in the world. In my opinion, it is lack of sincerity.

M. B. HALL: It is salesmanship.

R. HALL: I may add that you find the same terminology exactly used in the controversy between Newton and Leibniz, the controversy between the "first inventor" of the calculus and the "second inventor." Which individual is in which role is exactly what is at stake.

GINGERICH: Let me ask Willy Hartner to discuss some technical aspects of Copernicus as an innovator.

HARTNER: Earlier Professor Nelson remarked that when something is new, you can reassure people by calling it old. The opposite is also true, namely, that in many cases, things are claimed to be brand new, and they are age-old, and nothing but a slavish copy of what predecessors have done.

I have here some results of my computations concerning the theory of Mercury according to Ibn ash-Shatir, the *Commentariolus*, and *De revolutionibus*. Although there still remain many subtle questions that require further research, it is evident —and nobody doubts it—that the real aim of both authors is not to correct and improve the Ptolemaic theory but, on the contrary, to reproduce it as accurately as possible with the aid of a different kinematic model strictly in accordance with the axiom of uniform circular motion. The same of course is true of the two men's theory of the Moon and of the other planets. In the case of Mercury, taking as a test the maximum deviations from Ptolemy's value of the radius of the epicycle ($22^r;30$), they are in the magnitude of half of one part in Ibn ash-Shatir, of a sixth of a part in the *Commentarolius*, and of a fifteenth of a part in *De revolutionibus*. Unless Ibn ash-Shatir's parameters should have emerged from new observations, this indicates a successive numerical improvement testifying to Copernicus' mathematical skill. From the point of view of observational astronomy, however, it marks no progress but rather stagnation.

Rosen: There is a dreadful error made by Copernicus in the *Revolutions*, I, 10, dealing with the greatest and least distances of Mercury, but we have to bear in mind what it is Copernicus relied on. I have, I think, definitely established that for this information, Copernicus did not rely directly on Ptolemy himself, but instead he used a passage in Proclus' *Hypotyposis*. Of course, Proclus was not available to Copernicus in Greek, but he did have access to Giorgio Valla's two-volume work, *De expetendis et fugiendis rebus* (Venice, 1501), in which Valla compresses large sections of Proclus.

Now Giorgio Valla, although he was related to that very great man, Lorenzo Valla, and was himself a scholar of some repute, unfortunately really did not understand astronomy. When he compresses this passage of Proclus dealing with the orbits of Venus and Mercury, whether sun-centered or earth-centered, he goes way off the beam and leaves out a crucial sentence, probably because he couldn't understand Proclus.

The result is that when Copernicus tackles Valla to find out about the greatest and least distances of Mercury, Copernicus makes a dreadful error. He takes Valla's number for Mercury's greatest distance from the center of the Earth to be Mercury's interapsidal distance, in other words, the difference between Mercury's greatest and least distances.

You may say, "How can a great astronomer make an error of this sort?" The answer is that it didn't concern him. He is not talking about his own theory; he is rebutting a certain argument. He simply wants to show by using the principle of the full universe that, if you assume a stationary central Earth, the greatest distance of Mercury from that point is the same as the least distance of Venus, and so on. Hence for his purposes at the moment, it didn't matter what the significance of these numbers was; but I was astonished to see that Copernicus, misled by Valla's cockeyed Latin translation of Proclus' Greek, did not really grasp the significance of this whole passage and went sadly astray.

SWERDLOW: I have looked at the passages in Valla, in Proclus, and in Copernicus, and I thought they all seemed faithful one to the other. The Proclus distances are peculiar because he ends up with the greatest distance of Venus reaching beyond the least distance of the sun. I don't know why that doesn't bother Proclus, but it doesn't seem to.

GINGERICH: You are going to have to settle this during the coffee break, because it is getting very technical.

HARTNER: I am sorry, but I have to go into technicalities because it is getting fascinating for me. Ed, you speak about Valla; does he refer to Proclus directly or not?

ROSEN: No, Valla conceals the fact that he is condensing Proclus. In general, the man was a crook.

EISENSTEIN: It upsets me to hear him called a crook, because I think Valla wrote before there was a clear picture of what is involved in plagiarism. He was writing in a period before literary property rights were clearly defined, when people didn't have a clear idea of how and when to cite authority.

ROSEN: When other men translate from the Greek, they say this is Theodosius, this is Plato, and so on, but not Valla.

SWERDLOW: Book 18 is partially a paraphrase and partially a translation of Proclus' *Hypotyposis*, and he does mention Proclus in Book 18 and also earlier in Book 16.

VENICE, SCIENCE, AND THE INDEX
OF PROHIBITED BOOKS

Paul F. Grendler

This is a report on my attempt to discover how effective in practice were the *Index of Prohibited Books* and the Roman Inquisition in 16th-century Italy. The focus is Venice because of its importance as a publishing center and because its Inquisition records have survived and are easily accessible. The various editions of the *Index* and laws of the Inquisition are generally known, but little is known of their enforcement. There is a Venetian proverb: *"Una legge veneziana dura una settimana,"* that is, "a Venetian law lasts only one week." My task has been to try to determine if a certain law was enforced, to what degree, for how long, and why.

Whoever sought to censor the Venetian press faced a formidable task. From obscure origins, it became one of the three or four largest in the 16th century. A few statistics will give us an

This is an abbreviated version of the paper read. For documentation, please see Paul F. Grendler, "The Roman Inquisition and the Venetian Press 1540-1605," to appear in *Journal of Modern History;* and a forthcoming monograph on the same subject.

idea of the size of the industry. To make a conservative estimate, the Venetian bookman printed about eight million individual volumes in the second half of the 16th century. This was the result of about 8,000 editions, new and reprints, an average of 160 a year. These figures do not include pamphlets, which did not receive privileges. In any given year, 50 or more publishers produced at least one title, and about 500 publishers appeared on the title pages of Venetian books during the 16th century.

A Venetian press run varied according to the anticipated demand and whether the publishing firm was large or small. The average press run of a title of ordinary sales potential was about 1000 copies. A major publisher with a title of assured high demand printed press runs of 2000 or 3000. By comparison to the English-speaking world of today, 16th century Venice, a city of 125,000 to 190,000, produced an enormous number of books for its time.

Neither Church nor State in the Renaissance, nor perhaps at any other time, believed in complete freedom of the press. The Church was interested in doctrinal censorship, the State was interested in political censorship, and both in moral censorship. In practice, however, until the 1540s, the Venetian press had a minimum of censorship. For several years, the papacy had pressured the Venetians to do something about heretics and heretical books within the Venetian Dominion. The Venetians resisted the papacy until political miscalculation put them in an awkward position necessitating some gesture toward Rome. As if to erase the memory of their sympathy for the Protestant cause during the War of the Schmalkalden League (1546-47) and to assure Pope and Emperor of their orthodoxy, the Venetians in the spring of 1547 established a new magistracy with particular competence in heresy, called the *Tre savi sopra eresia*, or the "three sages concerned with heresy." Their task was to assist the Venetian Inquisition in every aspect of its activity.

From that point through the end of the century, the Venetian Inquisition functioned in the following way: The tribunal consisted of six members, three clerical and three lay. The Inquisitor, the Patriarch or his representative, and the Apostolic Nuncio or his auditor were the ecclesiastical components. The three Venetian nobles made up the civil component. It met every Tuesday, Thursday, and Saturday, month after month, year after year, with very few days missed. All six could be present at the trials, but only the three clergymen handed down sentences; the three laymen then authorized the execution of these sentences. But there can be little doubt that the laymen were the dominant component. The clerical members had to be aware of their sentiments; otherwise the tribunal could not function. The Inquisition could not issue a warrant for arrest without the concurrence of the three lay assistants. After all, the *Tre savi* were the government. They were named from the most important members of the patriciate, men who were frequently Procurators of St. Mark and held seats on the Council of Ten; I have found three members of the *Tre savi* who later became doge. On the whole, however, those nobles who were more sympathetic toward Rome tended to be named to the Inquisition. Notorious anti-papists like Niccolò Da Ponte and Leonardo Donà were seldom selected.

From a slow start in the 1540s, when the nobility were little concerned with heresy and heretical books, inquisitorial activity grew as the patriciate became Counter-Reformation minded. The rhythm of the development of the Inquisition's activity against heretical literature is as follows: In the 1540s, the machinery was set up, but the level of prosecution was low. In the 1550s the Venetians started to become inquisitors, although they still resisted jurisdictional initiatives from the papacy. In the 1560s, climaxing in 1569 to 1571, the Venetians prosecuted heretical books with as much zeal as even St. Pius v could want. Over the next 20 years, enforcement continued, but the fervor gradually waned until the papacy and the Republic quarreled over books in 1596, as they disputed other matters.

If book censorship is to be effective, some kind of list or index of banned titles is necessary. In January 1549, the Council of

Ten ordered the Inquisition to draw up such a list. The list or catalogue was to include "all the heretical books," "other suspect books," as well as books "containing things against good morals." The list was completed and printed in May. No sooner was it printed, however, than the Venetians drew back for a variety of reasons. By the end of June 1549, the battle was lost, and the catalogue was suppressed.

The 1550s witnessed a prolonged effort by the papacy to get an *Index* adopted by the Venetians, while the bookmen fought tenaciously with the limited weapons at their disposal to stop them. In March 1555, Rome sent a new *Index* to the Venetian Inquisition. The bookmen were given three months to comment before it would go into effect. They took advantage of this to present three written protests in which they made the following points: Many authors had their *opera omnia* prohibited despite the fact that most of their books had nothing to do with religious matters. Second, they pointed out the financial losses that the book industry would suffer. Third, they argued that the Church had tolerated the works of such pagan authors as Lucian for 1400 years; such titles were of great importance to humanistic studies and should not be banned. Fourth, they used the argument of 1549 that the Inquisition wanted to subject the bookmen of Venice to an *Index* not in effect in Rome.

Rome heeded the protests. It modified and then suspended entirely the 1555 *Index*. The bookmen resisted each new *Index*, but each time they had to concede some part of their freedom. The Venetian government was not very concerned with their plight, and, unfortunately, each new *Index* listed more titles than its predecessors.

Paul IV issued the next *Index* in early 1559. From January through March, the Venetian bookmen refused to obey. If they were to give up some of their books to be burned, they demanded financial compensation from the papacy. Nevertheless, the papacy discovered a weapon to enforce compliance. All major Venetian publishers owned bookstores outside the Venetian state, usually all over Italy, including the papal dominion. The papacy threatened to seize the stores and their contents within the Papal State. In the face of this threat, the bookmen began to comply. From April through August, they made their submissions, offering inventories and some books to be burned. They did not, however, give up all their prohibited books. From the inventories, it appears that they yielded northern Protestant books, but did not yet give up such Italian authors as Machiavelli and Aretino.

By 1560, the intellectual atmosphere had changed considerably. A generation of free, mocking, anticlerical authors had died, or had found the climate uncongenial to their writing, and had gone into retirement. Machiavelli's name was disappearing from books. Intellectuals were noticeably more cautious. At the same time, a genuine religious revival under the leadership of a reformed papacy occurred.

The bookmen were businessmen attuned to the intellectual atmosphere. They clearly saw what was happening and reacted like good merchants. They began to publish many more religious books and much less secular vernacular literature. By secular vernacular literature is meant poetry, drama, collections of letters, dialogues on various topics, courtesy books, Italian grammars, and Italian classics like Dante, Petrarch, Boccaccio, and Ariosto. Into this group fall most of the works of the most popular and prolific 16th-century authors like Pietro Aretino, Anton Francesco Doni, and many others.

I have analyzed the *imprimaturs*—that is, the permissions to publish new books—from 1550 through 1605 and found the following: From 1550 through 1560, secular vernacular literature accounted for 27 percent of the new titles published while religious titles accounted for 14 percent. During the 1560s, the figures began to change as vernacular literature dropped and the number of religious books rose. Then, from 1570 through 1605, the figures reversed. Secular vernacular literature accounted for 20 percent of the total and religious titles 31 percent. These

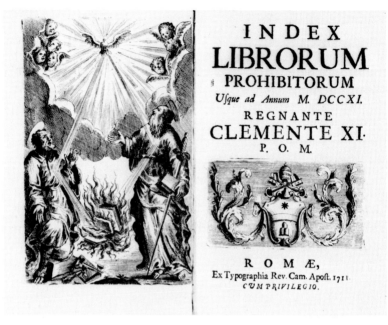

INDEX
LIBRORUM
PROHIBITORUM
Ufque ad Annum M. DCCXI.
REGNANTE
CLEMENTE XI·
P. O. M.

R O M Æ,
Ex Typographia Rev. Cam. Apoft. 1711.
CVM PRIVILEGIO.

The light of the Holy Spirit operating through St. Peter and St. Paul, the patrons of the Church, kindles the prohibited books in this frontispiece to a 1711 edition of the *Index*.

figures are so clear that they do not need comment. In short, Italians were "getting religion," and the bookmen were supplying them the books. The publishers simply switched from supplying a secular mass market to a comparable devotional one. There is no evidence that they lost money or that their presses were idled by the switchover.

The political climate changed as much as the intellectual atmosphere. For the Venetians, the major threat was the Turk, and to meet it, the Republic had need of papal assistance. Indeed, because of the Turkish menace, the Venetians were on friendlier terms with the papacy than at any other time in the century. In the 1560s, Church and State cooperated to enact new censorship legislation. A number of decrees erected various tedious legal hurdles before an author or publisher could obtain the necessary license to publish a book. These regulations were not new, but they were more extensive and better enforced. The government tightened the inspection of imported books at the customs house, giving the Inquisition power to have a man on the spot to inspect the books. All this was summarized in an omnibus law of the Council of Ten of 28 June 1569. For its part, the Council of Trent issued a new authoritative *Index of Prohibited Books* with extensive rules for authors and publishers. In 1564, the Venetian government accepted this *Index* along with all the other Tridentine decrees without a murmur, but the bookmen ignored it.

With the passage of the law of 28 June 1569, the Inquisition began enforcement, obviously with the approval of the government. It demanded inventories and the consignment of prohibited titles to the tribunal. The Inquisitor's men began making personal visits to the shops and storehouses of the bookmen. These visits were new. In the past, the Inquisition had no such authority. They caught a number of bookmen by surprise. They then carried off the prohibited books to the Inquisition and punished the bookmen.

From 1569 through 1571, a large number of the books were confiscated, this time Italian books of Machiavelli and Aretino, as well as those of northern Protestants who wrote on non-religious topics. The usual procedure was to burn the books and fine the bookmen from a few ducats up to fifty, depending on how guilty they were or how able they were to pay the fine. Twenty-two bookmen were fined or warned at this time.

There were limits to the Venetian acceptance of the war against heresy, however. Although the Venetian government demonstrated little sympathy for the bookmen, they did protect the German Protestant scholars at Padua. Of a total of 1000 to 1500 students in any given year, the Germans, by far the largest group of foreigners, numbered from 100 to 300. Most of them were Protestants, and a good number of the French scholars there were Huguenots.

As scholars and students, these foreign Protestants brought prohibited books into the Venetian territory and were good customers for the prohibited titles of Erasmus, Melanchthon, and others. The papacy wanted to keep these Protestant scholars and their books out of Padua, but the Venetians turned a deaf ear. They gave three reasons: (1) They did not wish to offend German princes by turning away German students. (2) The Venetians did not wish to lose the 25,000 to 30,000 ducats that the German scholars spent every year on food, accommodation, clothing, books, and so on. (3) The greater the number of scholars, especially foreign ones, the greater the reputation of the Studio di Padova. The papacy tried to answer these arguments, and the Venetian government agreed that every scholar had to make a profession of faith before receiving his degree. But the law was not enforced, and the papacy knew it.

From 1573 until the early 1590s, the *Index of Prohibited Books* was strongly enforced on the surface. Certainly, very few prohibited books were published, but there are many signs that there was a great deal of violation in the sale and importation of prohibited titles. Certain kinds of heretical and prohibited books could be found without difficulty if one knew where to look. Humanistic works by northern Protestant authors like Melanchthon, Protestant editions of the Bible like that of Antonio Brucioli, the works of Erasmus, Italian titles like the works of Aretino and Machiavelli, and the works of Ochino, Calvin's *Catechism*, and a few titles of Luther could be purchased at Venetian bookstores. Some bookstores made a practice of selling these books under the counter.

Scholars in the humanities particularly resented and disobeyed the regulations of the *Index* and Inquisition. Northern humanists, many of them Protestants, wrote a great deal and very well in those disciplines. If Italian humanists were denied access to these books, they would be shut off from the European world of learning. It was practically impossible, however, to obtain an Inquisition permission to hold these titles, so they disobeyed the laws and tried to be discreet about it. They did not run afoul of the Holy Office unless they were flagrant violators. To cite a case of 1580, one monk was not denounced for heretical books and opinions until he had been in the monastery for six years, had tried repeatedly to persuade his fellow monks to accept heretical views, and had shown himself to be an outspoken, irascible man, completely lacking in tact. One wonders how many others, more discreet and easier to live with, held heterodox religious books and views without being disturbed.

Throughout the 1570s and 1580s, the Venetian bookmen continued to smuggle prohibited books from Germany and Switzerland into Venice. On their regular journeys from the Frankfurt bookfairs, they acquired prohibited works, especially in Basel, and brought them into Venetian territory. The bookmen then found ways of eluding the Inquisition check at the customs house. One can only speculate on how they did it. Bribery is a possibility, either of the customs officials or the Inquisition agent, although it must have been difficult to bribe the latter. False title pages were, of course, used. Most likely, in my opinion, is that the sheer cumbersomeness of the process, which involved several people and minute lists of the titles, produced shortcuts and carelessness. Once through the customs, it was practically impossible to trace the books.

To give an idea of what impact the *Index* and Inquisition had on the availability of prohibited books to the scientific community of Padua, the library of Gian Vincenzo Pinelli provides an example. A Neapolitan nobleman born in 1535, he came to study in Padua in 1558 and stayed for the rest of his life. Endowed with money, love for scholarship, and a gift for hospitality, he opened his house to scholars of the most diverse interests and backgrounds. He is best known as Galileo's friend, host, and patron in the latter's Paduan career, within the years 1592 to 1610; but Torquato Tasso, Paolo Sarpi, Sperone Speroni, and Justus Lipsius also enjoyed his hospitality, while his correspondents included such scholars as Joseph Scaliger and Paolo Manuzio. He was acquainted with major church figures as Cardinals Caesar Baronius, Robert Bellarmine, Charles Borromeo, and Ippolito Aldobrandini, who became Pope Clement VIII. He was equally familiar with a number of Venetian patricians more interested in learning than politics.

Scholars came to Pinelli not just to enjoy his company, but to use his library, one of the famous collections of its day. You may know its sad fate. Upon Pinelli's death in 1601, it was first burglarized by a servant and then purged of about 200 books on politics by the Venetian government. The bulk of the collection was then loaded on a ship to be sent to Naples, but the ship was taken by pirates who, disappointed not to find treasure, dumped eight chests of books, two of portraits, and one of mathematical instruments into the sea. Eventually, part of the Pinelli collection was found and purchased by Cardinal Federigo Borromeo, and deposited in the Ambrosiana Library in Milan.

Luckily, an inventory, dated 7 October 1604, of the books and manuscripts was made, probably by order of the Council of Ten. It is now in the Marciana Library in Venice, unpublished and, so far as I can tell, unnoticed by scholars.[1]

The inventory catalogues a truly magnificent library of about 6500 printed volumes and manuscripts, ranging from incunables to volumes published the year before Pinelli's death. The titles are in Latin, Italian, Greek, French, Spanish, Hebrew, and Arabic, and the subject matter covered everything that interested Renaissance men. Among much else, Pinelli had a copy of the first edition of Copernicus' *De revolutionibus*.

The inventory listed at least 90 prohibited titles by about 44 different banned authors. That is roughly 1.5 percent of the total. The vast majority of the banned titles were works on topics other than religion, but were authored by scholars who happened to be Protestants. Although these Protestants had sometimes written a tract or two attacking Catholicism or propounding Reformed doctrine, they wrote many more volumes in areas such as law, mathematics, rhetoric, and so on, and are chiefly remem-

Part of the inventory of Pinelli's library, Biblioteca Marciana (Venice), Mss. Italiani, Classe X, 61 (6601), f. 100r. The second entry from the bottom is the 1543 Nuremberg edition of Copernicus, here mistitled *de Revolutionibus Motuum Celestium*, folio.

bered today for these other works. As the Jesuit *ratio studiorum* put it, their *opera omnia* had to be banned, because a book usually created affection for the author that might lead the respectful reader to accept the author's bad ideas as well, and because it was very rare that "a breast full of poison" could write anything uncontaminated.

The discovery of prohibited books always prompted the suspicion of heresy by the Inquisition. Should we suspect Pinelli of religious unorthodoxy? A closer look at the prohibited titles and his life dispels the suspicion. His correspondence and his library demonstrated that he was interested in religion, for he had many Catholic religious works but no prohibited titles by such northern heresiarchs as Luther, Calvin, Zwingli, Bucer, Bullinger, and Beza. Pinelli was a man of the broadest intellectual curiosity, but not a heretic. He was spiritually *catolicissimo* and intellectually most catholic. There is no sign that he was anything other than a loyal son of Rome, but he ignored the *Index of Prohibited Books* in many areas. This was not a question of ignorance, for he owned at least four copies of the *Index*.

Pinelli violated the Church's rules for the sake of learning. He is a vivid reminder that loyal Italian Catholics had the same commitment to the universal world of learning as their Renaissance predecessors, and that they honored that commitment in the middle of the Counter-Reformation.

My research suggests three conclusions: First, the *Index* and Inquisition became effective when the Venetian patriciate decided to support them. This took a number of years, and the joining together of political and religious motives. Once the ruling class made its decision, however, neither economic and jurisdictional reasons nor the pleas of the bookmen moved it.

Second, once supported by the government, the censorship was very effective for most, but not all groups of the reading public. The prepublication censorship insured that very few banned titles were printed in Venice. Halting the clandestine importation of prohibited titles and the resale of older banned volumes was more difficult, but again the Inquisition had notable success. For the majority of the reading population, the banned books were unavailable. But for the determined few, enough loopholes existed so that any title could be found if the reader were willing to bear the necessary difficulty, risk, and cost to get it. This was particularly true for the scholarly community. The evidence from the libraries of Pinelli and others argues that scholars did obtain banned nonreligious works from the north. It is reasonable to assume that they read them as well. Even if Italian scholars were hesitant to cite the works, they could still read, ponder, and discuss among themselves the forbidden titles. This was far from complete freedom of inquiry, but it was a good deal better than nothing.

Third, with the passage of time, State support for the *Index* waned. Just as a combination of circumstances generated this support earlier, so jurisdictional, political, and to a lesser extent economic motives eroded it in the 1590s. Eventually, in the 17th century, some banned titles were printed again in Venice.

Note

[1] Mss. Italiani, Classe X, 61 (6601), "Inventario della libreria di Giovanni Vincenzo Pinelli ereditata da Francesco Pinelli." My wife, Marcella, is preparing a critical edition of the inventory.

DISCUSSION OF P. F. GRENDLER'S PAPER

CHRISTIANSON: Why censorship? Professor Grendler asserts that suppression was the traditional response to heresy when the Venetians became convinced that Protestantism really was making inroads. The diplomatic failures of Catholicism with respect to the Schmalkalden League and in England, and the Ottoman threat were all part of this picture, of course. But if the motivation were suppression of Protestant heresy, then why were the Jews the first to suffer from censorship in Venice?

The 16th century was a period of intellectual and of religious conflict. It was also a period in which men tended to see all knowledge as comprising an interrelated whole. Certainly those factors in the intellectual milieu contributed to censorship.

But Professor Grendler keeps coming back time and time again to a question that I think is worthy of more comment: the scholar's relationship to the sources of power and wealth in his society. This is always a matter of utmost importance. In many ways, it is the factor by which all cultural activity, including science, is connected to the broader realities of the society within which it exists. It is a two-way street. It may be expressed in a positive form through patronage or support of scientific activity of one kind or another. It may be expressed in a negative form through various forms of censorship, scorn, or simple neglect.

The criteria for action on the part of those who have power are never simply cultural. Social responsibility is too broad for that. They have to fit cultural activity of various kinds into what they see as its proper place in society at large. Consequently, the scholars and the scientists are constricted; the limits of their activity are set by powerful and wealthy people whose motivations and points of view are not by any means wholly scientific or wholly intellectual.

As far as this repressive dimension of the relationship of scholarship to the sources of power and wealth is concerned, there might be a number of questions that we could ask:

First of all, Professor Grendler said that it was possible for a determined few to evade the ban of censorship. Did it make any difference who the person was, socially speaking? Could a patrician, for example, or a well-placed cleric, get and use banned scientific works with impunity whereas the possession of those same books might be incriminating to a Jew or a poor craftsman? Such social factors might determine which social groups were allowed to engage in activities such as science in the late 16th century.

Concerning the economic realities of publishing, could an individual author, either in Venice or elsewhere, bring pressure to allow his books to circulate even though they might otherwise be banned? I know that Tycho Brahe did this. His social position, his political and other connections allowed him to get copyrights to allow his books to circulate. He did so on the basis of individual negotiation through diplomatic channels.

In the sphere of technology, some questions also might be raised. The writings of Mrs. Eisenstein and others have made us aware that the letter press and associated developments constituted a technological revolution that had a tremendous social and intellectual impact on the 16th century. Professor Grendler and others have made us aware of the fact that the Venetians were as quick as anybody in exploiting the intellectual and economic advantages of that technical innovation. But like any important innovation with social consequences, the letter press had to be controlled and bridled in some way or another. It had to be worked into the fabric of the society which it was in the process of helping to change. Censorship in this context might be seen as an attempt to make the letter press work for social order rather than against it. Printing could be used to standardize and compartmentalize and regiment patterns of thought, as well as it could be used to open perspectives and to juxtapose ideas from various spheres of knowledge.

The question that we ultimately come to, in the face of all of these powerful diplomatic, social, economic, and technological forces, is, Where does science fit into the pattern? Is it simply a kind of a game, obscure and esoteric, being played by a few people who are really subject to the beck and whim of powerful social forces that they cannot hope to control? Are scientists a group who are increasingly isolated from other intellectuals and from other areas of intellectual activity as a result, or do they have a social feedback of some kind or another?

It would be foolish for students of 16th-century science to overlook the social forces, oppressive and otherwise, that worked upon scientific activity during that era. Professor Grendler's paper has been a demonstration of the impact of something like censorship. He has discussed the question of gaps in the walls of censorship. But how did scientists or their spokesmen plead the cause of science, bring their own influence, their own interest, to bear upon those powerful individuals who ran society and who administered the laws of censorship?

In the face of censorship of printed materials, how did scientists open up and use other channels of communication to bypass the censorship? In the period we are talking about, some of the other relevant media would be conversation, lecture, debate, correspondence, and manuscript treatises of various forms. These things are more or less hard to get at, but it is necessary to pin them down. In my own work, I have been interested in Tycho Brahe's relationship to a large number of co-workers. And it seems to me that a good deal of that relationship was expressed in ways that were not printed, through conversation or work on common projects, sometimes also expressed in manuscripts, sometimes not.

Here, then, is a final problem. What part of the action of 17th-century or late 16th-century science occurred in these non-printed forms and how do we as historians go about locating it?

GRENDLER: Did a patrician or a well-placed cleric have easier access to prohibited books than a commoner or layman? Yes, although the differences were not great. To my knowledge, however, few patricians and high clergymen had sufficient scientific curiosity to search for banned titles.

Could an individual author, editor, or publisher bring pressure to bear to obtain free circulation of a prohibited title? He might appeal to the Congregation of the Holy Office to change an absolute ban to permission for the title to circulate in expurgated form. This became quite possible in the 1590s with the elaboration of rules for expurgation, and some Italian titles that had not been printed for many years were again reprinted. But if the title remained on the *Index*, all he could do is try to secure its publition and distribution in some clandestine way. He might publish

An early printing and typesetting scene from B. Chasseneux's *Catalogus gloriae mundi* (Lyons, 1529). (Permission of Harvard College Library.)

it anonymously and/or arrange for it to be smuggled into the city.

Indeed, the state did use the press to protect the political and social order. The government was very concerned with political censorship. If someone authorized a treatise critical of the French king at a time when the Venetians considered him a friendly power, the state would move quickly and decisively. A civil tribunal, not the Inquisition, would confiscate the book and prosecute the author and, perhaps, the printer as well. There are examples of this in 16th-century Venice.

How did scientists plead the cause of science in order to stop censorship? To my knowledge, scientists in Venice and Padua did not make such appeals in the 16th century. The times were not propitious. Instead, men of learning tried to ignore or disobey the censorship machinery without drawing attention to themselves.

WESTMAN: Looking through Professor Grendler's list of Pinelli's library, I noticed not only the name of Philipp Melanchthon, but also several authors who published astronomical works —Simon Grynaeus, Caspar Peucer and Michael Neander. I noticed also several authors who published scientific treatises, but the books on this particular list are not the scientific ones; they are historical, theological, etc. Works on mathematics, astronomy, astrology, and physics really constitute a very low percentage of censored items compared to other subjects. The issues to which people were reacting were of a religious or political nature —which is not surprising, given the sensitivity of those areas particularly in our period of consideration. These are the areas where censorship strikes home. No one is censoring tables of numbers, such as Erasmus Reinhold's *Prutenic Tables*.

GRENDLER: These are only titles on the 1596 *Index* or prohibited individually between 1596 and 1604 by decree. It doesn't say that Pinelli doesn't own many, many other titles. I don't have the entire inventory here.

ROSEN: Even the first prohibition of Copernicus' book is not until 1616.

GRENDLER: Pinelli had Copernicus, but I haven't included in this list books that were later banned.

EISENSTEIN: It is not accidental, I think, that Pinella owned four copies of the *Index*. It suggested to him some of the avant garde, interesting titles that might be collected. Protestants often used the *Index* as a way of suggesting titles for publication. Just as today you find booksellers actually advertising books "banned in Boston," in 17th-century Holland books that weren't even on the *Index* were advertised as if they were.

R. HALL: How far did this principle of guilt by association extend? I haven't got the sort of imagination as to what can be so heretical about Conrad Gesner's *De lacte et operibus lactariis* or Sebastian Münster's *Cosmographia*. Was it normally the case that if a man published one little work that was very bad, they banned everything?

GRENDLER: This often happened, and I think that the reason was bureaucratic rather than religious. At first, with the highest intentions, the censors working for the Congregation of the Index in Rome intended to examine all suspect works. They very soon discovered that this was totally impossible. So, as the century moved on and new authors and titles appeared, if they found an author who had written one anti-Catholic work, as Gesner had, they would ban everything rather than read through all his works. The basic principle at work was stated in the Jesuit *ratio studiorum*: "How can a breast full of poison write anything uncontaminated?"

ROSEN: Reusch's book* on the *Index* in the 16th century shows that all the publications of certain authors were banned, and this is where you find the most prominent names, whereas in other cases only selected titles were prohibited. I think that Professor Grendler's remark is absolutely correct, since nobody had the time to read all this stuff and weigh the implications. So if they found anything that was objectionable, to be on the safe side they put that author in the first class, namely, those who were completely prohibited.

WALLACE: When I first started to teach 20 years ago—in a Catholic seminary, a Dominican House of Studies—I had to obtain permission to read books that were on the *Index*. At that time, it seems to me, a distinction was made between books that were prohibited because they were salacious and corruptive of

* Heinrich Reusch, *Die Indices librorum prohibitorum des sechzehnten Jahrhunderts* (Nieuwkoop, 1961 [reprint of 1886 ed.]), esp. p. 177.

morals and those that were prohibited because they were doctrinally suspect in matters of theology or philosophy. I had no difficulty obtaining permission to read books in the second category, but had I asked, I probably would not have been given approval to read those in the first. Thus some books were effectively prohibited even for professors in seminaries and universities, and this up to the time of the Second Vatican Council.

THE QUEST FOR CERTITUDE AND THE BOOKS OF SCRIPTURE, NATURE, AND CONSCIENCE

Benjamin Nelson

The presence of Edward Rosen in our midst today encourages me to begin by explaining that some pages in Rosen's lively and learned book on *Three Copernican Treatises* served as one of the main spurs that encouraged me to quicken my pace in a research program that has now carried me to new borders of the comparative historical *differential* sociology of culture and science—*in civilizational perspective*.[1]

Having had a rather extensive involvement in the study of the jurisprudence and moral theology of the 16th and 17th centuries, and having worked rather intensively on crises over disputed logics of variant casuistries of conscience and probabilities of opinion, I found myself stirred to curiosity over theological,

I am grateful to our editor for allowing me to dedicate this collegium statement to *Paul Oskar Kristeller* of Columbia University, friend and understanding mentor over many years. I offer this "chip from workshop" with the sincere hope that Professor Kristeller will accept it as fulfilling a promise to offer a contribution to the *Festschrift* in his honor now happily on its way through the press. It was in the Columbia Seminar on the Renaissance that I have been privileged for many years to enjoy the papers and discussions by Professors Kristeller and Edward Rosen.

legal, and forensic philosophical backgrounds of critical documents and episodes in the history of science. I focused especially on controverted texts central to issues at stake in the struggles associated with the teachings of Copernicus and Galileo:

1. The actual and the claimed intention of Copernicus' masterpiece.
2. The identification of Osiander as the author of the anonymous preface.
3. The interpretation of the meaning of Osiander's preface.
4. Bellarmine's proposal to Foscarini and Galileo.
5. The 1616 decree against Copernicanism by the Congregation of the Index.
6. The cultural contexts and outcomes of the 1633 sentence imposed on Galileo.

Early in my investigations I found myself noting that many available accounts of the prohibitions and the injunctions and the trial did not seem to me to explain what was happening. I could not grasp, for example how it was possible for Galileo to think of publishing his *Dialogue Concerning the Two Chief World Systems* after such injunctions as those of 1616, nor could I understand how less than a quarter century after the sentence against Galileo it would be possible for a theologian-scientist of such eminence as Cardinal Caramuel y Lobkowitz to say that the sentence against Galileo really was not in any sense a condemnation on the ground of heresy, but that it was basically a sort of practical interdiction. Least of all could I grasp why so few present-day scholars had detected and clearly set forth the powerful workings in these developments of such influential ideas as probabilities of varied strengths, certainty, and certitude of conscience.

These developments brought into sharp focus some of the different cultural, theological, historical and sociological backgrounds for the development of science, and, as it were, the histories of how people actually experienced and ordered their worlds as they went about "sciencing."

Let me turn now to the main issue of the present discussion. Edward Grant has called our attention to *"the great quest for reality initiated by Copernicus."* He explains:

> Modern science has shown a greater affinity with the XIVth century than with the century of Galileo and Newton. In the judgment of Pierre Duhem medieval scholastics had a truer conception of science than did most of the great scientists of the Scientific Revolution. Duhem even saw the Parisian nominalists as Christian positivists— forerunners of the positivist movement of his own day. He could not hide his scorn for the naïveté of some of the greatest figures of xviith century science who confidently believed they could—and should—grasp and lay bare reality itself. Most of their basic errors, Duhem insisted, derived from their delusive search for reality which served only to corrupt the theoretical structure of science.
>
> Duhem is, in general, quite right. Scholastics were most sophisticated and mature in their understanding of the role which an hypothesis must play in the fabric of science. They were not, as we have seen, deluded into believing that they could acquire indubitable truths about physical reality. But it is an historical fact that the Scientific Revolution occurred in the xviith century—not in the Middle Ages under nominalist auspices. Despite the significant achievements of medieval science—which Duhem himself did so much to reveal— it is doubtful that a scientific revolution could have occurred within a tradition which came to emphasize uncertainty, probability, and possibility, rather than certainty, exactness and faith that fundamental physical truths—which could not be otherwise—were attainable. *It was Copernicus who, by an illogical move, first mapped the new path and inspired the Scientific Revolution by bequeathing to it his own ardent desire for knowledge of physical realities.*[2] [My italics.]

As I see it, Copernicus was not the first to get men started on the *quest for Reality.* In my view, the central issues of the quest in which the pioneers of early modern scientific revelation were engaged had another name in the days of Copernicus and Galileo. It was the quest for simultaneous (objective) certainty and (subjective) certitude of conscience, and this double-edged quest for truth had been insistent through the medieval era and, indeed, recurrently throughout the history of Western thought.[3] Once this has been admitted, two linked questions instantly come to the fore: Why did the quests for certitude and certainty have the great thrust and significance they reveal in the 16th and 17th centuries? And why did the thrusts in this direction prove to have so much more significance in the West than they proved in China? In asking the last question I give expression to my conviction that few—if any—developments in the way of sciencing and science in the world are "natural" (in the sense of invariant). Rather, these shifts in orientation occur within civilizational contexts that have their own particular complexities.

References to the two kinds of certainties, (objective) *certainty* of proof and the (subjective) *certitude* of conscience, regularly recur in discussions among Catholic thinkers on the distinctive features of the works of the speculative and practical intellect and the mathematical and physical demonstration. Finding themselves unable to acquiesce in counsels of prudence and accommodation, the pioneers aspired to a knowledge of truth that was secure and that guaranteed access to reality. They wished to know the order of the world. They wished to uncover the *machina mundi*. They wished to be able to see the design and laws of the physical universe.

I would hope that these preliminary suggestions will not lead anyone present to suppose that I am putting forward a variant of the familiar hypothesis that has become all too familiar concerning the Protestant ethic and the enormous implication it had for the development of science. The fact is that many of the principal pioneers in the so-called revolution in science and philosophy in the 16th and 17th centuries were reared in Catholic cultural areas and were not as such stimulated by the Protestant ethic as is often claimed.

Another issue is very familiar to those who have read the writings of Joseph Needham. He argues that Chinese science and technology—and there I think he is mistaken in too often treating the two as though they were the same—were far ahead of the

West until the era from Leonardo to Galileo, when the great quantum jump to Galileo's mathematical physics occurred. The turn came in the West, Needham insists, because of the breakthroughs of mercantile capitalism in the 16th century.[4]

As far as the answer to Needham's challenge goes, it seems to me quite clear that he fails to realize the extraordinary amount of deep cultural preparation for science that developed within the West, certainly from Greek antiquity forward and again from the 12th century forward. The era of the "*Medieval* Renaissance" was a time of seedbedding of future scientific development and that occurred within natural theology, logic, and "physics." Culturally and in many ways that run deeper than those ordinarily dealt with in the historiography of science, the 12th and 13th centuries crystallized a strong background base for this thrust that occurred in the West and has never fully occurred in the East.

Many were the differences of orientation, outlook, organization at all levels, which made it unlikely that a science of a universalistic character could openly establish itself in China. You will recall that Matteo Ricci was the principal figure in one of the greatest missions in the history of Christendom, that which brought a number of very distinguished Jesuit scientists and engineers to China.[5] In his journal, Ricci gives a critical clue about some of these questions:

It was during this time, when they had settled down in their new residence, that the Fathers undertook a work which at first sight might not seem to be wholly in keeping with the purpose of their mission but, once put into practice, proved to be quite beneficial.

Dr. Hsu Paul had this one idea in mind. Since volumes on faith and morals had already been printed, they should now print something on European sciences as an introduction to further study in which novelty should vie with proof. And so this was done.

But nothing pleased the Chinese as much as the volume on the Elements of Euclid. This perhaps was due to the fact that no people esteem mathematics as highly as the Chinese, despite their method of teaching in which they propose all kinds of propositions but without demonstrations. The result of such a system is that anyone is free to exercise his wildest imagination relative to mathematics without offering a definitive proof of anything.[6]

I consider this passage an exceptionally revealing illustration of the wider horizons of what I have insisted on as the "quest for certitude."

I turn to another distinction that I have already sought to clarify elsewhere. It names the two key—but different—targets of the innovators: *the fictionalist theory of hypothesis or "saving the phenomena" and the probabilist casuistry of conscience and opinion.*

"Hypothesis" and "saving the phenomena" marked the approach that Osiander sought to impute to Copernicus. Cardinal Bellarmine later urged this approach upon Foscarini and Galileo. The resistance to this fictionalist concept of hypothesis by Copernicus, Galileo, Kepler, and Newton is now well known.

The second tradition is no less important in understanding the details of the various actions centering around the condemnations of Copernicanism and, of course, Galileo. It originates and develops independently of the status of mathematical schemas in astronomy. I refer now to the neglected tradition of the probabilist logic and epistemology loosely called probabilism. Probabilist theses arise less among mathematicians or mathematically oriented physicists or philosophers than among speculative natural philosophers, logicians, and, most prominently, moral theologians and philosophers.

I shall not here offer a detailed story of the probabilists and probabilism. Let me simply remark that in the 16th century, thanks to a very critical work by the Dominican, Bartholomeus de Medina, there was formulated a special kind of probability called *minus probabilismus* that altered the cast of all issues embraced under the rubrics, conscience, opinions, demonstration, proofs. It constituted one of the most critical contexts of many of the developments of the 16th and 17th centuries.[7]

The great 16th-century Jesuit missionary to China, Matteo Ricci, with Hsu Paul, who suggested the translation of Euclid into Chinese. (From Athanasius Kircher, *China monumentis*, Amsterdam, 1667.)

We need not be surprised that the quests for certainty and certitude are variously expressed by Copernicus, Galileo, Descartes, Pascal. After all, each had different stresses. In every case, however, objective certainty and subjective certitude of conscience were indispensable aims of all who felt a need to struggle against the accommodating conjecturalism of the learned. There was no passing forward, in fact, toward a mathematically grounded physical science without continuous effort to go beyond conjectural opinions. As I elsewhere argued:

> Whatever their hopes—whether they wished to advance pure science, logic, secular progress, true or "natural" religion—the pioneers of early modern science and philosophy sought to establish new foundations for exact and compelling knowledge and belief. The prevailing probabilist dialectic that inhibited the quest for new data, the recourse to well-designed experimental research, strict logical proof, and mathematical formalization inevitably became a major target of the 17th century writers.[8]

In this spirit I have claimed it was not so much Everyman who was upset by the scientists, but the people who were at the peak of ecclesiastical hierarchy and, of course, many, many very carefully disciplined thinkers.

The excitement that was generated by the works of innovating scientists and philosophers was great, but the battle was not felt to have been drawn until forthright claims to truth or certitude occurred, claims that openly challenged entrenched (received) doctrine threatening the very foundation upon which all vested authority rested: The claim that there could be some set of truly trustworthy assurances based upon the evidence of the senses concerning the plan and pattern of the "Book of Nature."

Andreas Osiander, a Lutheran official and theologian, the author of the false preface, did not need to reject the *De revolutionibus* as long as he could avow that Copernicus made no pretense of the truth of his hypothesis. The Roman Church would have been content to let astronomy make progress within its own area so long as neither physicists nor philosophers claimed to have demonstrative knowledge of anything in the physical or moral realms.

The time has now come to admit that I was myself startled one day to realize that I did not truly know the full contexts of two of Galileo's notable passages, which I had incorporated into the following statement:

> The founders of modern science and philosophy were anything but skeptics. They were, instead, committed spokesmen of the new truths clearly proclaimed by the *Book of Nature*, which they supposed revealed secrets to all who earnestly applied themselves in good faith and deciphered the signs so lavishly made available by the Author of Nature. Nature's Book, in their view, was written in numbers and never lied, whereas the Testaments were written in words which were easy and tempting to misconstrue. Men like Galileo and Descartes were vastly more certain about the truth revealed to them by number than they were by the interpretations placed upon Scriptures in the commentaries of theologians.[9]

The two passages which now intrigued me were statements offered by Galileo at critical moments in the controversies with his opponents, especially the theologians. The first statement had to do with what was usually called the "Book of Nature"; the second related to what was usually called the "Book of Revelation."

1. The Universe, that is, the Book of Nature, Galileo explained (in *Il Saggiatore*), was written in numbers and as such was a sort of "universal manuscript" available to those who knew numbers, a book not readily falsified by any for their own interests.[10] Numbers, as we would say today, were the medium and numbers were the message—a medium and message so constituted deserved more credence than the conjectural and often contradictory interpretations of Scriptural texts, the texts of the so-called Book of Revelation on the part of rival theologians. All who knew numbers were free to read the Book of Nature; they required no

special theological credentials and could test each other's readings. This was not the case with the Book of Revelation.

As everyone knew, any text was open to multiple constructions and yielded up its meanings only when these multiple constructions were put into play. Any phrase in Holy Scriptures or the Book of Revelation admitted of meanings at every level of interpretation, e.g., the literal, historical, analogical, anagogical, tropological, etc. All agreed that mistakes were bound to occur whenever an inappropriate meaning was imposed upon a text by a theologian insufficiently versed in the modes of interpretations.

2. The second passage in Galileo that now drew me forward into more intensive research was ascribed to Cardinal Baronius in his *Letter to the Grand Duchess Christina*.

Cardinal Baronius was correct, said Galileo, when he declared that the Holy Spirit had given us Scriptures—that is, the Book of Revelation—not to tell us how heaven goes, but how to go to Heaven.[11] On first seeing the passage, I became convinced that it represented the sophisticated point of view of a theologian who was already possessed of a rationalized structure of consciousness. Being so persuaded, I decided to look into the earlier life of the images to discover how such a distinction had been formed.

With luck, I soon found that an important step on the way to this destination were some passages in the *Natural Theology* or *Book of Creatures* by a Catalan theologian, teaching at Toulouse, generally known as Raymond de Sebonde (Ramon de Sibiude) on whose behalf Montaigne was to write an *Apology* and whose work Montaigne undertook to translate at the behest of his own father.[12] On closely examining the *Natural Theology* by Raymond de Sebonde, I discovered that the main discussion in the body of Raymond's text did not precisely say what Cardinal Baronius had inferred. Indeed, there was a distinction between the constructions we were to derive from two books, the Book of Nature and the Book of Revelation, but Raymond de Sebonde seems to have given the palm in the body of his work to the Book of Revelation, because the Book of Revelation did offer help in getting to Heaven. Raymond considered this knowledge to be of higher form and purpose than the less certain, more ambiguous and equivocal teaching of the Book of Nature. It was hard to say what was the message of that book. Its main message seems to have been to establish the degree of our dependence on God's power and will.[13]

A quite different sense of the relation between the two books is conveyed in the Prologue and in the first sections of Book One of Raymond's work. In the Prologue Raymond talks about the Book of Nature as absolutely *knowable with complete certitude* and embracing everything that anyone really needs to know for his salvation. What he means here by the Book of Nature is what might be called the combined "Books of Nature and Conscience" —that is to say, the notion of nature and the self.[14]

Raymond speaks about the Book of Nature in a manner that might easily have served as a critical context for those disposed to find revealed truth in the Book of Nature. It is not claimed here that Raymond's discussion is the source of Galileo's remark.[15]

Being convinced that a "provincial" theologian of the 15th century was hardly likely to have initiated this cluster of notions, I proceeded to look into what must have been his ground sources, the authors of the High Middle Ages, when the nuclear metaphors of pre-Reformation Christian thought achieved their classical form. I turned to Grosseteste, Roger Bacon, Bonaventura, and other writers of the 12th and 13th centuries. I was not surprised when I found that these images were simply everywhere, especially among the Franciscan writers.

Everything in the world was seen as being the work of God's hand. Everything in the world was seen as a "book." All actions, images, and artifacts comprised books. As we have noted, there was not only the Book of Nature or Creatures, there were also the Book of Conscience, the Book of Revelation, the Book of Life, and other Books as well.[16] Everything in the world (not evidently the work of a human hand) was somehow aided by the creative spirit. As the work of God's hand, it was directly revealed as incarnate Nature. It conveyed its own image directly.

Therefore, the notion that the Book of Nature was somehow more ultimate and available than the Book of Revelation was not the expression of a wholly new idea in Galileo's day but a new accent placed upon an old idea. There was a shift, so to say, in some of the theological notions and philosophical notions.

Let me mention briefly a noteworthy related case of the linking of theological and philosophical ideas. At the top of the frontispiece of Riccioli's *Almagestum novum* in 1651 we see the outstretched hand of Jehovah. A nearly concealed reference at the ends of Jehovah's fingers to the apocryphal Wisdom of Solomon 11:20 explains that the created world was all made and disposed in number, weight, and measure. The explicit use of Wisdom 11 in artistic representations of the creation intending

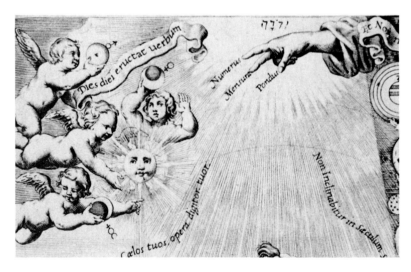

God's hand with "number, measure, and weight" are shown in this detail from the frontispiece of G. B. Riccioli's *Almagestum novum* (Bologna, 1651). (Permission of Harvard College Library.)

to stress the ordered creation of the world and its disposition in number, weight, and measure goes back quite a long way in the Western world, at least as early as the year 1000. A most striking representation of the theme of the Geometer God is found in the Winchester Gospel of about that date in which God is seen with compasses in hand.[17] It is quite remarkable that Riccioli should have thought to place a variant of this symbol as his frontispiece. Was it not a way of responding to the Galilean emphasis on number, weight, and measure? Very interestingly, it is exactly the symbol of Urizen, the Geometer God-Demon of the Compasses which Blake developed in his depiction of "The Ancient of Days," his attack upon mathematics, Newton—and Galileanism.[18]

In the same breath I allow myself to say that the idea of a "machine of the world" was by no means new in the 17th century. The 17th-century notion of a new *machina mundi* had new elements, new stresses, new force, new impetus, new proofs, but it was not a new idea. The notion of *machina mundi* is classical in origin and it acquires new meanings when there occurs a peculiar junction between the Greek and Christian ways of thinking about the Creation and order of the world.[19] More than one medieval philosopher-scientist spoke of unravelling the *machina mundi* and the *machina universitatis*.[20]

I close this story abruptly but allow some final words of clarification and caveat. I shall not be surprised if colleagues and scholars who are expert in the histories of physics and mathematics from the 13th century forward doubt that any such notions could have had any effect in the development of science. Had not the professional theologians and philosophers of the late Middle Ages passed beyond responding in any significant way to symbols and images coming out of these theological and religious traditions?

Even after calm reflection, I discover that I cannot agree with the views of disbelieving friends. To me the evidence seems overwhelming that many of these theological-religious symbols were experienced in different ways, given different kinds of

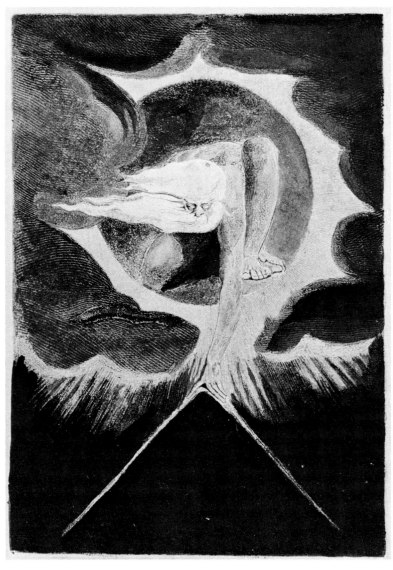

William Blake's "Urizen" with compasses, commonly called "The Ancient of Days" from his *Europe, a Prophecy* (1794). (Handcolored copy by Blake, permission of Harvard College Library.)

stress and emphasis, but they constitute the backgrounds of even the most highly disciplined theologians and philosophers who in their technical work chose to confine themselves to proofs that were solid in philosophy and in physics; nonetheless, they had these symbols as a ground. A twofold claim is implied here: (1) Modern science has not only metaphysical foundations of Greek origin; it also has *theological and religious*—Hebrew, Christian, and probably Islamic—foundations; (2) a full recovery of the cultural foundations of modern science requires greatly widened access to root images in the substrate of these traditions.

The men of the Middle Ages looked everywhere in the hope of seeing the alphabet, the signature, the characters of the message of the universal manuscript. They were confident that by continuous inspection of Nature they could read the numbers therein. For various reasons and with various outcomes men like Roger Bacon and Raymond Lull looked for numbers everywhere in hopes of achieving a kind of universal language which might, in the view of some, be a vehicle for the conversion of mankind. It was, therefore, I repeat, not altogether a new idea that identified the Book of Nature with numbers.

To summarize: The quests for subjective certitude of conscience and objective certainty of proof were powerful forces in the 16th and 17th centuries. The notion that there could have been anything else seems very unlikely to me. My other contention is that the thrust, the intensity of the two, certainty and certitude, had much to do with the fact that the West placed so great a stress upon the notion that the world was the work of God's hand and as such it constituted a Book of Nature, itself a revealed work, revealed to us by God's design. The Book was given to us and it was our task to decipher the pattern and prove the logic and character of the design. My last contention is perhaps a peculiar one whose discussion must be reserved for another day. The world was beyond any doubt a *machina mundi* and it was our responsibility to participate in *co-creation* through the development of mechanisms.

Notes

[1] My principal papers on these topics include " 'Probabilists', 'Anti-Probabilists' and the Quest for Certitude in the 16th and 17th Centuries," *Actes du Xme congrès internationale d'histoire des sciences 1* (Paris, 1965 [original draft, 1962]), pp. 267-273; "The Early Modern Revolution in Science and Philosophy: Fictionalism, Probabilism, Fideism, and Catholic 'Prophetism'," *Boston Studies in the Philosophy of Science*, edited by R. S. Cohen and M. Wartofsky (Dordrecht, 1968), vol. 3, pp. 1-40; "Communities, Societies, Civilizations: Post-Millennial Views on the Faces and Masks of Time," *Social Development: Critical Perspectives*, edited by M. Stanley (New York, 1972), pp. 105-133; "Sciences and Civilizations, 'East' and 'West': Joseph Needham and Max Weber," *Boston Studies in the Philosophy of Science*, edited by R. S. Cohen and M. Wartofsky (Dordrecht-Boston, 1974), vol. 11, pp. 445-493.

[2] Edward C. Grant, "Late Medieval Thought, Copernicus and the Scientific Revolution," *Journal of the History of Ideas*, vol. 23, pp. 197-220, especially pp. 219-220, 1962. A posthumously published book by N. R. Hanson offers a stimulating alternative way of characterizing Copernicus' central contributions; see pp. 175-187, 200, 220-235 in *Constellations and Conjectures*, edited by W. D. Humphreys (Dordrecht-Boston, 1973).

[3] The wide discussion of *certitudo mathematicarum* in 16th-century Italy is reported in Charles B. Schmitt, "The Faculty of Arts at Pisa at the Time of Galileo," *Physis*, vol. 14 (1972), pp. 243-272 (especially p. 620, note 92).

[4] Joseph Needham, *The Grand Titration: Science and Society in East and West* (Toronto, 1969), especially chapter 6.

[5] See Nathan Sivin, "Copernicus in China," *Colloquia Copernicana II: Études sur l'audience de la théorie héliocentrique* (Wrocław, 1973), pp. 62-122.

[6] *China in the Sixteenth Century: The Journals of Matthew Ricci: 1583-1610*, translated by L. J. Gallagher, S.J. (New York, 1953), pp. 476-477.

[7] For the literature and controversies related to the theses of *minus probabilismus* see Benjamin Nelson, "Self-Images and Systems of Spiritual Direction in the History of European Civilization," *The Quest for Self-Control: Classical Philosophies and Scientific Research*, edited by S. Z. Klausner (New York, 1965), pp. 49-103. An exceptional analysis of the relation of probabilism to "certitude" will be found in the very first treatise in Prospero Fagnani, *Commentaria super quinquos libros decretalium*, five books in seven volumes (Rome, 1661).

[8] Benjamin Nelson, "Comments on Edward Grant's 'Hypotheses in Late Medieval and Early Modern Physics'," *Daedalus*, pp. 612-616, Summer, 1961 (issued as *Proceedings of the American Academy of Arts and Science*, vol. 91, no. 3).

[9] Nelson, "The Early Modern Revolution in Science and Philosophy" [note 1], p. 12.

[10] Galileo writes: "Philosophy is written in this grand book, the universe, which stands continually open to our gaze. But the book cannot be understood unless one first learns to comprehend the language and read the letters in which it is composed. It is written in the language of mathematics, and its characters are

triangles, circles, and other geometric figures without which it is humanly impossible to understand a single word of it; without these, one wanders about in a dark labyrinth." The *Assayer* (1623), in Stillman Drake, *Discoveries and Opinions of Galileo* (New York, 1957), pp. 237-38.

[11] Galileo's words on this score are memorable: "From these things it follows as a necessary consequence that since the Holy Spirit did not intend to teach us whether heaven moves or stands still, whether its shape is spherical or like a discus or extended in a plane, nor whether the earth is located at its center or off to one side, then so much the less was it intended to settle for us any other conclusion of the same kind. And the motion or rest of the earth and the sun is so closely linked with the things just named, that without a determination of the one, neither side can be taken in the other matters.

"Now if the Holy Spirit has purposely neglected to teach us propositions of this sort as irrelevant to the highest goal (that is, to our salvation), how can anyone affirm that it is obligatory to take sides on them, and that one belief is required by faith, while the other side is erroneous? Can an opinion be heretical and yet have no concern with the salvation of souls? Can the Holy Spirit be asserted not to have intended teaching us something that does concern our salvation? I would say here something that was heard from an ecclesiastic of the most eminent degree: 'That the intention of the Holy Spirit is to teach us how one goes to Heaven, not how heaven goes.' " See Galileo (1615) with slight modification from the translation in Stillman Drake, op. cit. [note 10], pp. 185-186.

[12] See "Apology for Raymond Sebond," in *The Complete Works of Montaigne: Essays, Travel Journal, Letters,* translated by D. M. Frame (Stanford, California, 1948), pp. 318-457.

[13] Ramond de Sabonde, *Theologia naturalis seu Liber creaturarum* (1436), Introduction by F. Stegmuller (Stuttgart-Bad Cannstatt, 1966), pars 2a, tit. ccxii, pp. 312-15.

[14] Ibid., Prologus, pp. 27, 37. When the main text of the *Theologia* was removed from the *Index* in 1564, the Prologue remained. On pp. 40-43, note especially the great stress on certitude.

[15] Much research would need to be done before we could establish the actual sources of this saying of Galileo and Cardinal Baronius. A certain similarity in tone and phrasing does seem to exist between the Prologue of the *Theologia naturalis* and Galileo's *Letter to the Grand Duchess Christina*. See, for example, Ramond de Sebonde, op. cit. [note 13], p. 36; Galileo, in Stillman Drake, op. cit. [note 10], pp. 185-186.

[16] See chapters 1-4 in G. H. Tavard, *Transiency and Permanence: The Nature of Theology according to St. Bonaventure* (New York, 1954). For an older survey of the range of reference to the "Books" see E. Curtius, *European Literature and the Latin Middle Ages,* translated by W. R. Trask (New York, 1953).

[17] See p. 189 in Lynn White, "Cultural Climates and Technological Advance in the Middle Ages," *Viator, Medieval and Renaissance Studies,* vol. 2 (1971), pp. 171-201.

[18] Anthony Blunt, "Blake's 'Ancient of Days,' The Symbolism of the Compasses," *Journal of the Warburg and Courtauld Institutes,* vol. 2 (1937-1938) pp. 53-63.

¹⁹ See p. 531 in Helmut Koester, "ΝΟΜΟΣΦΥΣΕΣ: The Concept of Natural Law in Greek Thought," in *Religions in Antiquity, Essays in Memory of Ervin Ramsdell Goodenough,* edited by J. Neusner (Leiden, 1968).

²⁰ See for example Robert Grosseteste, *Commentarius in VIII Libros Physicorum Aristotelis,* edited by R. C. Dales (Boulder, Colo., 1963); M. D. Chenu, O.P., *Nature, Man and Society in the Twelfth Century. Essays on New Theological Perspectives in the Latin West,* edited and translated by J. Taylor and L. I. Little (Chicago, 1968).

DISCUSSION: PHILOSOPHICAL AND
THEOLOGICAL BACKGROUNDS

GINGERICH: Let us turn to Professor Rupert Hall, who has agreed to give some commentary, but who is at something of a disadvantage because he did not have an advance text.

R. HALL: I'd like to turn to a point that always interests me a great deal and almost makes me feel that when I turn my mind to it, I am capable of philosophic thought. I know that is purely a subjective delusion, but when I turn to it I do get this deceptive inner confidence. [Laughter]

I think that there is a difference between reality and words like objective certainty and subjective certitude. I don't think science is at all concerned with subjective certitude or ever has been, if I may say so. If I understand the words "subjective certitude" in a sort of simple and naïve, nontechnical sense, it seems to me subjective certitude is something that anybody may have. It is incommunicable. It is one's personal confidence. Of course, we all seek this. None of us wishes to live continually in some sort of schizophrenic state.

But I think one has to look beyond this to something else. All right. Objective certainty. But objective certainty again is a concept to be distinguished from reality, because if we believe that there is a reality, if we believe that the universe has a definite structure quite independent of ourselves as observers, then we may, I think, postulate this belief even though we say to ourselves that we haven't got access to this reality and perhaps even that we never will have access to this reality.

Now, objective certainty I put less than this, because we can be objectively certain of something and still not hold that that is reality. Objective certainty simply means something that is demonstrable to all intelligent beings. It is still something distinguishable from reality.

So there are three steps, it seems to me:

1. Subjective certainty, which is sort of a psychological condition, almost a psychological euphoria.

2. Then there is objective reality. These are the properties of the world around us that we can demonstrate to all men. If you don't believe that a crystal of niter is always shaped in such and such a way, then communication between us two is impossible. That's what I understand by objective certainty about things.

3. But then there is the idea of ultimate reality. I think that Edward Grant was right to talk about reality; I'm not claiming any sort of originality at this point at all.

If I may, I'd like to say a little on the subject of science and religion. I think that here one has got to beware a simple tautology. If the history of science in the 20th century were studied, I suppose one might assume that a prime condition for the study of science and history of the 20th century is that one should live in a country governed by a President and Congress whose capital is on the Potomac River, and so on. We must be very careful not simply to say that science has prospered in a particular society and period because the society in which it prospered has those particular characteristics. I agree entirely with Professor Nelson: Naïve correlations between religious faiths and scientific success, scientific achievement, are only to be drawn with very, very great caution.

I would say that attempts to pick out the Protestant ethic as an essential dynamic social factor for the development of science are not only erroneous but by now they are largely very arid. They no longer tell us anything that is very interesting about the development of science. In any case, "ethics" are temporally relative. To give a different example:

Islam as far as I know hasn't changed very much in its religion or ethics. Mecca is in the same place. The Koran is still the same text. But if that is what produced science between the 7th century and the 13th century of our era, it doesn't work after that time. If it was your fate to be a great scientist, you ought to be born in Islam if you were living at any time between roughly the 7th century of our era and the 12th or 13th century. But if you were going to be born as a great scientist after the 12th or 13th centuries, it was a misfortune to be born in Islam.

Let me now turn to Matteo Ricci and Euclid in the Chinese context. Of course, the Chinese had a very effective and complex mathematical tradition of their own. But it is obvious from Needham's writings, as well as others, that the Chinese were not geometers. You could not expect the Chinese to make sense of Euclid or to be other than astonished by the fact that something so obviously sophisticated was put before them that they didn't understand.

The Chinese mathematics is to me utterly incomprehensible. Even when transliterated into English I find it is still incomprehensible, but apparently it works, whereas Euclid I think I can understand. I don't think we should lay too much weight on that. This is just a cultural difference, it seems to me.

And surely again Professor Nelson is right in saying that the development of science in the West—and in a different, perhaps lesser way, as Needham would agree, since China never went through the scientific revolution stage—was due to cultural differences between the two societies.

I would like to applaud Professor Nelson's emphasis on a point he made right at the end, about the Christian belief that the universe is that created by God. If you have an anthropomorphic

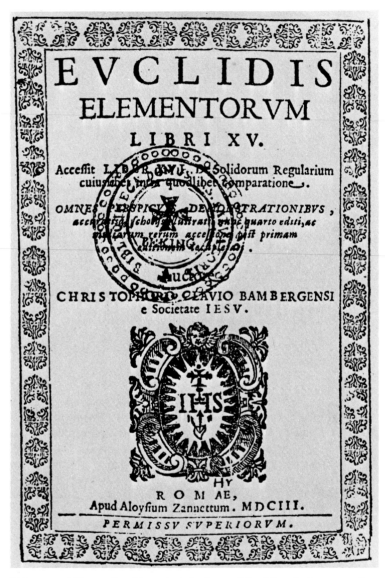

EVCLIDIS
ELEMENTORVM
LIBRI XV.

Acceffit LIBER XVI. De Solidorum Regularium
cuiuslibet intra quodlibet comparatione.

OMNES PERSPICVIS DEMONSTRATIONIBVS,
accuratisque fcholiis illuftrati, nunc quarto editi, ac
multarum rerum acceffione post primam
editionem locupletati.

Auctore
CHRISTOPHORO CLAVIO BAMBERGENSI
e Societate IESV.

IHS

ROMAE,
Apud Aloyfium Zannettum. MDCIII.

PERMISSV SVPERIORVM.

Title page of Christopher Clavius' 1603 edition of Euclid with the stamp of the Jesuit Library in Peking. (From F. Bortone, *P. Matteo Ricci S.I.*, Rome, 1965.)

idea of God, what God has done man can understand. It would be very little help if you said the universe is a deifying artifact if you then emphasize in a religious tradition the absolute incomprehensibility of God's purposes and actions in the creation of the universe. If this leads you to emphasize the fact that the universe is unintelligible to man, it doesn't take you anywhere. But if you look at it the other way and take the view that what God has done is intelligible to men, because, after all, man himself possesses a small spark of the Divine nature, then this gives you reason to hope.

Let me now make one distinction about the Book of Nature and the Book of Numbers. When Galileo spoke in the *Saggiatore* about the Book of Nature being written in numbers, I do not believe he meant the same sort of thing that Bacon and Raymond Lull might have said. To put the discussion on a different level, the sort of difference here is between the Platonic view of nature and mathematical physics, a distinction which is not yet clearly made in the history of science. So I am not quite happy about supposing that people like Raymond Lull and Bacon could be in a tradition that is directly a precursor of Galileo's appeal to this particular analogy.

NELSON: I chose to emphasize the notions of certitude and certainty because those were the terms in which all forensics in respect to conscience and opinion were, in fact, put in the medieval era and the early modern era. I ought to have made very much clearer than I did that absolutely every act and every opinion was in some way answerable to the grammars, logics, and forensics of the conscience of that time. All questions that were associated with statements made by people in natural philosophy would somewhere or other come up under the headings of probability and certitude.

As the issue would be phrased in the 16th and 17th centuries, the case was as to whether it was allowable to contend for probability in respect to the Copernican theory. Some here have perhaps noticed that on three occasions in the sentence against

Galileo it is clearly stated that he, Galileo, rendered himself vehemently suspect of heresy because he accorded probability to a hypothesis that would claim none, presumably because it had been condemned as opposed to Scripture. Moreover, subjective certitude is not a mere psychological requirement or disposition or whim, but subjective certitude of conscience and objective certainty of proof do get locked in together in the very mode of perceiving and experiencing these questions within the Christian world. My emphasis is upon the fact that these two notions were bound together in relation to the disciplined knowledge of physical reality. This claim does not admit of informed doubt.

Turning to another point at issue: The considerable debates about the Books of Nature and Revelation were, in fact, the testing ground of who was to have what legitimation to read the "Book" and which "Books" were to be conceived as being the work in which God's design can be understood. In the 17th and 18th centuries we have a fuller development of this idea, and we have the notion that nature is a universal manuscript that lies exposed to all and that it simply tells its story and it indeed is revelation. An exceptionally interesting use of the two books (of nature and creatures) is found in Sprat's *History of the Royal Society* (1667, pp. 370-71).

Finally, I would like to say just one word to Professor Hall in respect to our views in what I would call the comparative history and sociology of science and "sciencing" in civilizational perspective. For there to be breakthroughs toward universal sciences, for there to occur crucial transformations at the level of culture, science, and so on, there have to occur a lot of other sorts of breakthroughs, and these occur at the social relations level and the social organization level. China was not disposed this way. It took a long time for such changes to occur. There were different sorts of structures that were operating there.

Euclid was not available in Ming China. What I was hoping to convey in emphasizing Ricci's statement was that there really didn't exist, so far as one can tell, the kind of sets of structures and axioms, the proofs, the emphasis on demonstration that one finds in the West. Without it there are vast numbers of questions that could not be answered definitively, absolutely, in one way or another.

HARTNER: What does probability mean at the time of Galileo? Is it the trite sense in which we use it today? Could someone tell me in a few words the transition, the evolution of that term from the 13th century on to the time of Galileo?

WALLACE: I would not be prepared to do that at the moment. Both *probabilitas* and *certitudo* are mentioned in the writings of Thomas Aquinas, who wrote in the latter part of the 13th century. Sometimes for him *probabile* has the meaning of something that can be proved, or is "provable," whereas at other times, and more commonly, it has the meaning of being merely probable, a matter of opinion as opposed to a certain truth.

HARTNER: In many of the writings of the 14th and 15th centuries a clear distinction is made between *probabilitas* and *veritas*. The latter term is applicable only to the revealed, religious, truth, which is susceptible of interpretation but not of discussion or of proof in the mathematical sense of the word. On the other hand, *probabilitas*, in the cases known to me, still has its full semantic weight, such as results from its etymology: *id quod potest probari*, "that which can be proved," i.e., by means of philosophical, mathematical, and physical arguments. In the course of time, the original meaning gradually got lost, as evident from our modern word "probability," which is just a synonym of "likelihood."

WALLACE: That may be true. But I do not think that what you identify as the "original meaning" was the more common understanding of probability, and certainly it was not the meaning adopted by theologians in the 16th century when they absorbed the term into the context of the debates over probabilism—the Latin term they used was *probabilismus*. The problem relates to the quality of the knowledge one requires in order to judge the morality of a contemplated action, on the supposition that a person should not act unless he has safe ground for judging that he is acting correctly. It is practically impossible, of course, to have demonstrative certitude in matters relating to human action. Most theologians held nonetheless that a person should have convincing reasons for thinking that he is in the right before placing the act. The probabilists, however, would require only a probable opinion (and not necessarily the more probable or most probable opinion) to certify a particular course of action, and they were generally accused on this account of leaving the door open for laxity in moral matters.

NELSON: The notion of probability refers to two spheres that get connected throughout this period. One is the sphere of action which Professor Wallace has mentioned. The other is the sphere of opinion. That is, every opinion has to have a measure of probability. The funny fact is that anything that has a probability has a certain truth value—that is, a certain truth claim—and that is part of the trouble that Galileo gets into when he contends in terms of probability in his *Dialogue Concerning the Two Chief World Systems.*

WESTMAN: This discussion really concerns the status of knowledge and truth, and when we speak on a metaphorical level—the Book of Nature versus real books and the actual procedures of science—we have really two very different levels. Metaphor is an interesting matter: On the one hand, it is an extremely powerful component in scientific thinking, and, on the other hand, it has a huge potential for ambiguity, as we heard yesterday in Professor Oberman's paper.

For a long time people used the Book of Nature to justify numerology, and that is how the famous passage in Wisdom of Solomon 11:20 ["But thou hast ordered all things by measure and number and weight"] was interpreted; in the 17th century it is used in an entirely different way. Here metaphor shows a tremendous range of uses, and for this reason we have to be careful in differentiating the tradition of the Book of Nature.

Professor Nelson, in spite of the fact that I have read both of your articles on probabilism, I'm still confused about what you mean by subjective certitude and objective certainty. A scientist needs to believe that he is looking for the truth. Whether he can find it or not is another story. The feeling must be distinguished from the reality. Can he really find objective certainty? There we come up against the rules of logic and criteria of truth.

If that is the distinction which you are making, then I find it an interesting one. However, I really do not see from any evidence that you have presented that all these discussions about probability in a theological context really penetrate scientific discussions. Do you have evidence of this? Or is it just a conjecture?

NELSON: In the case of Galileo, the evidence it seems to me is quite clear. There is no doubt that the terms and phrases that I have emphasized figured prominently in the struggles among theologians and physicists as to the authority of different ways of arriving at the order of nature and the design of the world. At the very least, it is a struggle over credentials of those equipped to define the situation.

If, of course, you start out with the notion of highly individuated sciences that are totally distinct from theology and that are not in any way to be contaminated with any matter of conscience, then you have one situation. When you start asking about what evidence there is that a contemporary scientist needs objective certitude, the answer is, he may not need objective certitude, but the fact is in the 15th, 16th, and 17th centuries there was no way for science to break through and to develop itself unless it could cut itself free from the immense structure of legitimations and so-called "probable conjectures" that completely blocked its passage. So there had to be the quest for (objective) certainty and (subjective) certitude.

WILSON: In the 17th century you find sharply different stands on the notion of certitude. With Gassendi there is the feeling that a probable thing is quite adequate. I think that later on when Barrow, Newton, and Boyle seek certainty, it's partly a reaction simply to the multiplication of different mechanical constructions for the world; there are too many of them, and one obviously needs something to cut through all this.

I think of Galileo as a great polemicist, probably conscious of his use of words like probable and so on, secretly convinced that Copernican theory is right because he has his proof from the tides, but arguing all the time on probable grounds. How much is he really aware of the idea of certitude? Is it just a kind of peripheral thing on the horizon?

NELSON: May I speak to this? There was recession from the claim for certitude that occurred after the sentencing of Galileo. Many people began to talk in terms of probability. It is my view that Galileo must have been informed by friends that he should put the matter in terms of probable rather than that it was altogether a supposition. Why wasn't Galileo prepared to accept the recommendation made to him? The answer is that he himself went for certitude, and if you read Galileo's remarks on Bellarmine's letter to Foscarini, you will see he absolutely rejects the possibility of alternative hypotheses, and he argues for strict certainty and certitude. My contention is that one needs to know these kinds of context in order to get at the whole environment, the cultural backgrounds and environment of science.

GRENDLER: There was a wide intellectual context for the quest for certitude in the second half of the 16th century in Italy, that of the polemic between *parole* ("words") and *cose* ("things"). It developed out of a rejection of rhetoric. It was argued that one should not accept handed-down definitions, but should look at the experiential, and examine something that could be touched here and now. The opposition to *parole* was most often found in the works of moralists, political writers, and others, rather than in philosophers, until Campanella. He tended to say, perhaps in a confused way, that if one wished to find *cose*, one should look at the Book of Nature. I cannot say that Galileo was influenced by this polemic, but it certainly was part of the Italian context from the 1560s onward.

NELSON: Thank you for reminding us of it. There was a growing insistence that if things are clearly established within the Book of Nature, then we have to modify our way of thinking about Revelation.

A perfect illustration is in Foscarini's extraordinary *Letter*,* which was attacked and declared prohibited in 1615. It declares the absolute congruity of Revelation and Nature. With Revelation having to be interpreted, since God gave us Nature and it was His, Revelation may have gotten distorted in the process, Nature not.

GINGERICH: This would be an appropriate time for Father Wallace to make some remarks. I have asked him also to present some commentary on the symposium papers yesterday afternoon, the ones by Professor Temkin and Professor Oberman.

WALLACE: Why were astronomers in Galileo's time not content to settle for a fictionalist view of hypotheses? We have difficulty answering this because of our own modern understanding of scientific method—we hypothesize about all sorts of things and are not particularly worried whether our speculations happen to be true or not. In the 16th century, however, there were those who adopted the fictionalist view and there were those who vigorously opposed it. What was the reason for the latter's opposition?

I think that Galileo was definitely against fictionalism, and I have been led to believe that Copernicus was against it also, although I am not sufficiently acquainted with his writings to form that judgment on my own. But I have been studying Galileo's early notebooks, and the portion of these that treats of astronomy, the *Tractatio de caelo*, gives evidence of a strong realist commitment to the Ptolemaic system. This portion of the notes also gives evidence of being based on, or actually copied from, another source; at first it was thought that they might be based on Regiomontanus, but subsequent checking revealed that they

---

* Paolo Antonio Foscarini, Lettera sopra l'opinione dei Pittagorici e del Copernico della mobilità della terra e stabilità del sole (Naples, 1615).

are taken almost verbatim from Christopher Clavius' *Commentary on the Sphere of Sacrobosco*. Now Clavius is first and foremost a realist. For him, the eccentrics and epicycles of the Ptolemaic system exist in the heavens and are really there. Clavius could not be indifferent to the Copernican system any more than he could be to the Ptolemaic—in his view neither was a mere fiction, they were both competing views of reality. And his reason for holding this is that he accorded to astronomy the same status as natural philosophy, which in his mind was a true science that could discover causes from their effects. If a person could not reason from effects or appearances in astronomy to their causes, that is, to the true arrangements and movements of the heavenly bodies, then he could not reason from effect to cause in natural philosophy, and philosophy would cease to be a science. Clavius is explicit on this in his commentary on Sacrobosco:

> If it is not right to conclude from the appearances that eccentrics and epicycles exist in the heavens, because a true conclusion can be drawn from false premises, then the whole of natural philosophy is doomed. For the same way, whenever someone draws a conclusion from an observed effect, I shall say, 'That is not really its cause. It is not true because a true conclusion can be drawn from a false premise.' And so all the natural principles discovered by philosophers will be destroyed.

For Clavius, therefore, the rejection of philosophy and its effect-to-cause reasoning that was apparently implied in the fictionalist view of astronomy was completely unacceptable, and that explains his vigorous rejection of fictionalism.

Now, although Clavius was quite wrong in according reality to Ptolemy's system with its eccentrics and epicycles, I think that methodologically he made a definite impression on Galileo in his youth, with the result that in his later researches Galileo was obsessed with the idea of demonstrating the truth of the Copernican system. Galileo, in my mind, was never a fictionalist; in his early teaching he taught the truth and reality of the Ptol-

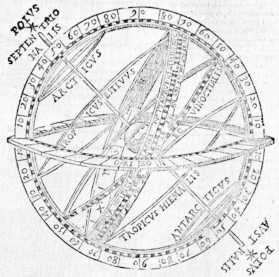

Title page of Christopher Clavius' *Commentary on the Sphere of Sacrobosco*. Clavius added a critical discussion of Copernicus to this 1581 edition.

emaic system, and then, he turned just as vigorously to arguing for the Copernican system. But neither, in his mind, was to be regarded as a mere fiction; the one or the other was a true representation of reality.

However, if we go back to the 14th century to Nicole Oresme, Oresme was certainly aware of the possibility of the earth's rotation accounting for the apparent diurnal revolution of the heavens. But, after evaluating the evidence, he concludes simply: "The truth is, the earth is not so moved but, rather, the heavens. However, I say that this conclusion cannot be demonstrated but only argued by persuasion." Oresme's truth-claims for reason were thus not as strong as Clavius' or Galileo's. And, like most orthodox thinkers of his time, he was perfectly willing to *believe* that a true picture of the structure of the heavens had already been given in Revelation, in the Book of Scripture. Moreover, while conceding the possibility of coming to *know* the true system of the world by reason alone, since he was actually able to offer only "persuasive" rational arguments in support of one system or other, he settled in favor of the scriptural account. In any event, he was not going to contradict what seemed to him the clear teaching of Sacred Scripture on the basis of only persuasive or probable arguments.

In this connection, I was much impressed with Professor Temkin's paper yesterday, and its balanced and very full presentation of the factors that bear on an understanding of 16th-century science. We must be extremely careful to avoid anachronisms of any kind, and particularly attributing the attitudes of the present day to the scientist of the 16th century. Professor Temkin gave due credit to the connections between magic and early science, between the hermetic tradition and the development of modern science. But he also made very clear that the age of Copernicus was characterized by a strong union of scientific thought with religious thought—the real world that scientists were investigating was equally the world of God's making, about which much had already been divinely revealed. This was, for me, the most characteristic and central idea of the 16th century concerning the universe. In its light, of course, one can readily understand why Osiander could maintain in his preface to the *De revolutionibus* that truth of the universe's structure is obtained from divine revelation alone, whereas Copernicus himself could be convinced that such truth was also accessible through scientific investigation. Compared to the problems created by this epistemological dualism, difficulties relating to the displacement of man from the center of the universe probably created fewer problems for people of the 16th century than later historians have claimed.

One point in Professor Temkin's presentation, however, I found arguable. In rightly maintaining that Aristotle's doctrine of the four causes still loomed large during this period, he seemed to place undue emphasis on the final cause. I have made an extensive examination of the use of Aristotelian causal analysis from the 13th through the 16th century, particularly in the study of the world of nature, and have found that here final causality is not regarded as a significant factor, particularly not in matters relating to scientific explanation. Otherwise Professor Temkin's paper gave us an extremely accurate and balanced picture of science in the 16th-century context.

I was not equally impressed, on the other hand, with Professor Oberman's account yesterday of the role of nominalism in the rise of the new science, for much of what he said went counter to what my own studies over the past five or six years have shown. I am quite willing to concede that Oresme and the Paris School generally played an important part in developing the ideas that made modern science possible, but my studies have shown that they did this only when they rejected nominalist theses in natural philosophy, such as that local motion does not require a cause, while retaining such theses in logic, particularly their logic of terms. On this account I would prefer to call them "terminists," rather than "nominalists." Unlike Ockham, the Paris terminists were willing to regard local motion as a real entity distinct from the object moved and requiring its own causes. Thus they were realists in their analyses of motion and in their application to the heavenly bodies of the dynamic principles that followed therefrom. It was therefore their realist attitude rather than their nominalist attitude that exerted the influence on Copernicus and Galileo, for this encouraged them in their search for the true causes that would not only explain the appearances but also lay bare the structure of the real.

Moreover, for a correct picture of the influence of Bradwardine, Swineshead, Buridan, Albert of Saxony, and Oresme on 16th- and 17th-century science, it simply will not do to take a text or a generally accepted position of the 14th century, and then extrapolate it to the 16th century, without taking account of the details of its transmission. That is one of the reasons why I have been spending so much time with Galileo's *Juvenilia*, or early notebooks, and the authors he cites therein, for these citations enable us to trace some of the lines of transmission from the 15th and early 16th century to the onset of the 17th. At present, however, there is too much of a gap in 15th- and early 16th-century studies, to talk meaningfully about the way in which 14th-century science directly influenced the founders of modern science.

ROSEN: Concerning the passage that you read from Clavius, the attack that he makes on those who deny the reality of the epicycles and eccentrics is directed against Averroes, who denied it. This is a realist position that Clavius shares, by the way, with Copernicus and Rheticus. In the Latin translation of Averroes the eccentrics and the epicycles turn out to be nothing at all. They are not even figments of the imagination. It is against that position that Clavius (in common with Copernicus and Rheticus) accepts the reality of the eccentrics and epicycles.

WALLACE: That is quite true. Galileo's training at Pisa was not in Averroist Aristotelianism, nor was it in the Greek Aristotelianism of the Alexandrinists. Rather it was definitely a scholastic Aristotelianism, somewhat eclectic, but essentially Thomistic with some Scotistic overtones. It seems to have derived from

Spain: Francisco Toledo and Benito Pereira are apparently his main sources, together with Clavius, and all three of these are Jesuits who were teaching at the Collegio Romano at the time Galileo was studying at Pisa.

The thesis I have developed in the first volume of my *Causality and Scientific Explanation* is that the mathematical component of modern science had its origin at Oxford with Ockham, Bradwardine, and the Mertonians, and that their work was then taken up and developed at Paris, with Jean Buridan and his group. The difference between Oxford and Paris, however, lies in the fact that Oxford was more nominalist in its natural philosophy, analyzing motion for example linguistically and in the imagination alone *(secundum imaginationem)*, whereas Paris was more realist, and thus could see this type of mathematical theorizing as directly applicable to the world of nature. The final component of modern method then came at Padua, for here the mathematical teachings of Oxford and the realist dynamics of Paris were combined in a way that suggested the possibility of experimental proof, and thus experimentation finally came into its own as an integral part of scientific method.

ROSEN: May I introduce a somewhat different aspect into this discussion? A great American astronomer, Otto Struve, said that astronomy has made remarkable progress since the Church lost its power to burn people for their astronomical activities and opinions. I think that we should face the facts bluntly and squarely. The attempt to remove the power of theologians over the expression of scientific opinion is greatly strengthened by Nicolaus Copernicus' dedication of his *Revolutions* to Pope Paul III.

I would like to remind you that in this magnificent and dramatic appeal for what is today called freedom of thought, Copernicus himself does not use this expression. Copernicus points out that there may be certain objectors whom he characterizes by the Greek word ματαιολόγοι. If you look at the two passages in the New Testament (1 Timothy 1:6 and Titus 1:10) that use this word, you see quite plainly that it means "those who babble nonsense."

Now, what does Copernicus say? "If there are those who babble nonsense and then dare to object to my statements, let me remind them that"—and then comes that famous expression which is so widely misunderstood—"astronomy is written for astronomers."

What he means is that the only persons who are qualified to judge an astronomical book are professionally trained astronomers, and this excluded the theologians. Half a century after Copernicus wrote this, an Italian *Protestant* refugee, Alberico Gentili, Regius Professor of Civil Law in Oxford University, in his *Law of War* (I, 12) said: "Theologians, keep quiet about a matter outside your field." (*Silente theologi in munare alieno.*)

If you will allow me to say so, we are here today by courtesy of Nicolaus Copernicus, Alberico Gentili, and their successors who said to the theologians, "Mind your own business, and stop burning Giordano Bruno and people like him and their books." Science has advanced since it was separated from theology and theology was confined to its own sphere. This very important point should never be obscured.

HARTNER: Would you include the modern salvation theories also, such as Marxism?

ROSEN: I would put it this way: I believe that as an activity, science is not unique. It stands with other things, and it flourishes only in a free soil. And wherever authoritarian control is exercised, for whatever purpose, it is obvious that intellectual activity ceases to flourish.

A simple question: Where are the great followers of Turgenev and Dostoevski? Does that answer your question?

HARTNER: Certainly.

GRENDLER: We should remember that theologians do not burn people or books. What happens is that political and ecclesiastical leaders sometimes burn them when they discover that new ideas seem to threaten the secular and/or spiritual order in some way.

Theologians may denounce a person or idea, but they seldom have the power to eliminate an opponent. Condemnations and prohibitions emanating from theologians are only effective if the leaders of society are willing to implement them. Possibly for this reason Copernicus addressed his letter to the Pope, a man of ecclesiastical and political power, but certainly not a theologian.

R. HALL: In fairness, it must be said that there are other censors besides theologians.

ROSEN: Of course. This was the group with whom we were immediately concerned. But as Professor Hartner indicates, as the states become secularized the seats of power are shifted out of their hands into the hands of other bigots.

NELSON: I think Ed Rosen is totally correct in saying that in the 16th century we have all through Europe a new trend, the demand that theologians not dictate any of the principles either in the sciences or in politics or in economics or anywhere else. However, I would not assume that because the outcomes were so fortunate in respect to, say, astronomy that they were equally fortunate in all other spheres. Sometimes it meant turning over authority to others who have exceedingly partisan narrow interests.

I have written about that in my book on usury.* Here I have reported the forwardness of 16th-century controversials who were saying: "We don't want theologians telling us how businessmen are supposed to run their affairs." The secularizing outcomes of the new propaganda were not as helpful as some might suppose on many critical questions.

R. HALL: To revert to those passages that Ed Rosen began to discuss, I have always wondered whether I was right in reading these references to the mathematicians and their competence in mathematics as also implying a sort of restriction.

When Copernicus says, "What I'm going to talk about is the business of mathematicians; only mathematicians are capable of

---

*Benjamin Nelson, *The Idea of Usury: From Tribal Brotherhood to Universal Otherhood,* second edition, enlarged (Chicago, 1969).

pronouncing on it," is he also saying, "I'm not going to talk about other matters that might also be of interest in which other people have their own competence"? For example, Aristotelian natural philosophy?

SWERDLOW: I have thought about exactly those implications to that passage. Is he excluding not only theologians but also philosophers, Aristotelians, you name it? Anything except a real astronomer? It is possible. The context, though, is completely theological.

ROSEN: There is a widespread misinterpretation of those three famous words, *Mathemata mathematicis scribuntur*—"astronomy is written for astronomers"—namely, that they express contempt for the common people. Of course, this is not what the passage means at all, and Noel Swerdlow is entirely right in saying the context is theological.

Does Copernicus mean to exclude the natural philosophers? The answer is contained in the *Revolutions*, Book 1, Chapter 8, where Copernicus is discussing the question whether the universe is finite or infinite. But what he says explicitly is this: "Let us leave the question whether the universe is finite or infinite to be discussed by the natural philosophers."

So obviously he is not excluding the natural philosophers. He is excluding himself from the ranks of the natural philosophers, but he turns over to them such ultimate questions, which he regards as beyond his own reach as a theoretical—

SWERDLOW: Unlike the heliocentric theory, which is within his reach.

ROSEN: Quite so.

Albrecht Dürer's engraving of Philipp Melanchthon. The inscription reads "Dürer was able to depict Philipp's features as if living, but the practiced hand could not portray the soul."

THE WITTENBERG INTERPRETATION
OF THE COPERNICAN THEORY

Robert S. Westman

Any major scientific theory consists of varied components ranging from very general, "deep" assertions about the nature of reality, through a level of more specific conceptual claims that articulate the initial presuppositions and refine their intelligibility, to a still more specific level where these claims become amenable to more direct confirmation and falsification.[1] At any given time, but usually during periods of transition from one major theory to another, some tension will exist between the most general presuppositions or regulative ideals of a theory and its more particular components. In the work of Copernicus, tension areas are most evident in his criticisms of the received

I am very grateful for research support received from the American Philosophical Society (Penrose Fund, Grant No. 6450) and the Academic Senate, University of California at Los Angeles in 1972-73. Subsequent versions of this paper were read before the Psychohistory Study Group of Los Angeles in July 1973 and at the University of Pittsburgh Philosophy of Science Seminar in January 1974. I wish to acknowledge with appreciation the criticisms and suggestions that I received at these sessions as well as from my UCLA colleagues, Peter Loewenberg and Amos Funkenstein; and also from Bruce Moran.

astronomical tradition: Ptolemy's extension of the Platonic axiom of celestial motion to include the equant; the system of purely homocentric spheres; the contemporary treatment of the lunar motions and the motions of the eighth sphere; the failure of traditional astronomers to satisfy the requirement that the periods of the planets increase in some uniform way as their distances from the earth increase; and, finally, the state of astronomical observations and the degree of acceptable error. It is not quite accurate to say, in agreement with Thomas Kuhn, that Copernicus' discontent reflected a "crisis state" in 16th-century astronomy.[2] It is evident, for one thing, that many of these problems were not new ones; they were endemic to the pre-Copernican, Graeco-Arabic astronomical tradition.[3] Unless one wishes to say that medieval astronomy was in a state of chronic crisis, which is clearly unreasonable, then we must admit that there existed no predecessor or contemporary of the young Copernicus, who shared such a general discontent with such a broad spectrum of issues, a discontent that was not only technical but also deeply philosophical. Copernicus, like all creative scientists, found things to be upset about that others took for granted.

The singularity of Copernicus' dissatisfaction with the astronomical tradition that he inherited helps to explain, in part, why the solutions that he proposed to problems uniquely perceived by him could not be fully appreciated by the great majority of astronomers living in the 25 years or so after the publication of *De revolutionibus orbium coelestium* (1543). My purpose in this paper, then, is not to dwell upon the interesting question of how Copernicus came to make his great discovery,[4] but rather to focus upon the factors—methodological, institutional, and emotional—which actually determined the early reception of the new theory.

A major theme in this study, one which is generally overlooked, is that contemporary assessment of a new scientific theory will depend upon how that theory is perceived. This may seem an obvious point, yet there exists an understandable tendency among historians and philosophers of science to assume that a later, well-supported version of a theory was the same version available to its earliest recipients. Moreover, the perception of a new, unfamiliar theory will be determined by other considerations, such as prior emotional expectations and procedural habits, by the reactions of close colleagues, as well as by the context in which the new ideas are first encountered. Learning the essentials of a new scientific program directly from its creator, for example, will undoubtedly constitute a different kind of experience than the once-removed occasion of reading a text. Yet, if it is almost inevitable that different persons will perceive the same theory in somewhat diverse ways, commonality of professional training and contact will generate a counter-force that encourages the sharing of perceptions.

The "Wittenberg Interpretation" is a phrase which I shall use to designate a common methodological outlook or style, a consensus on how to "read" the newly published *De revolutionibus,* which was shared by a group of young astronomers at the University of Wittenberg gathered under the fatherly tutelage of the famous Protestant reformer, Philipp Melanchthon (1497-1560). There were several features of this "reading" of Copernicus' innovation and, within this common orientation, certain shades of variation. By and large, the principal tenet of the Wittenberg viewpoint was that the new theory could only be trusted within the domain where it made predictions about the angular position of a planet;[5] and, in some cases, Copernicus' predictions were taken to be an improvement over those of Ptolemy.

Beyond this basic attitude, some members of the Wittenberg Circle believed that certain Copernican models, such as that which replaced the Ptolemaic equant with epicyclic devices, were to be preferred. An important plank of the Wittenberg program, though one not to be realized until the succeeding generation, was the goal of translating these equantless devices into a geostatic reference frame. The least satisfactory Copernican claim was the assertion that the earth moved—and that it moved with more than one motion. In some public summaries of the theory, this claim is explicitly denied and taken to be refuted by the Aristotelian postulate that a body can have only one simple motion. Osiander's anonymous preface to *De revolutionibus* provided the slightly more moderate view, however, that all celestial motions (including the earth's) were hypothetical. But what the Wittenberg Interpretation *ignored* was as important as that which it either asserted or denied. In the writings, both public and private, of nearly every author of the generation that first received the work of Copernicus, the new analysis of the relative *linear* distances of the planets is simply passed over in silence. This ignoring of the relative ordering of the planetary spheres does not mean, of course, that astronomers were disinterested in such matters. The familiar 16th-century woodcut of the geocentric cosmos with its four sublunary elements was graphic testimony to the unquestioned consensus on the true number and order of the planets. The Wittenberg Interpretation of the Copernican theory, therefore, was no mere repetition, as Pierre Duhem implied,[6] of the ancient methodological formula, "to save the phenomena"; for it represented more than a position of epistemic resignation with regard to what one could know about actual, celestial motions. It also maintained that certain parts of the new theory were to be adopted and preferred *if* interpreted in a framework where the earth was at rest and it rejected or ignored other aspects of the theory that it regarded either as irrelevant or as possessing little truth content.

The social context of this early response to the Copernican innovation was an informal circle of scholars drawn together under the leadership of Melanchthon, a generation of men who had been born in the period ca. 1495-1525. The development of informal academies, especially in Italy, had begun in the 15th century. Structurally, they were composed either of a patron with a surrounding circle of intellectuals or of a charismatic intellectual about whom gathered a group of scholars or, finally, of a group of intellectuals coming together for informal discussions.[7] Melanchthon's circle is an example of the second type, but unlike the Italian academies it evolved *within* the walls of the university. Lacking the symbols of autonomy and power, the bureaucratized organizational structure and tight membership criteria, as well as control over publication (which would characterize such a later, professionalized scientific society as the Paris *Academie des Sciences*), Melanchthon and his disciples yet exercised considerable influence on the discipline of astronomy by staffing many of the leading German universities with their pupils and by writing the textbooks that were used in those institutions. The effect of this informal scientific group on the early reception of the Copernican theory cannot be underestimated. Thanks to its efforts, the realist and cosmological claims of Copernicus' great discovery failed to be given full consideration. There was but one notable exception to this methodological consensus: Georg Joachim Rheticus (1514-1574).

For reasons that will be made clear, it was Rheticus alone who departed from the "split" interpretation of the innovation of Copernicus, an interpretation that generally characterized its earliest reception in the German universities and in many parts of Europe.

THE ROLE OF PHILIPP MELANCHTHON

Melanchthon and the University of Wittenberg. It was Johannes Kepler, writing some 65 years after the publication of *De revolutionibus,* heir to a half century of critical commentary and discussion of the Copernican theory and consciously committed to the construction of a new *kind* of physical astronomy, who first clearly expressed the existence of two Copernicuses: the one, author of a truly sun-centered cosmology, which was supported by a few physical arguments and the discovery of a new, simpler argument for the ordering of the planets; the other, an inventor of planetary models calculated on the assumption that the planets revolve about a sun, which is eccentric to the true center of the universe and in which the equant is excluded as a candidate for saving the phenomena.[8] It is hardly surprising that Kepler should have possessed such an accurate historical perspective, for he was not only the recipient of a bifurcated Copernican tradition but he was also self-conscious of his own role in bringing to it a new unity.[9] While the origins of this dissociated viewpoint may be traced back to an earlier split in Western medieval astronomy and to Osiander's unique role in affixing his anonymous letter to *De revolutionibus,* the origins of its *institutional* entrenchment and promulgation must be sought in Melanchthon's Wittenberg Circle.

Son of a sword-cutler from Bretten in the Palatinate, a grandnephew of the great Hebraic and Cabalistic scholar, Johannes Reuchlin (1455-1522), Melanchthon was early on disposed toward humanistic and Greek studies, although he also developed a deep interest in astrology and astronomy under the tutorship of Johann Stoeffler (1452-1531). He was an active leader in the humanistic movement at the University of Tübingen but left there in 1518, at the age of 21, to assume a professorship of Greek at the newly formed University of Wittenberg.[10] There is evidence that he was the object of a vigorous recruiting campaign from two other universities—Ingolstadt and Leipzig—but he turned down these offers in spite of a higher salary and great faculty banquets in his honor. His biographer, Joachim Camerarius, adds that Melanchthon was "not so lusty a drinker as the Leipzig professors were."[11] At Wittenberg, the violent, popular energy of the Reformation was beginning to assert itself. Some of the more zealous followers of Luther demanded not only a return to the simplicity of the mass as in the early years of the Church, but the abolition of all education. Christ and his apostles had not been educated, it was argued, and the Gospel was intended for the simple, not the wise.[12] Added to these conflicts was the very serious social disruption brought about by the Peasants' War (1524-1525). Furthermore, in the years between 1521 and 1536, attendance figures at the German universities generally tended to decline quite considerably and, in some cases, drastically.[13]

It was in this context that Melanchthon launched a vigorous and far-reaching campaign of educational reform, which was to have a profound effect on the structure and content of German learning. He involved himself in establishing and reforming the principles of organization for the Protestant universities, which, beyond Wittenberg, included Tübingen, Leipzig, Frankfurt, Greifswald, Rostock, and Heidelberg. The newly founded universities of Marburg (1527), Königsberg (1544), Jena (1548) and Helmstedt (1576) also reflected the Melanchthonian spirit of education. He wrote innumerable textbooks in a wide variety of areas ranging from the classical trivium to physics, astronomy, history, ethics, and, of course, theology. His prefaces appeared in editions of the *Sphere of Sacrobosco*, Erasmus Reinhold's Commentary on Peurbach's *New Theories of the Planets* (1542), Euclid's *Elements*, and many other scientific treatises.[14]

In short, Melanchthon's ideals were not merely contemplative. He was a dedicated teacher who was always close to his pupils. As he once remarked to his friend Camerarius, "I can assure you that I have a paternal affection for all my students and am deeply concerned about everything that affects their welfare."[15] In practice, Melanchthon's precepts and personal example inspired several generations of pupils and teachers to the most signal achievements in astronomy and the other sciences. Indeed, it might be suggested, although space does not allow a full treatment here, that the relatively high state of the astronomical art in Germany in the 16th century, when compared with England, France and perhaps even Italy, was stimulated in large measure by the reforms and charismatic leadership of Melanchthon who, in his own lifetime, was rightly called *Praeceptor Germaniae*.

Melanchthon and the Copernican Theory. It is one thing to *encourage* expansion in a particular discipline, to recruit outstanding talent and to solicit substantial financial support. These excellent qualities, however, do not necessarily guarantee a completely open and receptive attitude toward innovation. Great humanistic scholar, administrator, and pedagogue that he was, Melanchthon could not properly be called a working astronomer. His lectures on physics and astronomy, first published at Wittenberg in 1549 (although written several years earlier), reveal a firm grounding in the original texts of Aristotle and Ptolemy, a clear systematization of arguments in support of various claims, but a woeful lack of diagrams.[16] More significantly, Melanchthon was hardly a man just entering an age of youthful exuberance when he first learned of the Copernican theory through Rheticus' *Narratio prima*, which was sent to him on 15 February 1540.[17] At the age of 43, he was certainly not old, but he had already been associated with the University of Wittenberg for more than two decades; he was the veteran of many academic battles, and he was at the height of his career.

His earliest reference to Copernicus, in a well-known letter to Mithobius, on 16 October 1541, is merely an incidental one and treats the new theory as a disturbance rather than as a threat.[18] Later statements, however, take a harder line. In his published lectures of 1549, he perceives the new theory as an old and absurd "paradox," which Aristarchus had once defended and which young students ought to stay away from since it conflicts with Scripture.[19] Further passages clearly argue against the earth's motion from the Aristotelian doctrine of simple motion.[20] Contrasted with these views, however, we find evidence of a more positive and favorable interpretation. Thus, he praises Copernicus' lunar theory because it is "so beautifully put together" (*admodum concinna*) although adding that students ought rather to learn the Ptolemaic viewpoint.[21] In several places, he uses Copernican data for the solar apogee[22] and for the apogees of the superior planets;[23] and, in the second edition of his *Initia doctrinae physicae*, as Emil Wohlwill first demonstrated, he tones down the negative allusions to Copernicus by deleting several phrases about those who argue that the earth moves "either from love of novelty or from the desire to appear clever."[24]

The customary dichotomy of "pro" and "anti" Copernican, then, becomes less than adequate as a description of Melanchthon's views and those of his disciples. To be sure, one can certainly discern a traditional component in Melanchthon's attitude toward the new theory. He recognizes that the earth's motion *could* be interpreted as a real one and he explicitly rejects this possibility as contrary to the divine testimony of Scripture. By contrast with the position of Osiander, then, Melanchthon's interpretation is slightly less moderate, for Osiander's anonymous introduction nowhere rejects the motion of the earth with Scriptural and physical arguments but simply awards to it an open hearing, namely, the same hypothetical status as the planetary theories of the ancients.[25] In the spirit of Osiander's preface, however, there is also a *pragmatic* component in Melanchthon's evaluation of the Copernican theory. He seems to be saying that there may be something of value in this new teaching that will be useful for students and professors. On a psychological level, Melanchthon's viewpoint appears as a benign paternalism—tolerant, perhaps even amused by the revival of an old paradox, willing to encourage some experimentation, but ultimately wedded to established points of view. Not surprisingly, it was this same capacity for compromise and flexibility which had already gained for the *Praeceptor Germaniae* his reputation as an effective arbitrator of controversies, whether theological or personal.[26] While the florid and frequently dogmatic Luther had taken a characteristically strong (and, in this case, negative) view of a new idea on which he had only bare, hearsay knowledge,[27] Melanchthon adopted a typically moderate position, which absorbed the criticism of Luther while allowing considerable freedom to explore new pathways. Limited though it was, then, Melanchthon's attitude would permit and even encourage some articulation of the conceptual and empirical components of the Copernican theory: the improvement of observations, discussion of the precessional problem, and consideration of the Copernican planetary models.

THE DISCIPLES OF MELANCHTHON

Of the group of astronomers who pursued their discipline in the course of Melanchthon's tenure at Wittenberg, from about 1530 to 1560, one might include at least ten members, but none were more important than three figures: Erasmus Reinhold (1511-1553), Caspar Peucer (1525-1602), and Georg Joachim Rheticus (1514-1574). Over the years, a strong father-son relationship had developed between them and Melanchthon, and under his aegis the science of astronomy flourished. Two of the three persons mentioned above, Reinhold and Peucer, eventually held high administrative positions as rector of the university; and each of the three, in his own way, helped to lay down a particular pattern for the reception of the Copernican theory.

Virtually nothing is known of the life of Reinhold.[28] Between 1541 and 1542, he first became directly acquainted with the theory of Copernicus through his colleague Rheticus, who had only recently returned from Frombork. Already in his Commentary on Peurbach's *New Theories of the Planets* (1542) a major source of his interest and curiosity about the new theory is discernible in his criticism of contemporary astronomy where we note a dissatisfaction with the planetary theories that reminds us at once of a similar discontent in Copernicus.[29] Somewhat later, he refers to a new hope for the restoration of astronomy:

> I know of a recent author who is exceptionally skillful. He has raised a lively expectancy in everybody. One hopes that he will restore astronomy. He is just about to publish his work. *In the explanation of the phases of the moon he abandons the form that was adopted by Ptolemy. He assigns an epicyclic epicycle to the moon . . . .* [My italics.] [30]

Now, we recall that Melanchthon had also mentioned Copernicus' lunar theory but there was only one printed source where Reinhold could have obtained this information by 1542: in the recently-published *Narratio prima* of Rheticus. Here, Rheticus had written of Copernicus' handling of the moon's second inequality: "He assumes that the moon moves on an epicycle of an epicycle of a concentric . . . ."[31] On the following page, we grasp the larger context in which Rheticus writes:

> Furthermore, most learned Schöner, you see that here in the case of the moon, we are liberated from an equant by the assumption of this theory, which, moreover, corresponds to experience and all observations. My teacher dispenses with equants for the other planets as well . . . .[32]

Liberation from the equant! How much pride Copernicus had taken in this achievement; and Reinhold leaves no doubt that it was this accomplishment that had initially impressed him so deeply. At the bottom of the title page of his own, personal copy of *De revolutionibus,* Reinhold wrote in beautiful and carefully formed red letters: *"Axioma Astronomicum: Motus coelestis aequalis est et circularis vel ex aequalibus et circularibus compositus"* ("The Astronomical Axiom: Celestial Motion is both uniform and circular or composed of uniform and circular motions").[33]

It was precisely this axiom, or boundary condition, which Copernicus had tried to satisfy, if not always successfully, in his construction of the planetary models and which Reinhold consistently singles out in his annotations. Thus, in Book V, Copernicus initiates his discussion of the planetary theories once more by attacking the ancients for conceding that "the regularity of the circular movement can occur with respect to a foreign and not the proper center; similarly and more so in the case of Mercury."[34] He then articulates this initial presupposition in terms of three particular models: a circle eccentric to an eccentric circle; an epicycle on an epicycle on a homocentric circle, and an eccentric circle carrying an epicycle.[35] Once again, Reinhold's pen boldly announces the point,[36] and with a great flourish reserved for the orbit of Mercury: *"Orbis Mercurij Eccentrj Eccentrus Eccentrepicyclus"* ("The Orbit of Mercury is an eccentric on an eccentric on an eccentric circle carrying an epicycle").[37]

It is no wonder that literati scorned the strange, seemingly pedantic activities of astronomers. For Reinhold, however, Copernicus had truly restored astronomy by attempting to clean up the morass of equants, which had cluttered the art.[38] It was not the revolutionary cosmological arguments of Book I in the *De revolutionibus*, therefore, that had brought such high esteem from Reinhold, for with the exception of a few underlinings and factual paraphrases, his own copy is practically devoid of interpretative comments.[39] This seeming disinterest is further confirmed by his unpublished commentary on Copernicus' work where he maintains what Aleksander Birkenmajer calls "the most perfect neutrality on the problem of geocentrism and heliocentrism."[40] There remained, in Reinhold's view, but one more major task: to systematize and to recalculate the motions according to the Copernican requirements. And this is precisely what he did. In an article that nicely complements our conclusions in this section, it has been shown with the aid of a computer, that the *Prutenic Tables* are based solely upon Copernican planetary mechanisms.[41]

Reinhold, thus, clearly became the leading figure in the Wittenberg Circle in the years after Rheticus' departure for Leipzig in 1542. His great work on the new tables of motion received the full moral and financial support of Melanchthon who wrote to Albrecht, Duke of Prussia, on his behalf.[42] And his tables soon rapidly showed their competitive strength against other compilations of planetary data.[43] Through Reinhold the image of "Copernicus the Calculator" seeped inexorably into astronomical discussions both inside and outside the walls of the universities.

The man largely responsible for consolidating and institutionalizing the Wittenberg Interpretation within the universities was a pupil of both Reinhold and Melanchthon. His name was Caspar Peucer. Born in 1525, the son of an artisan, he showed such remarkable talents as a young man that he began to attend the University of Wittenberg at the age of 15 where he stayed at the home of Melanchthon.[44] In 1550, this informal relationship with

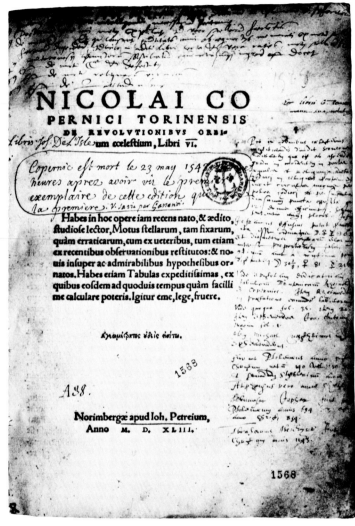

Title page of Caspar Peucer's copy of *De revolutionibus*, with his notes at right and his name at upper right. The notice below the title, "Copernicus died on 23 May 1543 a few hours after having seen the first example of this edition, which is the first," was added by the 18th-century French astronomer Joseph-Nicolas Delisle. From Delisle the book went to Dépôt de la Marine, whose anchor stamp appears in the upper center, and thence to the Observatoire de Paris, with whose permission this page is reproduced.

the great reformer was formalized when he married Magdalena, the daughter of Philipp Melanchthon. A poem at the beginning of Peucer's *Elements of the Doctrine of the Celestial Circles* (1551) endearingly, and not surprisingly, dubs Melanchthon with the title of "Father" and a chronological list of "*Astrologi*," starting with the creation and proceeding up to 1550, ends with Erasmus Reinhold, "*Praeceptor mihi carissimus*."[45] When Reinhold died of the plague in 1553, Melanchthon's son-in-law succeeded him. In 1559, he was named to the chair of medicine and in 1560 he succeeded finally to the rectorate of the University upon the death of Melanchthon.

The Wittenberg Interpretation, as reflected in Peucer's work and the writings of other Wittenberg professors, firmly echoes the views of Peucer's two mentors, Melanchthon and Reinhold. It is important to recognize, however, that there were different pedagogical levels at which the theory might be considered: introductory and advanced. In 1545, Melanchthon had reorganized the Faculty of Arts and therein defined the subject matter for the various curricula. Students in the natural sciences, at the lower levels, were to be instructed in arithmetic, physics, the second book of Pliny's *Natural History*, Aristotle's *Ethics* and the *Sphere* of Sacrobosco, while those seeking the Master's degree were to have lectures on Euclid, planetary theory (Peurbach) and Ptolemy's *Almagest*.[46] It is hardly surprising, then, that the introduction of the Copernican theory into the scientific curriculum should follow the limits already prescribed by the existing system. Peucer, in his widely used introductory textbook, mentions the new theory briefly and cites quantitative material from it to illustrate various aspects of the celestial motions; for example, he uses Copernican data on the absolute distances of the sun and moon from the earth in his discussion of eclipses and on the definition of the length of the day.[47] Just as in Melanchthon's lectures, Copernicus is portrayed as the reviver of the theory of Aristarchus[48] from which, of course, students could learn nothing of the extensive arguments and demonstrations set forth by Copernicus on the planetary theory. The only arguments to which beginning students were exposed were those *against* the motion of the earth—that it is contrary to Holy Scripture and the laws of simple motion—and the usual claim that astronomers use many different kinds of devices to save the phenomena.[49] Just as today, when we do not include advanced and very recent research in introductory texts, so Peucer and Melanchthon drew the line, perhaps a bit too strictly, at the treatment of planetary theory.

At the Master's level, the handling of planetary motions became far more sophisticated. Here the basic text was not Sacrobosco, but Peurbach's *Theoricae novae planetarum*. There can be little doubt that Wittenberg students used the version of Peurbach to which was affixed the commentary of Reinhold, at least until the middle of the 1550s.[50] Many later Wittenberg professors, such as Johannes Praetorius (1537-1616), Andreas Schadt (1539-1602), and Caspar Straub, who were all probably pupils of Peucer, continued to use the Peurbachian framework for their lectures on advanced planetary theory; and it was in this context that several aspects of the Copernican theory were accorded more extensive and serious consideration than in the introductory courses.

Space does not permit us to examine in detail the views of Peucer and the subsequent articulation of the Wittenberg Interpretation after about 1560, but at least two generalizations may be attempted here. In the first place, the status of Copernicus' work was clearly ranked on the same level as Ptolemy's. Peucer, Schadt, Straub, and others advised their students to consult *De revolutionibus* directly and often recommended the comparison of corresponding topics in the *Almagest*.[51] In a particularly influential and representative source of the Wittenberg viewpoint, Peucer's *Hypotheses astronomicae*, it was suggested that Copernicus' precessional model could be "transferred," in principle, to a geostatic reference frame if a ninth and tenth spheres were added.[52]

A second basic feature of the evolution of the Wittenberg Interpretation was the conspicuous absence of urgency about the issue of cosmological choice. In the vast majority of advanced treatments of planetary theory we do not encounter the Scriptural and physical arguments so prominent in debates about the Copernican theory after 1616. Questions of the actual motion of the earth and the order of the planets were rarely considered. The passages in Book I of *De revolutionibus*, extolling the newly discovered harmony of the planets and the eulogy to the sun, with its Hermetic implications, were usually passed over in silence. Copernicus was seen, in general, as the *reformer* of Ptolemaic astronomy—not in a revolutionary sense, however, as later thinkers such as Kepler would believe, but in an essentially conservative sense, as the admired inventor of new planetary hypotheses and an improved theory of precession.

While the Wittenberg reading of *De revolutionibus* persisted at least until the end of the 16th century, it was already being questioned and reformulated by a new generation of astronomers in the 1570s. For men such as Johannes Praetorius, Tycho Brahe, and Michael Maestlin, it seemed that their predecessors had overlooked the serious cosmological claims of Copernicus.[53] Of the men of the earlier generation, only one member had praised Copernicus for asserting an "absolute system" of the planets. Where Peucer and Reinhold had been dutiful "sons" of Melanchthon, the third "son," Rheticus, had not followed the cautious path laid down by the "father" of the Wittenberg circle. Rheticus had found a different kind of father to idealize.

Rheticus: Personality and Theory-Choice. There was something different about Rheticus. Karl Heinz Burmeister, his most recent and comprehensive biographer to date, suggests that he had an "abnormal personality," an insane genius characterized by "bionegative elements": celibacy, childlessness, and homosexual tendencies.[54] Considerably more sympathetic and perceptive than Burmeister's artificial pigeon-holing, Arthur Koestler writes with the impressionistic flair of the novelist:

Rheticus, like Giordano Bruno or Theophrastus Bombastus Paracelsus, was one of the knight errants of the Renaissance whose enthusiasm fanned borrowed sparks into flame; carrying their torches from
one country to another, they acted as welcome incendiaries to the Republic of Letters . . . . [Rheticus was] an *enfant terrible* and inspired
fool, a *condotierre* of science, an adoring disciple and fortunately,
either homo- or bi-sexual, after the fashion of the time. I say 'fortunately' because the so afflicted have always proved to be the most
devoted teachers and disciples, from Socrates to this day, and History
owes them a debt."[55]

The comparisons with Paracelsus and Bruno are interesting
although undeveloped. Rheticus certainly possessed a kind of
Brunian-Paracelsian *Wanderlust*. Between 1538, when he temporarily left his post at Wittenberg, and 1554, when he settled in
Cracow, his life was one of almost constant peregrinations. Unlike the former individuals, however, he was not fleeing the authorities, because, like Paracelsus, he had thrown the works of
Galen and Avicenna into a public bonfire;[56] nor like Bruno, because he was attempting to reconcile Protestants and Catholics
in a Hermetic religion of the world.[57] The only journey that he
undertook involuntarily, so far as we know, was his hurried departure from Leipzig in 1550, which was probably due to his
having engaged in illegal sexual acts.[58] Unfortunately, there is
scarcely enough evidence for us to conclude anything about the
general pattern of his sexual behavior.

Koestler is closest to the mark when he writes of the "enthusiasm" of Rheticus. If indeed one were to point to a single most
prominent trait in Rheticus' personality, based upon the tone of
his writings, the testimonies of his contemporaries and his own
life activities, one would have to seize upon his great energy and
intensity—whether in the vitality of his work, in his widespread
travels, or in his evident pursuit to lay to rest something inside
of himself. Where one observes for any individual an unusually
great investment of psychic energy in a particular object or goal,
there one can often expect to find powerful, conflicting emotions.
Rheticus, in particular, seems to have had an *inordinate* need to

associate himself with famous men, to be pleasing (or displeasing) to them and, in short, to be loved by them. Consider his decision to leave Wittenberg in 1538, at the age of 24, after he had been teaching at the University for two years and was very well established. In a letter to Heinrich Widnauer in 1542, he recalls being "attracted by the fame of Johannes Schöner [1477-1547], in Nuremberg, who had not only accomplished much in scientific subjects but had excelled in all the best things of life as well."[59] From Nuremberg, Rheticus then journeyed to Tübingen where he visited the "famous pupils" of Johannes Stoeffler (1452-1531)[60] and thence to Ingolstadt where he met with Peter Apianus (1495-1552). "And finally," Rheticus writes,

> I heard of the fame of Master Nicholas Copernicus in the northern lands, and although the University of Wittenberg had made me a Public Professor in those arts, nevertheless, I did not think that I should be content until I had learned something more through the instruction of that man. And I also say that I regret neither the financial expenses nor the long journey nor the remaining hardships. Yet, it seems to me that there came a great reward for these troubles, namely, that I, a rather daring young man [*iuvenili quadam audacia*] compelled [*perpuli*] this venerable man to share his ideas sooner in this discipline with the whole world.[61]

Beyond the manifest purpose for visiting Copernicus, namely, to learn about his theory, lay a deeper personal motivation. The "great reward" for Rheticus was to be found in his *relationship* with the older man, in causing him to share with the younger a treasure that no one else had been able to pull from him. This interpretation is reinforced by an interesting event that occurred near the end of Rheticus' life. A young mathematician named Valentine Otho (b. 1550), who had studied with the successors of Rheticus at Wittenberg, Caspar Peucer and Johannes Praetorius, came to visit him in 1574 in the town of Cassovia, located in the Tatra Mountains of Hungary.[62] In Otho, the aging Rheticus saw himself again as he had once been, almost as though he were looking into a mirror of the past. Otho reports:

We had hardly exchanged a few words on this and that when, on learning the cause of my visit he burst forth with these words: "You come to see me at the same age as I was myself when I visited Copernicus. If I had not visited him, none of his works would have seen the light.[63]

Rheticus' personal relationship with Copernicus, therefore, was singular; and it was a relationship that no one at Wittenberg would ever fully comprehend. In his conversion to the Copernican theory—and, in this case, we can use the word with its religious and emotional overtones—the subjective meaning, which the new ideas had for Rheticus, and his sense of sharing in their birth, would provide the emotional energy that fueled the rational arguments adduced in its favor.

The *Narratio prima* itself, the little work written by Rheticus in 1540 as a kind of "trial balloon" for Copernicus, was not merely a "report" of Copernicus' ideas as the title seems to suggest. It contains ideas that were not to be found in *De revolutionibus* or in any of Copernicus' other writings;[64] it contains at least one conceptual error born undoubtedly of excessive zeal[65] and, most importantly, it gives selective emphasis to one Copernican argument in a way that goes well beyond the public position of its author. In short, as one reads the *Narratio prima* today, one gets a strikingly different impression of the Copernican theory than that reflected in the Wittenberg textbooks.

The work opens with a consideration of problems that were certainly of great interest to the Wittenbergers: the calendar, lunar motion, and the rejection of the equant. This is clearly the work of a man who has full command of the leading astronomical issues of his time. Yet there exists a quality of enthusiasm in his writing that, not surprisingly, is distinctly absent from the university textbooks. Even more than its general tone, however, is the unmistakable concentration upon one feature of the Copernican theory that had been consistently ignored by the Wittenberg astronomers: the demonstration of a system in the necessary interconnexity of the relative distances and periods of the

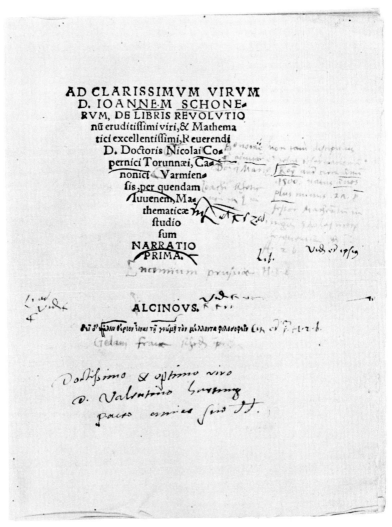

Title page of Rheticus' *Narratio prima* (Danzig, 1540) with Rheticus' inscription to his friend Valentine Gasser. (Photograph by Owen Gingerich, permission of Yale Medical Historical Library.)

planets. In the old hypotheses, Rheticus wrote, "there has not yet been established the common measure [*mensura communis*] whereby each sphere may be geometrically confined to its place" and where "they are all so arranged that no immense interval is left between one and the other . . . ."[66]

In these claims we find three assumptions about the planetary models: (1) that each planet is carried by a uniformly revolving sphere;[67] (2) that there are no gaps between the spheres (principle of plenitude); and (3) that the relative planetary positions are to be measured with respect to a common unit. Underlying these assumptions lay Copernicus' strong preference for unified explanations, which he had invoked metaphorically when he criticized the Ptolemaic hypotheses on the grounds that they had not been able to demonstrate the "fixed proportionality" of the universe.

With them it is as though one were to gather the hands, feet, head and other members for one's images from diverse models, each excellently drawn, but not related to a single body, and since they in no way match each other, the result would be a monster rather than a man.[68]

The metaphor of the human body was only one of several metaphors with a long intellectual tradition in the West that expressed the architectonic unity of the parts in an integrated whole.[69] The human soul, the world, number, musical harmony, the well-proportioned dimensions of churches—each of these embodied the principle of unified, well-ordered wholeness. Copernicus had endowed this abstract value with a powerfully new concrete import by showing that the order of the planets was in agreement with the most general requirement of any good theory,[70] while Rheticus goes even further in his development of this theme. There are at least ten references to the harmony of the Copernican system in the *Narratio prima,* a number far greater than in *De revolutionibus* itself. Added to this, Rheticus ends his treatise with a conspicuously Platonic-Pythagorean passage on the harmony of the soul.[71]

None of Rheticus' Wittenberg colleagues had been struck by this deeper vision of the universe. None could write with the conviction of a man who had not only countenanced a new intellectual system but had personally known its creator. Amplifying Copernicus' criticism of the Ptolemaic hypotheses and appealing to the criteria of unity and necessity, Rheticus wrote:

> . . . my teacher was especially influenced by the realization that the chief cause of all the uncertainty in astronomy was that the masters of this science (no offense is intended to divine Ptolemy, the father of astronomy) fashioned their theories and devices for correcting the motion of the heavenly bodies with too little regard for the rule which reminds us that the order and motions of the heavenly spheres agree in an absolute system. We fully grant these distinguished men their due honor, as we should. Nevertheless, we should have wished them, in establishing the harmony of the motions, to imitate the musicians who, when one string has either tightened or loosened, with great care and skill regulate and adjust the tones of all the other strings, until all together produce the desired harmony, and no dissonance is heard at all.[72]

The emotionally heightened, Platonic-Pythagorean imagery of Rheticus's *Narratio prima,* fully infused with the rhetorical brilliance of the classical Renaissance style, and a sense that he has truly "seen the light" (a quality reminiscent of Paracelsus and Bruno), lies in stark contrast with the neutral feeling tone of the Wittenberg scientific establishment. In a remarkable and revealing passage, where Rheticus explains his decision to adopt the Copernican theory, we see the link between the intellectual and the emotional basis for his conversion.

> I sincerely cherish Ptolemy and his followers equally with my teacher, since I have ever in mind and memory that sacred precept of Aristotle: "We must esteem both parties but follow the more accurate." And yet somehow I feel more inclined to the hypotheses of my teacher. This is so perhaps partly because I am persuaded that now at last I have a more accurate understanding of that delightful maxim which on account of its weightiness and truth is attributed to Plato: "God ever geometrizes";[73] but partly because in my teacher's revival

of astronomy I see, as the saying is, with both eyes and as though a fog had lifted and the sky were now clear, the force of that wise statement of Socrates in the *Phaedrus:* "If I think any other man is able to see things that can naturally be collected into one and divided into many, him I follow after and 'walk in his footsteps as if he were a god'." [74]

On the one hand, Rheticus finds that the statement attributed to Plato takes on a new and deeper significance within the context of the Copernican theory. Copernicus had shown a necessary ordering of the planets to emerge from the correlation of heliostatic periods with the magnitudes of the planetary orbits and that the earth-sun-planets triangles could be compared with one another by the commonly shared earth-sun radius. The mathematics of proportionality now seemed to have achieved a deeper meaning in the sacred heavens themselves. On the other hand, Copernicus' very act of unification seems to have had a liberating, almost intoxicating effect on Rheticus. In some way, by being allowed to share the theory of the great man, a veil that had covered an inner personal turmoil was lifted from the young man.

Rheticus' Conflict: A Possible Explanation. Let us pause for a moment and ask a new question: What was the nature and cause of Rheticus' unconscious conflict? Can we offer an adequate psychodynamic explanation for Rheticus' behavior? Such questions of underlying motivation are seldom raised in the historiography of science, which has been dominated by a highly successful, but thoroughly intellectualist, methodology. [75] Yet, having broached the question, one is faced immediately by the kind of unfortunate paucity in the data with which historians of our period are all too familiar. A satisfactory psychodynamic explanation, which assumes the importance of early childhood experiences and perceptions, of fantasies, unconscious wishes and dreams, must be able to draw, at the very least, upon sufficient information about significant *patterns* in an individual's early relationships with others so as to establish accurate *parallels* with later behavior. Conflicts that remain unresolved in early years will tend to reappear later on in different guises; and the historian's skillful identification of repetitive patterns can help to explain the cause of later actions which seem truly irrational—i.e., inappropriate, exaggerated, unexpected, and even destructive in a particular individual. In addition to the general models of relating to mother, father, brothers and sisters, *specific* events of traumatic severity can sometimes intensify a low-lying conflict or, perhaps, initiate a new one. The response to the devastating experiences of rape, warfare, or the observation of violent death, for example, will depend upon factors such as age, basic ego-strength, and external support, but the chances are that these events will be upsetting even for a normal individual. In the case of Rheticus, we have virtually no information on the general patterns of his early familial relationships, but we do have evidence of one very important event in his life: the death of his father. From this small piece of evidence we may form a possible account of what might have occurred in Rheticus' youth.

In 1528, when Rheticus was 14 and had, therefore, just reached adolescence, his father, Georg Iserin, a doctor in the town of Feldkirch, Austria, was convicted on a charge of sorcery and beheaded.[76] Witch trials were not uncommon during this period.[77] Whatever had been the specific crime of Rheticus' father, however, the effects of his antisocial behavior (suspected or real) must have been difficult for the family. At least one painful result was that the alleged sorcerer's name could no longer be employed legally after his execution, and hence, the unfortunate widow had to revert to her Italian maiden name "De Porris" and her son, Georg, later took on the additional name of "Rheticus" because he had been born within the old boundaries of the Roman province of Rhaetia.[78] Georg Joachim Iserin, Jr., had lost not only his father but also part of his old identity. What must it have been like? We do not possess Rheticus' direct testimony, but we can imagine what a painful experience it must have been for the young lad to have lost the man whom he admired more than anyone else in the world. Indeed, we would predict that his later relationships with men, particularly with older men, would be marked by a deepened and intensified ambivalence. On the one hand, the manifest horror and grief over the power of aggression that had led to the violent death of his father would later appear in his need to atone for the least vaguely hostile feelings. Together with this, we should expect to find determined efforts to repair unconsciously the damage earlier wrought on his father by the search for wholeness and strength. On the other hand, there must have been an unconscious sense of liberation—at last he was freed from the tyranny of the old man—a feeling which was fully consonant with his later identification with intellectual rebellion in the persons of Copernicus and Paracelsus.[79]

Let us now see if we can find some support or refutation for these hypotheses in Rheticus' adult writings. We find evidence of ambivalence in his relationship with Johannes Schöner. The *Narratio prima* starts with the following dedication: "To the Illustrious John Schöner, as to his own revered father, G. Joachim Rheticus sends his greetings."[80] The motto on the title page is, however, of an entirely different tenor, a sort of revolutionary manifesto embodied in a Greek quotation from Alcinous, which reads: "Free in mind must be he who desires to have understanding."[81] Reverence and rejection of authority: the ambivalence is announced at the beginning of the work and the themes reappear clearly in the text itself. Thus, he says to Schöner:

> Most illustrious and most learned Schöner, whom I shall always revere like a father, it now remains for you to receive this work of mine, such as it is, kindly and favorably . . . . If I have said anything with youthful enthusiasm (we young men are always endowed, as he says, with high rather than useful spirit) or inadvertently let fall any remark which may seem directed against venerable and sacred antiquity more boldly than perhaps the importance and dignity of the subject demanded, you surely, I have no doubt, will put a kindly construction upon the matter and will bear in mind my feeling toward you rather than my fault.[82]

The tone is cautious, apologetic, and deferential as if to implore his father unconsciously not to beat him for his aggressive sentiments. In the next sentence, we learn that Copernicus was also very respectful of ancient authority.

. . . concerning my learned teacher I should like you to hold the opinion and be fully convinced that for him there is nothing better or more important than walking in the footsteps of Ptolemy and following, as Ptolemy did, the ancients and those who were much earlier than himself.

Then Rheticus describes the rebellious, innovative act almost as though it were beyond Copernicus' control, but ending with the quotation from Alcinous that had appeared on the title page.

However, when he [Copernicus] became aware that the phenomena which control the astronomer, and mathematics compelled him to make certain assumptions even against his wishes, it was enough, he thought, if he aimed his arrows by the same method to the same target as Ptolemy, even though he employed a bow and arrow of a far different type of material from Ptolemy's. At this point, we should recall the saying: "Free in mind must he be who desires to have understanding." [83]

The imagery of the bow and arrow, symbolizing the attack on ancient ideas, suggests at least part of the appeal which Copernicus' work held for the young Rheticus, for he had shared in the attack as well by influencing Copernicus to publish his new theory after so many years. In Copernicus, Rheticus had found a kind and strong father with a streak of youthful rebellion in him—a man who was different, as Rheticus' father had been; a father who could attack ancient authority with his intellectual weapons without himself being destroyed; a father who, *like the system he created*, had a head and a heart that were connected to the same body. With such a man Rheticus was not only eager to identify, but to idolize and deify. [84]

It is unfortunate that we do not have a detailed account of what transpired upon Rheticus' return to Wittenberg in September 1541. From the fervor of his writings and, by contrast, from the relative pallidness of his reception, however, it is clear that others could not appreciate the nature of the personal experience that he had undergone. Still, this is perhaps understandable. We can empathize with a person who has undergone a terrible experience; yet few of us are conscious of the emotional scars that others have carried secretly with them since childhood. When Rheticus came back from Frombork, he returned not only to report to Melanchthon, as Hans Blumenberg has written, but he returned full of fire, a believer and a convert.[85] As we might say, he came back "overheated." It is no wonder, then, that Rheticus' great enthusiasm was not shared by the kindly and paternal Melanchthon, although his confidence in the young man's abilities continued unabated.[86] He could have had little way of knowing why the experience for Rheticus had been so relieving, nor was Melanchthon predisposed to go against tradition and the status quo of which he was such an important part. As Melanchthon wrote to his friend Camerarius on 25 July 1542:

I have been indulgent toward the age of our Rheticus in order that his disposition, which has been incited by a certain enthusiasm [*quodam Enthusiasmo*], as it were, might be moved toward that part of philosophy in which he is conversant. But at various times, I have said to myself that I desire in him a little more of the Socratic philosophy, which he is likely to acquire when he is the father of a family.[87]

But the Socrates whom Rheticus followed was the God-like Copernicus, the man who had lifted the fog for him and shot his arrows true to the target—not the quiet, moderate Socrates of Melanchthon. In the end, the vision of Rheticus was not censured; it was merely ignored and repressed from public consciousness.

Conclusions. Let us now try to formulate some general conclusions. Broadly speaking, this paper has addressed itself to the problem of theory-choice from three basic perspectives: historical, philosophical, and psychosocial.

First, we have tried to demonstrate the historical fact that the University of Wittenberg, by far the most important mid-16th-century German university, in no way supports the view of scholars such as Stillman Drake who say that, "the doors of 16th-century universities were closed to new scientific ideas from the outside."[88] While one would not wish to describe the 16th-century universities as hotbeds of scientific revolutionary activity, the opposite view is equally misleading. Recognition of the great range of intermediate responses to innovation helps to make the historical landscape less rigidly compartmentalized.

A second important conclusion is closely related to the first. One needs to exercise great caution in defining just what it is that scientists are choosing between when they make what philosophers call a "theory-choice." For it is clear from this study that a scientist's perception of a theory will affect the decisions that he makes about it. A beginning student at Wittenberg in the 1550s, for example, would have a rather misleading conception of the Copernican theory if he saw it only as the revival of Aristarchus' schematic picture of the universe. The Master's candidate, like his professors, admired the theory of Copernicus for what it had to offer by way of new parameters and alternative models that might be translated into a geostatic framework. This helps to explain why commonly employed dichotomies, such as "pro-Copernican" and "anti-Copernican," "Copernican paradigm" and "Ptolemaic paradigm" fail to provide an accurate description of the choice situation in the earliest phase of the theory's reception. Theory- or paradigm-choice simply was not *seen* as a major problem by the majority of astronomers who constituted the Wittenberg Circle. They did not treat the Copernican theory *as though* it were a paradigm but merely made use of those parts of it where they were in agreement.

If we were to look at the early reception of Copernican astronomy at Wittenberg through Kuhnian spectacles, therefore, we should have to make the paradoxical statement that it had been welcomed respectfully into the fold of Ptolemaic normal science—a situation which should never occur by Kuhn's reckoning. Professor Kuhn himself actually gives different assessments of the historical situation depending upon whether he writes as an historian or as a philosopher. In *The Copernican Revolution,* when he is working with his historical spectacles, he is less apt to dichotomize although the analysis is not always satisfactory. Hence, he can write of the reception of the *Prutenic Tables:* "Every man who used the *Prutenic Tables* was at least acquiescing in an implicit Copernicanism."[89] Here the notion of an "implicit Copernicanism" at least suggests a transitional process even if the interpretation itself is misleading, since use of the *Prutenic Tables* in no way committed one to the paradigmatic features of the Copernican theory. On the other hand, Kuhn's philosophical lenses in *The Structure of Scientific Revolutions* tend to blind him somewhat to the intermediate steps between paradigms. Here he writes: "Copernicus' innovation was not simply to move the earth. Rather, it was a whole new way of regarding the problems of physics and astronomy, one that necessarily changed the meaning of both 'earth' and 'motion'."[90] Kuhn's reading of Copernicus' innovation is actually a later, more fully rationalized interpretation, for even Copernicus himself barely glimpsed the new meaning of "earth" and "motion" implied by his theory. "The whole new way of regarding the problems of physics and astronomy" was itself subject to the later interpretations of men such as Kepler, Galileo, and Descartes. In short, we might venture to suggest that since the full implications of a new theory are never evident from the outset, attitudes toward old and new programs of research, which develop early in the game, will *always* be conditioned by some interpretation. As in the case of the Wittenberg Interpretation, the reading may be so conservative that, while it may have the positive effect of increasing the exposure of parts of the theory to discussion and further use, it will do so at the price of ignoring its most fundamental presuppositions.

This brings us to a final general conclusion. It concerns the role of scientific institutions in affecting choices between research programs. Social units control their members by a variety of sanctions.[91] As Warren O. Hagstrom has shown, the reward system in scientific groups may be viewed as analogous with gift-giving in certain primitive societies, a system governed by reciprocity in which social recognition is exchanged for information.[92] Conformity with methodological standards in such groups can be enforced by the *withholding* of recognition. When scientists disagree with one another, they do not usually lose their jobs except under the extreme conditions of a Lysenko Affair.[93] But they do run the risk of forfeiting full recognition for their contributions. By never challenging the paternalistic, familial framework established by Melanchthon, Reinhold and Peucer helped to institutionalize the moderate, pragmatic interpretation of the Copernican theory advocated by the *Praeceptor Germaniae*. The case of Rheticus suggests that recognition of his discovery of the cosmological power of the new theory was consciously or unconsciously withheld because he had deviated from the methodological consensus and because his identification with the revolutionary side of Copernicus may have seemed threatening, or, at least, a childish enthusiasm that ought not to be accorded serious consideration.

While it has not been our purpose here to evaluate the rationality of Rheticus' decision to become a Copernican in the context of recent theories of rationality, we are somewhat closer to the Kuhnian account of conversion. Kuhn allows that the factors that *actually* motivate an individual's shift from one research program to another are not necessarily rational; and he is quite correct when he writes: "I do not expect that, merely because my arguments are logical, they will be compelling." [94] As historians, we need to be able to explain why the same arguments will be evaluated in diverse ways by contemporaries no matter how we might judge their rationality from the advantage of retrospect.

We have tried to argue that one important component in Rheticus' assessment of the Copernican innovation was unconscious: The overwhelming wish to put the pieces of his murdered father back together again, to restore wholeness and unity; and this wish was manifested on the intellectual level by his strong identification with a man who had no son and whose theory restored the fragmentation of the Ptolemaic monster into a whole and complete organism. If our psychohistorical reconstruction is correct, then it helps to explain why Rheticus was so strongly persuaded by the argument from harmony; and it also helps to explain why he was not drawn into the Wittenberg consensus. Indeed, one wonders how often correct choices—correct, that is, in light of a later theory of rationality—are made for reasons that are *felt* but not consciously understood, and for reasons that become much more understandable to us in light of a psychoanalytic theory of irrationality. As our study shows, there is a place for both conscious and unconscious factors in the history of science.

Notes

[1] On the structure of scientific theories, see Gerald Holton, "Presupposition in the Construction of Theories," in Harry Woolf, editor, *Science as a Cultural Force* (Baltimore, 1964), pp. 77-108; Gerd Buchdahl, "History of Science and Criteria of Choice," in Roger Stuewer, editor, *Historical and Philosophical Perspectives of Science* (Minneapolis, Minn., 1970 [*Minnesota Studies in the Philosophy of Science*, vol. 5]), pp. 204-230; "Methodological Aspects of Kepler's Theory of Refraction," *Studies in History and Philosophy of Science*, vol. 3 (1972), pp. 265-298; *Metaphysics and the Philosophy of Science* (Oxford, 1969); Paul K. Feyerabend, "Problems of Empiricism, II," in Robert Colodny, editor, *The Nature and Function of Scientific Theories* (Pittsburgh, 1970 [*University of Pittsburgh Studies in the Philosophy of Science*, vol. 4]), pp. 275-353.

[2] Thomas S. Kuhn, *The Copernican Revolution* (Cambridge, Mass., 1959), p. 141; *The Structure of Scientific Revolutions*, second edition (Chicago, 1970), p. 69. Kuhn now believes that crisis, "the common awareness that something has gone wrong," is not an "absolute prerequisite" for a revolution but rather "the usual prelude, supplying, that is, a self-correcting mechanism which ensures that the rigidity of normal science will not forever go unchallenged. Revolutions may also be induced in other ways, though I think they seldom are." (ibid., p. 181).

[3] The literature on this subject is too vast to summarize here, but some fundamental studies include: Willy Hartner, "Medieval Views on Cosmic Dimensions and Ptolemy's Kitab al-Manshurat," *Oriens-Occidens*, pp. 319-348, (Hildesheim, 1968) [first published in *Mélanges Alexandre Koyré*, (Paris, 1964), vol. 1, pp.

254-282]; "The Mercury Horoscope of Marcantonio Michiel of Venice," *Oriens-Occidens*, pp. 440-495 [first published in Arthur Beer, editor, *Vistas in Astronomy* (London and New York, 1955), vol. 1, pp. 85-138]; "Nasir al-Din al-Tusi's Lunar Theory," *Physis*, vol. 11 (1969), pp. 287-304; E. S. Kennedy and Victor Roberts, "The Planetary Theory of Ibn al-Shatir," *Isis*, vol. 50 (1959), pp. 227-235; and Jerome R. Ravetz, *Astronomy and Cosmology in the Achievement of Nicolaus Copernicus* (Warsaw, 1965).

⁴Two recent studies that shed new and important light on the problem of Copernicus' discovery are Noel Swerdlow, "The Derivation and First Draft of Copernicus' Planetary Theory: A Translation of the *Commentariolus* with Commentary," *Proceedings of the American Philosophical Society*, vol. 117 (1973) pp. 423–512, and Curtis A. Wilson, "Rheticus, Ravetz and the 'Necessity' of Copernicus' Innovation," in R. S. Westman editor, *The Copernican Achievement* (Berkeley and Los Angeles, in press).

⁵This does not mean that, on occasion, Copernicus' values for the absolute distances of the sun and moon were not used in discussions of matters such as eclipses.

⁶Pierre Duhem, *To Save the Phenomena*, translated by E. Dolland and C. Maschler (Chicago and London, 1969 [first published in 1908]), pp. 87–91.

⁷See Joseph Ben-David, *The Scientist's Role in Society* (Englewood Cliffs, N.J., 1971), p. 63 ff.

⁸Johannes Kepler, *Astronomia nova*, in Max Caspar, editor, *Johannes Kepler Gesammelte Werke* (Munich, 1937), vol. 3, p. 237.

⁹For the unity in Kepler's thought, see Robert S. Westman, "Kepler's Theory of Hypothesis and the 'Realist Dilemma'," in F. Krafft, K. Meyer, and B. Sticker, editors, *Internationales Kepler Symposium Weil der Stadt 1971* (Hildesheim, 1973), pp. 29-54, and also in *Studies in History and Philosophy of Science*, vol. 3 (1972), pp. 233-264; Gerald Holton, "Johannes Kepler's Universe: Its Physics and Metaphysics," *American Journal of Physics*, vol. 34 (1956), pp. 340-354; Jurgen Mittelstrass, "Methodological Elements of Keplerian Astronomy," *Studies in History and Philosophy of Science*, vol. 3 (1972), pp. 203-232.

¹⁰See Clyde L. Manschreck, *Melanchthon, the Quiet Reformer* (New York and Nashville, 1958), pp. 19-26; Friedrich Paulsen, *German Education, Past and Present*, translated by T. Lorenz (New York, 1908), p. 60 ff.; John Dillenberger, *Protestant Thought and Natural Science* (New York, 1960), p. 28 ff.

¹¹Joachim Camerarius, *Corpus Reformatorum, Philippi Melanthonis opera quae supersunt omnia*, edited by C. G. Bretschneider (Halle, 1834), vol. 1, p. 42; *De vita Melanthonis narratio*, edited by G. T. Strobel (Halle, 1777), p. 26.

¹²Manschreck, op. cit. [note 10], p. 76.

¹³See Franz Eulenberg, *Die Frequenz der Deutschen Universitäten von Ihrer Gründung bis zur Gegenwart* (Leipzig, 1904 [*Abhandlung der Philologisch-Historischen Klasse der Königliche Sächsischen Gesellschaft der Wissenschaften*, No. II].), p. 288. Of the 14 universities surveyed by Eulenberg in this period, Wittenberg consistently had the largest enrollment figures and began to recover its earlier losses by 1528.

[14] See William Hammer, "Melanchthon, Inspirer of the Study of Astronomy; With a Translation of His Oration in Praise of Astronomy (*De Orione*, 1553)," *Popular Astronomy*, vol. 59 (1951), pp. 308-319.

[15] Melanchthon, ed. cit. [note 11], vol. 3, p. 562; quoted and translated in Manschreck, op. cit. [note 10], p. 152.

[16] Philipp Melanchthon, *Initia doctrinae physicae, dictata in Academia Witebergensi* (1549), in Melanchthon, ed. cit. [note 11], vol. 13, p. 216 ff. A manuscript in Melanchthon's hand, "De Supputatione Motus Solis," MS PR 5974 in the Huntington Library, confirms this view.

[17] Edward Rosen, *Three Copernican Treatises*, third edition, (New York, 1971), p. 394.

[18] Melanchthon, ed. cit. [note 11] vol. 4, p. 679. This is the interpretation given to the passage by A. Bruce Wrightsman in his unpublished doctoral dissertation ("Andreas Osiander and Lutheran Contributions to the Copernican Revolution," University of Wisconsin, 1970, pp. 345-346). Cf. Rosen's view: ". . . when Rheticus returned from Frombork to Wittenberg, on 16 October 1541 Philipp Melanchthon, Luther's principal lieutenant, harshly condemned 'that Polish (Sarmatian) astronomer who set the earth in motion' " (Rosen, op. cit. [note 17], p. 400).

[19] Melanchthon, ed. cit. [note 11], vol. 13, p. 216.

[20] Ibid., p. 217.

[21] Ibid., p. 244.

[22] Ibid., pp. 225, 241.

[23] Ibid., p. 262.

[24] The change in tone from the 1549 edition to the 1550 edition was first pointed out by Emil Wohlwill in "Melanchthon und Copernicus," *Mitteilungen zur Geschichte der Medizin und der Naturwissenschaft* (Hamburg and Leipzig, 1904), vol. 3, pp. 260-267.

[25] In his copy of *De revolutionibus*, Michael Maestlin (1550-1631) wrote a critical marignal note by the anonymous preface, confirming Osiander's "neutral" position toward the Copernican theory: ". . . the author of this letter, whoever he may be, while he wishes to entice the reader, neither boldly casts aside these hypotheses nor approves them but rather he imprudently squanders away something which might better have been kept silent." Maestlin's heavily annotated copy is located in the Stadtbibliothek, Schaffhausen, Switzerland.

[26] In 1525, for example, Melanchthon intervened on behalf of a group of nuns who were threatened with expulsion from their convent. Mother Charitas Pirkheimer wrote in her memoirs: "He [Melanchthon] was more moderate and modest in his speech than any Lutheran I had yet heard. That the people were being subjected to force was very displeasing to him . . . . Thus he brought it about that the people desisted somewhat from hostility towards us and were no longer so violent concerning us. . . ." *An Heroic Abbess of Reformation Days: The Memoirs of Mother Charitas Pirkheimer, Poor Clare, of Nuremberg*, introduction by F. Mannhardt, S.J. (St. Louis, Mo., 1930), p. 26.

[27] See Wilhelm Norlind, "Copernicus and Luther: A Critical Study," *Isis,* vol. 44 (1953), pp. 273-276.

[28] See Owen Gingerich, "Erasmus Reinhold," in *Dictionary of Scientific Biography* (New York, in press).

[29] See Duhem, op. cit. [note 6], p. 71.

[30] Ibid., p. 72. I have modified Duhem's sometimes-loose translations where necessary. In all cases, I have reinterpreted the passages cited.

[31] Georg Joachim Rheticus, *Narratio prima,* translated by E. Rosen, in Rosen, op. cit. [note 17], p. 134.

[32] Ibid., p. 135.

[33] Reinhold's copy of the 1543 edition is presently located in the Crawford Library of the Royal Astronomical Observatory in Edinburgh. I should like to thank Owen Gingerich for generously allowing me to use his microfilm of the work which permitted me to check and supplement the notes, which I took from the original in November 1972. See Owen Gingerich, "The Role of Erasmus Reinhold and the Prutenic Tables in the Dissemination of the Copernican Theory," in *Studia Copernicana,* vol. 6, *Colloquia Copernicana II* (Warsaw, 1973), pp. 43-62, especially pp. 56-58.

[34] Nicolaus Copernicus, *On the Revolutions of the Heavenly Spheres,* translated by Charles Glenn Wallis, in *Great Books of the Western World* (Chicago, 1952), vol. 16, p. 740 (from book 5, ch. 2). I have checked all translations against the 1543 edition.

[35] Ibid., p. 742 (from book 5, ch. 4).

[36] Reinhold's copy of *De revolutionibus* [note 33] f. 142: "Primus Modus per Eccentrepicyclum"; f. 142v: "Secundus Modus per Homocentrepicyclos."

[37] Ibid., f. 164v; Copernicus, op. cit. [note 34], p. 785 (from book 5, ch. 25).

[38] Erasmus Reinhold, *Prutenicae tabulae coelestium motuum* (Tübingen, 1551), p. 21, "Praeceptu calculi motuum coelestium"; see Duhem, op. cit. [note 6], pp. 72-73.

[39] In Book I, the longest note occurs in chapter 6, f. 4v on the diurnal motion of the earth.

[40] Aleksander Birkenmajer, "Le Commentaire Inédit d'Erasmus Reinhold sur le *De revolutionibus* de Nicholas Copernic," in *La Science au seizième siècle,* (Paris, 1960 [first published in 1957]), pp. 171-177. The neutrality with respect to cosmology in astronomical tables such as Reinhold's *Prutenic Tables* had been understood by Copernicus according to his friend Bishop Tiedemann Giese, who predicted that if Copernicus were to publish only his tables, without the hypotheses and principles underlying them, scarcely anyone would seek out those principles. As Rheticus reports Giese's views: "There was no place in science, he asserted, for the practice frequently adopted in kingdoms, conferences and public affairs, where for a time plans are kept secret until the subjects see the fruitful results and remove from doubt the hope that they will come to approving the plans" (Rheticus, ed. cit. [note 31], p. 183).

[41] Gingerich, op. cit. [note 33], pp. 48-50.

[42] Hans Blumenberg, *Die kopernikanische Wende* (Frankfurt am Main, 1965), pp. 95-96.

[43] Gingerich, op. cit. [note 33], pp. 51-55.

[44] The following biographical details are based mainly upon the articles on Peucer in *Nouvelle Biographie Générale*, pp. 767-770, and Thomas-Pope Blount, *Censura celebriorum authorum* (Geneva, 1710), pp. 735-737.

[45] Caspar Peucer, *Elementa doctrinae de circulis coelestibus, et primo motu, recognita et correcta* (Wittenberg, 1553), Preface.

[46] Walter Friedensburg, editor, *Urkundenbuch der Universität Wittenberg* (Magdeburg, 1926), vol. 1, selection 271, pp. 256-257. Walter Friedensburg, *Geschichte der Universität Wittenberg* (Halle, 1917), pp. 216-217.

[47] Peucer, op. cit. [note 45], f. 54; P2.

[48] Ibid., f. E4v; f. G2v.

[49] Ibid., f. E4v.

[50] Editions appeared in 1542 and 1553. There were a few references to Copernicus in the edition of 1553.

[51] See, for example, Caspar Straub, *Annotata in Theorias planetarum Georgii Purbachij*, f. 2, Erlangen Universitätsbibliothek MS. 840, 2 November 1575; Andreas Schadt, *In Theorias planetarum Purbachij annotationes Vitebergae*, f. 72, Erlangen Universitätsbibliothek MS. 840; [Caspar Peucer], *Hypotyposes orbium coelestium (Hypotheses of the Celestial Orbits, Which Are Called the Theories of the Planets: Congruent with the Tables of Alfonso and Copernicus, and even the Prutenic Tables: Published for Use in Schools)*, (Strasbourg, 1568), p. 301.

[52] Peucer, op. cit. [note 51], pp. 516-517. For a discussion of various publications of this work, see Ernst Zinner, *Geschichte und Bibliographie der Astronomischen Literatur in Deutschland zur Zeit der Renaissance* (Leipzig, 1941), pp. 35, 244, 250, and Gingerich, op. cit. [note 33], pp. 59-60.

[53] For a more complete discussion of this generation, see Robert S. Westman, "Three Responses to the Copernican Theory: Johannes Praetorius, Tycho Brahe, and Michael Maestlin," in R. S. Westman, editor, *The Copernican Achievement* (Berkeley and Los Angeles, in press).

[54] Karl Heinz Burmeister, *Georg Joachim Rhetikus 1514-1574: Eine Bio-Bibliographie* (Wiesbaden, 1967), vol. 1, p. 190.

[55] Arthur Koestler, *The Sleepwalkers* (New York, 1963; first published in 1959), pp. 153-154.

[56] Cf. Walter Pagel, *Paracelsus: An Introduction to Philosophical Medicine in the Era of the Renaissance* (Basel and New York, 1958), p. 20 ff.

[57] See Frances A. Yates, *Giordano Bruno and the Hermetic Tradition* (London, 1964).

[58] Cf. Ernst Zinner, *Enstehung und Ausbreitung der Coppernicanischen Lehre* [*Sitzungsberichte der physikalisch-medizinischen Sozietät zu Erlangen, 74*], (Erlangen, 1943), p. 259. Zinner is Koestler's source, op. cit. [note 55], p. 188.

[59] Burmeister, op. cit. [note 54], vol. 3, p. 50.

[60] Stoeffler had been succeeded by Philipp Imser.

[61] Burmeister, op. cit. [note 54], vol. 3, p. 50.

[62] Ibid., vol. 1, p. 175.

[63] Quoted in Leopold Friedrich Prowe, *Nicolaus Coppernicus,* vol. 1, pt. 2, (Berlin, 1883-1884), p. 387; quoted and translated in Koestler, op. cit. [note 55], pp. 189-190.

[64] Particularly striking is his millennial prophecy based upon the motion of the sun's eccentric (Rheticus, ed. cit. [note 31], p. 122).

[65] As Curtis A. Wilson shows in a forthcoming article, Rheticus claimed erroneously that the Copernican account of precession could not be explained on the geostatic theory ("Rheticus, Ravetz and the 'Necessity' of Copernicus' Innovation," in Westman, op. cit. [note 4]).

[66] Rheticus, ed. cit. [note 31], pp. 146, 147.

[67] Rheticus speaks *as though* the spheres were solid although he nowhere calls them "material" or "solid." On the other hand, he nowhere says that they are merely an aid to the imagination, as does Melanchthon. His concern with the problem of plenitude and the force exerted by one sphere on another, however, suggests that he *did* think of them as solid (Cf. esp. Rheticus, ed. cit. [Note 31], p. 146).

[68] Nicolaus Copernicus, *De revolutionibus orbium coelestium,* in F. Zeller and C. Zeller, editors, *Nikolaus Kopernikus Gesamtausgabe* (Munich, 1949), vol. 2, p. 5.

[69] For further discussion of this metaphor, see Westman, op. cit. [*Studies,* note 9], p. 249 f.; also, "Johannes Kepler's Adoption of the Copernican Hypothesis," (University of Michigan: unpublished doctoral dissertation, 1971), pp. 138-177; Leo Spitzer, *Classical and Christian Ideas of World Harmony,* ed. A. G. Hatcher (Baltimore, 1963; first published in 1944-45).

[70] See note 1.

[71] With the exception of the gods, Rheticus says, only the human mind can understand harmony and number; diseased souls are cured by musical harmonies and happy republics are governed by rulers who possess harmonious souls. Earlier, he speaks of the connection between Copernicus' celestial harmony and the number six, the latter "honored beyond all others in the sacred prophecies of God and by the Pythagoreans and the other philosophers." (Rheticus, ed. cit. [note 31], pp. 196, 147).

[72] Ibid., p. 138.

[73] As Edward Rosen points out, this question is not found in the Platonic corpus. The source is Plutarch's *Moralia: Questiones conviviales,* Book VIII, Question 2. Ibid., p. 168.

[74] Ibid., pp. 167-168.

[75] For a helpful survey of some issues in the psychohistory of science, see J. E. McGuire, "Newton and the Demonic Furies: Some Current Problems and Ap-

proaches in the History of Science," *History of Science,* vol. 2 (1973), pp. 21-48.

[76] Burmeister, op. cit. [note 54], pp. 14-17.

[77] See A. D. J. MacFarlane, *Witchcraft in Tudor and Stuart England* (New York, 1970), and H. C. Erik Midelfort, *Witch Hunting in Southwestern Germany, 1562-1684* (Stanford, 1972).

[78] See Edward Rosen, [Review of Burmeister], *Isis,* vol. 59, (1968), p. 231.

[79] Rheticus had met Paracelsus in 1532 and was impressed by him. It is interesting that he did not then decide on a career in medicine, a vocational choice that would have identified him directly with his father. But in 1554, at the age of 40, and after his travels had ended, in effect, Rheticus became a practicing physician in Cracow and his interest in Paracelsus revived. Cf. Burmeister, op cit. [note 54], pp. 35, 152-155.

[80] Rheticus, ed. cit. [note 31], p. 109.

[81] Ibid., p. 108. A facsimile of the title page is printed in Burmeister, op. cit. [note 54], vol. 2, pp. 58-59.

[82] Rheticus, ed. cit. [note 31], p. 186.

[83] Ibid., pp. 186-187.

[84] In 1557, Rheticus wrote of Copernicus "whom I cultivated not only as a teacher, but as a father. . . ." (*"quem non solum tanquam praeceptorem, sed ut patrem colui . . ."*). This letter (Rheticus to King Ferdinand I, Burmeister, op. cit. [note 54], vol. 3, p. 139), which shows no animosity toward Copernicus, would seem to weigh against Koestler's interesting thesis that Rheticus felt betrayed by Copernicus when the latter failed to mention him in his great work.

[85] Blumenberg, op. cit. [note 42], p. 109.

[86] Melanchthon promoted him to the Deanship of the Faculty of Arts upon his return and later recommended him highly for other positions. Cf. op. cit. [*Corpus,* note 11], vol. 4, p. 812; Burmeister, op. cit. [note 54], vol. 1, p. 67 ff.; Wrightsman, op. cit. [note 18], p. 351.

[87] Melanchthon, ed. cit. [note 11], p. 847.

[88] Stillman Drake, "Early Science and the Printed Book: The Spread of Science Beyond the Universities," *Renaissance and Reformation Journal,* vol. 6 (1970), p. 49.

[89] Kuhn, *Copernican Revolution* [note 2], p. 188.

[90] Kuhn, *Structure of Scientific Revolutions* [note 2], pp. 149-150.

[91] Cf. Amitai Etzioni, *Modern Organizations* (Englewood Cliffs, 1964), p. 58.

[92] Warren O. Hagstrom, *The Scientific Community* (New York, 1965), pp. 12-22.

[93] David Joravsky, *The Lysenko Affair* (Cambridge, Mass., 1970).

[94] Thomas S. Kuhn, "Reflections on My Critics," in Imre Lakatos and Alan Musgrave, editors, *Criticism and the Growth of Knowledge* (Cambridge, 1970) p. 261.

DISCUSSION: THE RECEPTION OF
HELIOCENTRISM IN THE SIXTEENTH CENTURY

GINGERICH: Thank you for a very interesting paper. Let us go straight to Jan Henderson, who has agreed to start out the discussion.

HENDERSON: As I see it, there are a number of different issues in the reception of Copernican theory.

There is a question of whether the assumptions on which the Copernican theory is based are to be considered as a physical reality or merely as a convenient mathematical device. This question can be applied to the solar, lunar, and planetary models of Copernicus, where one is concerned with the issues of uniform circular motion, the equant, and "saving the phenomena." It is also applied to the question of geostatic versus a heliostatic system.

For a mathematical astronomer at this time, the question of geostatic versus heliostatic is not necessarily a burning issue, and Reinhold, whom I have studied, is a good example of this. A Copernican versus Ptolemaic model simply represents a change of reference system, which has no effect on the astronomical results, that is, the predicted longitude of the celestial body in question.

These are issues on which we have touched during our discussion in the past few days; it's not easy to determine the viewpoint held by an individual. For example, in the case of Reinhold, does he believe in the geostatic or heliostatic system of the universe? In considering comments on Copernican lunar theory, one should keep in mind that (1) this issue is independent of the question of a heliostatic versus geostatic system, and (2) the Copernican lunar model does represent an improvement over the Ptolemaic model aside from the question of uniform circular motion, an improvement of which Copernicus was, I think, quite proud.

The Copernican lunar model, identical to that of Ibn ash-

Shatir's, does solve the problem of the great variation in lunar parallax and diameter in the Ptolemaic model. By adding a secondary epicycle rather than varying the distance of a single epicycle, Copernicus is able to maintain the same mean lunar distance at quadrature as at syzygy. It seems to me this is a feature that could be appreciated by the Wittenberg astronomers without necessarily being all that controversial, at least to an astronomer at that time as opposed to a philosopher.

I would like to interject one further issue as far as the reception of Copernican theory goes, a nonphilosophical one; namely, the question of prediction and observation, and Copernicus' derivation of parameters and construction of tables. This was obviously an important issue for astronomers at that time. Again, this is a case where the question of a geostatic versus heliostatic system is not relevant.

There was considerable dissatisfaction in Copernicus' time with the inadequacy of almanacs and ephemerides for predicting celestial phenomena. Copernicus' contribution, as I see it, is not only the introduction of heliocentrism and the restoration of uniform circular motion, but for the first time since the *Alfonsine Tables* someone had attempted to derive new parameters and construct new tables, with the hope that this would produce greater agreement between prediction and observation.

As I see it, this whole question can be treated independently of the Copernican models, in a sense, for it is not the particular choice of models that is responsible for whatever improvement there may be, but parameters. I am not going to attempt to answer the question of whether either the Copernican tables or the *Prutenic Tables* produce better results than those obtained from the *Alfonsine Tables*. Rather, I would like to raise two questions.

There is another man in the Wittenberg Circle at this time about whom I would like to know something more, and this is Mathias Lauterwalt. Now, the only connection in which I have come across him is a letter written to Rheticus in February of 1545. Maybe he never actually did very much because he only lived to 31 years, but I would like to know if he was really an astronomer or if he was just a student of Rheticus. He was born in 1520, came to Wittenberg in 1540, was appointed to the faculty in 1550, and died in 1551. I was curious about him because the contents of this letter are striking. He is writing to Rheticus, who by this time is in Leipzig; Rheticus had observed a lunar eclipse in December of 1544 and Lauterwalt has a note about this. Lauterwalt, who is about six years younger than Rheticus, and whose tone in this letter is very respectful, says "I don't doubt your honesty but there could be something wrong here because I observed this eclipse and found it occurred in Wittenberg a half of a quarter of an hour before four." He then gives the calculations of Copernicus, which are used to predict the time of the eclipse, and he goes on and gives details of his observation. Then he cites another observation that he had made, a conjunction of Mercury and Jupiter, and here he makes the point that he really appreciates Copernicus. He is very satisfied with the agreement of predictions and observations. In this connection he mentions that if you use the *Alfonsine Tables*, the conjunction would be predicted to occur the day before. In order to account for this, those people against Copernicus say they are actually seeing Jupiter when, in fact, it is Mercury.

I would personally like to know if anyone has an example of people using the actual tables of Copernicus before 1551, and what they thought of them.

WESTMAN: I think that you agree with one of the general points that I argued: namely, that the question of "geostatic versus heliostatic" cosmologies was not a "burning issue," as you put it, in the first three decades after the publication of Copernicus' theory.

It is fairly clear, then, that the discussion of planetary models by themselves does not enter into the matter of theory-choice. I certainly agree with this; but, on the other hand, it does give us an important clue as to what it was that led certain people like Reinhold to make friendly "noises" about Copernicus. I think that some of those vague eulogies that he writes are really praise for Copernicus' abolition of the equant. You do not need to commit yourself to the Copernican theory in order to get rid of the equant.

As for Lauterwalt, you said that he was quite respectful of Rheticus; I would say that Lauterwalt was not alone. Melanchthon and everyone else say nice things when Rheticus returns to Wittenberg. In other words, there is no personal animosity; it is just that people ignore what he is saying—and this, for both social and psychological reasons. I would interpret Melanchthon's dismissal of Rheticus' "childish" views as an expression of the Wittenberg scientific community's "consensus mechanism," which rejects the idea of using the Copernican theory as anything more than a new *calculational* scheme.

ROSEN: The view that there was any such thing as a unified Wittenberg Interpretation is in conflict with the obviously intense disagreement in the University of Wittenberg on how this whole matter should be handled. Reinhold and Rheticus did not see eye to eye, and this is undoubtedly part of the reason why Rheticus left the university. When Rheticus came back from Frombork in the autumn of 1541, the reaction of the faculty to him was plain, because he was instantly elected dean. This was an elected position. There is no ambivalence about this at all: Our boy has come home and he has made good. Nevertheless he couldn't get permission to print the new astronomy in Wittenberg. This was for ideological reasons, grounds already previously expressed by Luther and Melanchthon, who were number one and number two in the town. Therefore, Rheticus had to leave Wittenberg University to get another job elsewhere, and he also had to leave Wittenberg in order to find a printer for the *Revolutions*.

The proof is a very simple one. The trigonometrical part of the *Revolutions* was published in Wittenberg in 1542. It was noncontroversial and purely technical, but Rheticus had to go to Nuremberg to his friend, Hans Peter [Petreius] to get the *Revo-*

lutions published under Catholic, not Lutheran, auspices, in 1543.

GINGERICH: Wait a minute. How can you say it is printed under Catholic auspices since Petreius is a Lutheran printer?

ROSEN: I know, but you must remember that there was a fierce controversy in Nuremberg at this time. In the early 1540s there was a militantly pro-Catholic as well as a militantly pro-Lutheran faction. The great bulk of the citizens said, "A plague on both your houses. We want Nuremberg as it is and we don't want the Pope and we don't want Martin Luther."

GINGERICH: I think you have only the flimsiest evidence to say that he was unable to get it published in Wittenberg, because you have no real idea if he tried there. After all, three of the five books Rheticus carried up with him to Frombork as gifts for Copernicus were published by Petreius. Petreius had a considerable reputation as a scientific printer. There was no comparable printing in Wittenberg at that time.

ROSEN: Hans Lufft published the *Trigonometry*.

GINGERICH: In Wittenberg there was a tremendous growing tradition of printing cheap university textbooks, about the leading place for this, and the *Trigonometry* fits perfectly into that kind of scheme. It is small, it is obviously the kind of thing that will be regularly used and it is something that's needed for local distribution.

The whole printing pattern of Petreius is another thing. He had a publishing house equipped to handle a long and complex scientific treatise, and this is not at all true of Lufft. In terms of the size and scope of the *Trigonometry*, it is a thin pamphlet compared to *De revolutionibus*. I suppose sales and financing would be another aspect of it. I will require you to have better evidence than that.

ROSEN: The facts speak for themselves.

HENDERSON: Why did Reinhold have the noncontroversial *Prutenic Tables* published in 1551 in Tübingen if there were an adequate printer in Wittenberg? That is certainly something one

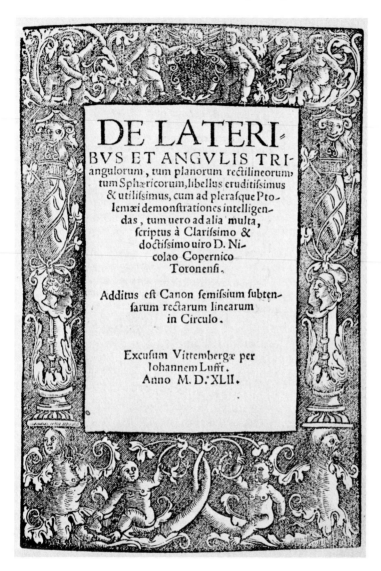

Title page of the trigonometrical part of Copernicus' work, separately issued in Wittenberg in 1542. (Photograph by Owen Gingerich from the Horblit Collection; now at Harvard College Library.)

would want to oversee. With so many numbers there is no way logically for a printer by himself to correct them.

ROSEN: Before returning to Wittenberg from his visit to Copernicus, Rheticus wrote to the Duke of Prussia, a Lutheran secular prince, saying "Give me two letters of recommendation, one to the University of Wittenberg and the other to the Elector of Saxony, asking for permission to print the book," and three days later, the secretary of Duke Albert of Prussia issued both documents. That is the last you ever hear of it; the Elector of Saxony never gave permission for the publication of the book, nor did the University of Wittenberg and that is why the *Revolutions* was not published in Wittenberg.

Technological differences are imaginary. It was ideological differences and perhaps also personal difficulties that explain Rheticus' departure from Wittenberg, and there is besides the purely Catholic dress in which the book appears. It is dedicated to the Pope, there is a covering letter from a cardinal, and among the people who are named by Copernicus as those who overcame his unwillingness to print is a personal friend who is a Roman Catholic bishop.

The situation is entirely clear. It has nothing whatsoever to do with manufacturing or paper or anything of that sort.

I am surprised that in this discussion of Erasmus Reinhold there has been no mention of the commentary in which this man sets forth his views.

HENDERSON: There is nothing very explicit on Reinhold's reaction to Copernicus in the commentary. I've been over the passages in which Reinhold asserts the alternative to the heliocentric view. Birkenmajer[*] pointed out a very important development, and that is that Erasmus Reinhold, in his rejection of sun-centered planetary theory, reached the stage normally at-

---

[*] Aleksander Birkenmajer, "Le commentaire inédit d'Erasmus Reinhold sur le *De Revolutionibus* de Nicolas Copernic," in *La Science au Seizième Siècle* (Paris, 1960), pp. 171-177; reprinted pp. 761-766 in Birkenmajer's *Études d'Histoire des Sciences en Pologne* (Wrocław, 1972).

tributed to Tycho 30 years later.

SWERDLOW: There are the two little notes on it in one chapter on Jupiter that do indicate sort of a Tychonic theory. It's a little messier than that. In Reinhold's commentary it is difficult to pin down what he thinks about a lot of things. For example, in deriving the eccentricity and direction of the apsidal lines of the superior planets, he goes into an enormous amount of detail in the whole process. In so doing he moves back and forth between using Ptolemy's model with the bisected eccentricity and a Copernican model with the 3/2 eccentricity and little epicycle. What does he think is the correct model?

ROSEN: Those are technical points and they are acceptable on either cosmological basis but there are other passages . . .

GINGERICH: No, just two little lines in the margin. It's such a small thing.

SWERDLOW: One is in the margin and one at the foot of the verso of the same folio.

ROSEN: Earlier, I was interrupted when I was talking about what Professor Westman calls the Wittenberg Interpretation. If I try to say it simply in my own way, what Reinhold did was to reject the conceptual foundation and accept the technical work; in other words, recognize Copernicus as a practical astronomer of the highest rank and reject the theoretical basis of his system. If that is what you are saying, then this is also the reaction all the way from Gemma Frisius to Father Clavius, the principal Catholic spokesman. You would have to call this the Wittenberg-Rome axis. There is no ignoring of Copernicus for twenty years; instead, there is a very active discussion. The normal reaction is recognition of Copernicus' technical proficiency. This is the standard view, I think, of qualified people, of whom Reinhold on the Protestant side and Clavius on the Catholic side can be taken as specific examples. The first great change is in England, not a backward country by any means at this time, namely in Digges, in Digges' effect on Bruno, and in Bruno's effect on Kepler. That is the next great phase in the improvement of

Folio 233 of Reinhold's *Commentarius in opus Revolutionum Copernici* (Berlin manuscript Latin 2°391), cited by Swerdlow in the discussion; the annotation hinting that Reinhold had considered a proto-Tychonic system is in the left center margin.

Copernicanism. At the same time this acceptance, of course, leads to the ecclesiastical objection.

Incidentally, Reinhold's preface to the *Prutenic Tables* says explicitly what his commentary implies, namely, what I call the standard reaction. Copernicus is proficient as a technical astronomer, and is the second since Ptolemy, but his conceptual basis has to be rejected.

SWERDLOW: But the whole point of Reinhold's commentary on Copernicus is redoing every numerical parameter in Books III, IV, and V of *De revolutionibus*, which seems, if anything, to show that Reinhold did not feel that Copernicus' work was technically proficient enough.

ROSEN: That is exactly Reinhold's point, but I think you draw a diametrically erroneous conclusion. According to Reinhold, Copernicus is a first-rate astronomer but he makes numerical mistakes, and these have to be eliminated in order to make the work even better than it is.

SWERDLOW: I am not saying Reinhold doesn't think Copernicus is first-rate. In the *Prutenic Tables* he gives more credit to Copernicus than he gives to himself. It's just that the commentary is a complex work. It has contradictory things within it and I would hesitate to generalize about just what Reinhold thought about all of this.

GINGERICH: Dr. Nelson has wanted to speak for a long time. We'll have Dr. Nelson, Dr. Christianson, Dr. Wallace in that order and then the coffee break. Please don't say anything too controversial.

NELSON: Apparently the knowledge of Osiander's authorship of the false preface was more general than I had supposed. Is there any work which specifies how many people know it? It is quite evident that until Kepler made it known, it was unknown to many people.

ROSEN: Part of the answer is this. In 1601, on the verso of the title page of Kepler's *New Astronomy* he reveals the story. Two hundred years later one of the greatest historians of astronomy,

J. B. J. Delambre . . .

GINGERICH: But that is irrelevant, Ed. He is asking a totally different question.

ROSEN: Hold it a minute, please. In his *Histoire de l'astronomie moderne*, vol. 1, p. 139, Delambre says that Copernicus is the author of the Foreword. Here, more than two hundred years after Kepler has published the *New Astronomy*, one of the masterpieces in the history of astronomy, Delambre still doesn't know that Osiander was fingered by Kepler as the man who inserted the false Foreword.

Now, to go back to the earlier aspect, between 1543 and 1609 it was known privately that Osiander had admitted having done this and having got away with it. Peter makes a very interesting remark when he says, "that Foreword was given to me with the rest of the treatise," which means that Osiander was a crook. He handed his own Foreword to Peter to give to the workmen, and Osiander did not indicate his own authorship, so that Peter was taken by surprise when the protest from Rheticus came. Later, Osiander, in a conversation with Philip Apian, confessed that he put this trick across. So the answer to your question is that in certain limited circumstances before 1609 it was known or rumored that Osiander was the author of the false Foreword.

CHRISTIANSON: Do you think the circles would have included the Wittenberg people we were talking about today?

HALL: Rheticus must have known, musn't he?

ROSEN: Rheticus blew his top, according to Tiedemann Giese's letter of July 26th, 1543. Rheticus must have screamed about this thing and he also said that he would handle that man, Osiander, in such a way that he would mind his own business and never damage astronomers again. Rheticus was absolutely purple on the subject, but the question is, who knew about the Giese letter?

WESTMAN: Well, the chances are that his "purple rage" must have spread and others heard about it. But it is not an unreasonable inference to suppose that there were other people who did not know about it. In other words, there was private knowledge

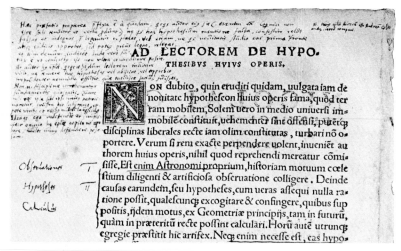

AD LECTOREM DE HYPO-

THESIBVS HVIVS OPERIS.

ON dubito, quin eruditi quidam, uulgata iam de nouitate hypothefeon huius operis fama, quòd terram mobilem, Solem uero in medio uniuerfi immobilé conftituit, uehementer fint offenfi, putétcq difciplinas liberales recte iam olim conftitutas, turbari nõ oportere. Verum fi rem exacte perpendere uolent, inueniét au thorem huius operis, nihil quod reprehendi mereatur cõmi fiffe. Eft enim Aftronomi proprium, hiftoriam motuum cœle ftium diligenti & artificiofa obferuatione colligere. Deinde caufas earundem, feu hypothefes, cum ueras affequi nulla ra tione poffit, qualefcunq excogitare & confingere, quibus fup pofitis, ijdem motus, ex Geometriæ principijs, tam in futurũ, quàm in præteritũ recte poffint calculari. Horũ autẽ utrunq egregie præftitit hic artifex. Neq enim neceffe eft, eas hypo-

Michael Maestlin, Kepler's astronomy teacher, identified Andreas Osiander as the author of the anonymous introduction in a miniscule handwritten note at the upper right in his own copy of *De revolutionibus*. (Courtesy of the Schaffhausen Stadtbibliothek.)

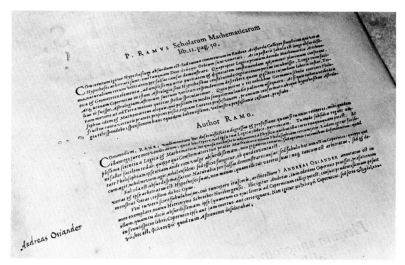

Johannes Kepler named Osiander as author of the introduction on the back of the title page of his *Astronomia nova*. The Swiss scholar Stephen Spleis emphasized it in his copy of Kepler's work. (Photograph by Owen Gingerich, from the Schaffhausen Stadtbibliothek.)

of the identity of the author of the anonymous *praefatiuncula* and this knowledge was communicated *orally* or through the private marginalia on copies of *De revolutionibus*.

CHRISTIANSON: I think Tycho fits very well into the whole context of what we are talking about here today. He was very definitely a Philippist [follower of Melanchthon] in his theological orientation. Philippism was pretty much squeezed out of the German universities beginning in the 1570s but it survived in Denmark where it had the protection of King Frederick II and, in fact, was the dominant interpretation of Lutheranism throughout Tycho's lifetime. It was attacked very severely in the early 17th century in Denmark and at that time former associates of Tycho were among its foremost defenders. They were eventually silenced by about 1616.

Another thing we should not pass over too quickly was your emphasis upon Melanchthon's theories of education. They were very important to the curricular structure of education in Lutheran universities. The quadrivium was given a good deal of interest as basic education for everybody that went to the university, and most students were being trained for the clergy. This meant that chairs of mathematics were established at those universities that were re-endowed according to the Philippist model; a couple of chairs at Wittenberg, a couple of chairs at most of them out of a total of a dozen to twenty chairs. Then because of the 16th-century practice of moving on up through a series of chairs, more than those two people who held the chairs of mathematics at any one time might have been in that chair at some stage in their academic career. It meant, in other words, that astronomy was an establishment activity in the Lutheran university. This ties in with the psychobiographical approach because it might have affected the attitudes of people who held those chairs of astronomy, and of their colleagues.

It also served to create an audience for mathematical writing, not only among the colleagues of the professors of mathematics, but also among their students and former students. I think this is a significant external factor that Bob Westman has brought out.

WALLACE: Did the Wittenberg interpretation differ from the Rome interpretation, and was there a difference between Clavius, say, and Melanchthon?

If I understand him correctly, Professor Westman is saying that the Wittenberg interpretation was generally based on a fictionalist view of scientific theories. Being mainly philosophers and theologians, the Wittenberg scholars did not take astronomical anomalies all that seriously; they really were not interested in calculations, because they did not feel these would affect their basic world view. So they were not greatly upset over astronomy. At the same time, they probably did not have the same zealous concern to preserve orthodoxy with regard to the interpretation of Scripture as did their Roman counterparts.

Much the same attitude towards calculations existed in the early days at the Collegio Romano, until the growing knowledge of astronomy reached the stage where one could no longer brush aside the realist claims. Then the fictionalist interpretation came to be increasingly questioned, and once the realist interpretation had emerged as a serious alternative to it, the problem of Scripture and its proper interpretation could no longer be ignored.

Would this be how the Wittenberg interpretation differs from later views that came into prominence as observational evidence began to grow? With such views, of course, there would be a much stronger thrust towards realism and a gradual rejection of fictionalism.

WESTMAN: First of all, I would not say that the "Wittenberg interpretation" can be characterized simply as a fictionalist viewpoint. I want to reiterate that there is an important difference, which seems to be followed fairly consistently at Wittenberg, between the Bachelors and Masters levels of instruction. On the lower level, the Copernican theory virtually means the simple Aristarchan scheme (which is, of course, highly misleading since it is a totally unarticulated model) and it is accompanied by explicit citations from Holy Scripture and Aristotelian natural philosophy against the earth's motion. Although one frequently finds a few references to Copernican parameters on the introductory level, it is in the Masters curriculum that one encounters a truly serious consideration of Copernican planetary models, the problem of precession, commutation of models into different reference frames and frequently—most notably with Reinhold—the rejection of the equant. This latter position, which also characterizes the methodological viewpoint adopted in private research, is really the stance of Osiander, although Osiander nowhere indicates a special preference for equantless models.

WALLACE: Is Osiander's different from the Wittenberg interpretation?

WESTMAN: Osiander, I would say, represents a more moderate dimension of the Wittenberg interpretation, because he is claiming that *all* motions, including the earth's motion, should be awarded hypothetical status, whereas, in the elementary texts, Melanchthon, Peucer, and the other Wittenberg astronomers say "no," we cannot let our students think that the earth has any real motion at all. Osiander's position is quite clear. It is a position of epistemic resignation.*

ROSEN: Are you saying that according to Osiander the sun doesn't rise in the morning and set in the evening? Osiander says we obtain truth by revelation; it is revealed in the Sacred Scripture that the sun rises in the east and sets in the west. Osiander is not a one-hundred-percent fictionalist. He holds that everything you say in astronomy is fiction but everything the Bible says is God's word and is true.

WESTMAN: But he nowhere explicitly rejects the motion of the earth, does he? Can you think of anywhere where he does?

EISENSTEIN: Does he hold to a double-truth doctrine? That

---

* ". . . the causes of the apparent unequal motions are completely and simply unknown to this art. And if any causes are devised by the imagination, as indeed very many are, they are not put forward to convince anyone that they are true, but merely to provide a correct basis for calculation." (Osiander's introduction, translated by E. Rosen, *Three Copernican Treatises*, third edition (New York, 1971), p. 25.)

would be another possibility.

ROSEN: No. Osiander holds that anything you say in astronomy may be dead wrong but if it is useful let's keep it as a device. However, what the Bible says was told to us by God and this is true. Now, God said (as they read the Bible then) that the earth stands still, the sun moves, the stars move. To say that all motion is fictional according to Osiander seems to me sheer nonsense.

WESTMAN: No, I don't quite go along with that. First of all, one difficulty in speaking about Osiander is that he wrote so little on astronomy. In astronomy we know him from that one famous little preface that he adds in the *De revolutionibus,* and it seems to me that if you examine it very carefully, you can only come to the conclusion that he is saying that all motions that one is about to read *in this book* are fictional. They are all hypothetical.

Osiander maintains the same position with respect to the Ptolemaic hypotheses: "Let us therefore permit these new hypotheses, *which are no more probable;* let us do so especially because the new hypotheses are admirable and also simple, and bring with them a huge treasure of very skillful observations."

ROSEN: First, you say all motions, and later, all motions in this book. Those two statements are not equivalent. An axiom is not a boundary condition. Your friend is not your enemy; black isn't white; top isn't bottom; male isn't female.

WILSON: In Osiander's preface, it seems to me that he has a fairly good argument for saying that the Ptolemaic and Copernican theories are fictional, in that the brightness of Venus as far as the naked eye can see remains fairly constant, and neither theory is satisfactorily in accord with this fact.

R. HALL: Yes, Tycho says this. Tycho makes it a positive objection against Copernicus, does he not?

WILSON: It is an equal objection to Ptolemy.

ROSEN: But remember, Curtis, Osiander makes the generalization: He who enters the door of astronomy thinking that he has found the truth will depart a greater fool than when he entered.

HOLTON: There is another aspect to what Bob Westman has been talking about. That is the psychobiographical one, and it is still a novel and daring kind of thing to do. It does seem that the Copernicans occasionally allowed themselves what I might call an indecent public exposure of their euphoria. This really is a component of Copernicanism, and one sees this in Rheticus, too; it is characterized by the message: "I have seen the True Light." If we would know more about the psychodynamics of the people involved, perhaps one could explain a little bit better what motivated them to throw themselves into that kind of a battle, and what distinguishes theirs from the style of the established scientists.

WESTMAN: Thank you for your comments. I realize that I am going out on a limb (a new limb) and I am glad that you took up the issue of psychobiography because I have a few doubts about it myself. I think that my interpretation of Rheticus' ambivalence stands up. I think that the objections, if there are any, should come from those who can say, "Well, look, there are other conditions here that would explain why people were ignoring or refusing to comment on this section of Rheticus." However, it seems to me there is a check on this: the private response. When you look at the notes people made on Rheticus' work, you see that all this business of harmony was not affecting people. You can appreciate some aspect of a new theory without necessarily committing yourself to it; that is the major point.

NELSON: Bob Westman, you seem to be saying that the strongest claims, the most rhetorical ones, are likely to be made by people who have certain sorts of life cycle crises or else who undergo unhappy experiences at a point in their lives in connection with their father through whatever sets of circumstances.

I would like to say a little bit more about what all of the culturally and psychologically relevant elements may be, not only in the circle that you are discussing but, say, within Roman Catholicism, including universities and scholars. What kinds of circumstances would impel some to make powerful statements, often in a rhetorical mode, that would not be purely scientific, in defense of the reality, the certainty of the Copernican claim? How would you put all that together? Suppose you thought of advancing your effort in a comparative historical way rather than in terms of the psychobiographies of these individuals?

WESTMAN: First of all, and this is an objection that really could have been raised, these metaphors about harmony and system and so forth have been around for centuries and also we have had a lot of unhappy people around for centuries as well.

Now, why is it that I go ahead and link these two together? The answer is because I find evidence in Rheticus that they are so linked. Rheticus personally experienced something in the Copernican system that reverberated with an earlier experience in his childhood. It seems to me that what appealed so deeply to Rheticus, both intellectually and emotionally, was Copernicus' restoration of unity and wholeness where before there had been a "monster": the feeling of fragmentation and guilt, on the one hand over his father's death and, on the other hand, the desire for wholeness.

Now, I am not contending that other people experienced precisely what Rheticus felt. Certainly not. What I am saying, however, is that these personality constellations do give us some sense of what it was that made Rheticus emphasize the harmony and symmetry and order in the new theory. Why does one put things together in a particular way? Why does one feel or hear or see or remember certain events with unusual intensity and with a selective perception?

NELSON: Would you use a similar explanation for Kepler, Riccioli, Campanella, and others who do have a good bit of rhetoric about innovation and new theory?

WESTMAN: I really have not studied their personalities in such great depth nor have I studied them in the context of their thoughts; but I would say, potentially, that my methodology is applicable there.

NELSON: Your mode of explanation has both the strength and the weakness of Erik Erikson's way of talking about very critical changes that do occur within civilizational settings; that is, his analysis explains the Lutheran theory and the vast cultural, social, political changes that were in process by reference to what goes on in the life of a single person. But it seems to me that we do need wider structures if we are going to make sense of all of those who express such changes or are interested in such changes. Also, it seems to me, we have to get more deeply (than you did today at least) into political-social issues that are in some way at stake in any alteration of central ideas upon which so much rests.

I mean, I would accept a lot of what you say about Rheticus, but I am not at all sure I would use that as a general explanation of those changes.

WESTMAN: I must reiterate that I too am not using it as a general explanation. I agree with you. If you are going to generalize, then you must be quite careful. I really think that the reaction to Arthur Koestler is, in a way, a reaction to some of the things that I have been talking about here. Koestler was outside of the establishment; he wrote a book * and he wrote it very well. It's an extremely provocative book. He made a lot of mistakes and was in many cases, not a good historian, but I think people threw Koestler out with the bath water. He had a lot to say about the psychology of creativity, and it seems to me that some of these things also carry over into the psychology of theory-choice. I feel that some of the exaggerated statements made in 1959 by Stillman Drake and Giorgio de Santillana in their *Isis* review completely missed his good qualities.

ROSEN: If you are going to use the facts of Rheticus' biography to explain why he became a Copernican, how do you use the same facts to explain his becoming a Paracelcist, as he was a good part of his life?

M. B. HALL: It could also be he was a man who liked extremely radical, disturbing theories.

---

* *The Sleepwalkers* (London, 1959).

Rosen: Rheticus is a curious and inquisitive man. "Here is an established astronomy and I hear there is some guy living somewhere who says it is wrong. I have got to find out about it." He takes a lot of trouble to find out and then when he is bounced out of his career as a professor for reasons I have already indicated, he has to make a living in another way. So he goes to study medicine at an advanced age and he becomes a doctor. Then he finds there are controversies in the field of medicine somewhat similar to those in the field of astronomy and he makes up his mind that Paracelsus is right and the people attacking him are wrong. Thus, he decides to become a devotee of Paracelsus, as he has previously been a devotee of Copernicus. It has nothing to do with the beheading of his father, the fact that his sister had marital difficulties, that his mother's will was involved in a lawsuit, etc.

Nelson: May I come back to Osiander again? I wonder if it isn't apposite to remark that the notion of the motion of the earth was treated as a plausible fiction long before Osiander, and that his statement of it doesn't represent anything very new?

Westman: I think that people are too ready to say, "Look, Osiander is just repeating a traditional viewpoint," but let me point out one fundamental difference. He is repeating the traditional viewpoint in a new situation, namely, where you have two theories competing with each other. Given this situation, you have a choice. You either fall back on tradition, or you struggle to come up with some new way of interpreting the theory. I do not see it merely as a party-line view. It is a decision to fall back on tradition.

The tone of Osiander's preface is one which attempts at once to reassure the reader that Copernicus' work will *not* upset traditional methodological assumptions in astronomy:

> Since the novelty of the hypotheses of this work has already been widely reported, I have no doubt that some learned men have taken serious offense because the book declares that the earth moves, and that the sun is at rest in the center of the universe; these men un-

doubtedly believe that the *liberal arts*, established long ago upon a correct basis, should not be thrown into confusion. But if they are willing to examine the matter closely, they will find that the author of this work has done nothing blameworthy.[*]

Notice also that Osiander recognizes the possible challenge or threat that the new work might offer to the university curriculum. The liberal arts will not be "thrown into confusion" by this treatise.

EISENSTEIN: When you come to this question of a choice of theory, it does seem to me one important dimension is that Copernicus, by spelling out a new theory in full mathematical detail, does present the next generation with a choice that preceding ones never had. In this sense he introduces a different dimension into astronomy. For the first time you have a community facing a choice between two theories and after the *Prutenic Tables* they also face a choice of two tables.

HENDERSON: The choice of tables, with the *Prutenic Tables*, doesn't imply anything about what theory you subscribe to. There are numerous people using them, but it really doesn't matter as far as . . .

EISENSTEIN: Right, but wouldn't it be that Tycho as a young man looked things up in these two sets of tables and decided they were both wrong and he should consult the book of nature? What I am trying to suggest is the novelty of a young student encountering two conflicting authorities. The experience was only possible after Copernicus; it was not present before. Tycho grew up in an environment where he almost had to make a choice.

ROSEN: May I interject a remark here? What you say relates to something extremely important in the biography of Copernicus. It wasn't suspected at all until a previously unknown letter[**] was published, that he was disposed to print part of his

---

[*] E. Rosen, *Three Copernican Treatises*, third edition (New York, 1971), p. 24.

[**] *Zeitschrift für die Geschichte und Altertumskunde Ermlands*, vol. 25 (1933), pp. 237-239.

work. It is quite clear from this letter that Copernicus was eager to issue his tables. Now you have a constant testing of all competitive tables at all times in the hands of competent observers and the facts decide which table is superior in a way that cannot be applied to conflicting beliefs. Thus, Copernicus was eager to publish, but he was eager to publish only nonincriminating material, namely, tables, and as Jan correctly says, you don't know what the tables are based on. When you get to theory that is another question. When his friends urged him to publish, they were prodding him to take a more dangerous course. Tables, yes, but more, and he says no, only tables. This proves the contention that the man's fear is dread of possible consequences of the publication of the theory, since he is eager to publish tables where no harmful consequences can possibly follow.

R. HALL: Can you very briefly explain a point that's long puzzled me about the difference between Reinhold's tables and Copernicus' own tables? Does a difference arise because the basic observational data used by Reinhold is different, because the methods of computation from the data are different, or because sheer computational errors are rectified or . . .

GINGERICH: It is extraordinarily difficult and frustrating to calculate a planetary position from *De revolutionibus* because so many of the numbers you need are buried in the text and not in convenient form. Basically what Reinhold has done is place things in a much more convenient fashion. Reinhold's *Prutenic Tables* bears the same relation to *De revolutionibus* that the *Handy Tables* of Ptolemy bear to the *Almagest*.

R. HALL: Thank you, that point is easily dealt with. Secondly, I have often seen allusions in the literature to the computations of absolute distances in the heliocentric system. I think Bob is the first person I have actually heard suggest that in the mid-16th century this particular thing was singled out by someone as a significant point that might in some hypothetical sense count in favor of the Copernican view.

GINGERICH: This is pointed out by Gemma Frisius in about

1555. In a letter to Stadius, and printed in Stadius' *Ephemerides,*
Gemma comments that the two great attractions of the Copernican hypothesis are the explanation of retrograde motion and
the spacing of the planets in a specific, fixed way.

SWERDLOW: That is what Copernicus says. That is what the
theory is all about.

GINGERICH: But it is another thing for somebody to pick it up
and notice that these are the two things.

R. HALL: The third point comes back again to the problem of
theory-choice. If you are going through Duhem's work, the point
strikes you that the ancient writers who discuss the fictionalism
of astronomy were making the general point that any geometric
model can have another geometric model constructed that is
equivalent to it. Ptolemy, I think, only illustrated this by the
rather trivial fact that a concentric with an epicycle would be
equivalent to an eccentric. However, it is obvious there is no
alternative system of mathematical astronomy competing with
the total Ptolemaic system. You might have knowledge of Mar-
tianus Capella, but as far as I know, there is no evidence that any
such alternative fits in this viable system.

SWERDLOW: Not true. In Indian planetary theory one has a
large body of transmitted Greek material. A small part of it is
Ptolemaic, but most is non-Ptolemaic.

Now, by comparison with Ptolemaic, the other is, quite hon-
estly, very primitive, very crude, and poorly done. Yet, it is evidently what a good many astrologers actually used. They had
tables based on these simple models. These are models with two
concentric epicycles, one of which controls the equation of center while the other controls the equation of the anomaly. They
seem to be based on a very strict preservation of uniform circular motion. There is never a bisected eccentricity. What we know
of this is out of the Indian treatises, but it is definitely Greek
material; we just don't have it from Greek sources.

Now, there is no heliocentric alternative. I mean, van der

PHÆRESEON SOLIS.   43

I Sexagena

Gradus	Adde Centri (par / //)	Dif.A.S	Scrupu Propor	Dif.S	Subtrahe Orbis (par / //)	Dif.A.S	Excessus	Dif.A.S	Gradus
30	7 19 58	0 55	31 55	31	1 50 38	0 5	32 39	3	30
31	7 20 53	0 47	31 24	31	1 50 40	0 2	32 41	2	29
32	7 21 40	0 39	30 53	32	1 50 41	0 1	32 43	2	28
33	7 22 19	0 31	30 21	31	1 50 40	1 0	32 44	1	27
34	7 22 50	0 23	29 50	31	1 50 36	0 4	32 44	0	26
35	7 23 13	0 16	29 19	31	1 50 31	0 5	32 44	0	25
36	7 23 29	0 7	28 47	32	1 50 24	0 7	32 43	1	24
37	7 23 36	0 1	28 16	31	1 50 14	0 10	32 42	1	23
38	7 23 35	0 9	27 45	31	1 50 3	0 11	32 40	2	22
39	7 23 26	0 17	27 13	32	1 49 49	0 14	32 37	3	21
40	7 23 9	0 26	26 42	31	1 49 34	0 15	32 34	4	20
41	7 22 43	0 34	26 10	32	1 49 16	0 18	32 30	4	19
42	7 22 9	0 42	25 39	31	1 48 56	0 20	32 26	5	18
43	7 21 27	0 51	25 8	32	1 48 35	0 21	32 21	6	17
44	7 20 36	0 59	24 36	31	1 48 11	0 24	32 15	6	16
45	7 19 37	1 8	24 5	31	1 47 45	0 26	32 9	7	15
46	7 18 29	1 16	23 34	31	1 47 18	0 27	32 2	8	14
47	7 17 13	1 25	23 3	31	1 46 48	0 30	31 54	8	13
48	7 15 48	1 33	22 32	31	1 46 16	0 32	31 46	8	12
49	7 14 15	1 42	22 1	30	1 45 43	0 33	31 38	10	11
50	7 12 33	1 51	21 31	31	1 45 7	0 36	31 28	10	10
51	7 10 42	2 0	21 0	30	1 44 29	0 38	31 18	10	9
52	7 8 42	2 8	20 30	31	1 43 50	0 39	31 8	12	8
53	7 6 34	2 17	19 59	30	1 43 8	0 42	30 56	11	7
54	7 4 17	2 25	19 29	30	1 42 25	0 43	30 45	13	6
55	7 1 52	2 35	18 59	31	1 41 39	0 46	30 32	13	5
56	6 59 17	2 43	18 20	20	1 40 52	0 47	30 19	13	4
57	6 56 34	2 52	18 0	29	1 40 2	0 50	30 6	13	3
58	6 53 42	3 0	17 31	30	1 39 11	0 51	29 51	15	2
59	6 50 42	3 10	17 1	29	1 38 18	0 53	29 37	14	1
60	6 47 32		16 32		1 37 23	0 55	29 21	16	0

| Subtrahe | S A | | | | Adde | S A | | S A | Gradus |

4   Sexagenæ

L 3

A page from Erasmus Reinhold's *Prutenicae tabulae* (Tübingen, 1551). (Photograph by Charles Eames from the Honeyman Collection.)

Waerden's booklet on the heliocentric theory is totally wrong.[*] Ptolemy had the best material, but we can't say that there was nothing else. Ptolemy's success made all this other material seem worthless.

ROSEN: There is a passage in Copernicus [*Revolutions*, Book 3, chapter 20], which you must not overlook here. In talking about alternative models he says, "Since so many arrangements lead to the same result, I would not readily say which one is real, except that the perpetual agreement of the computations and phenomena compels the belief that it is one of them."

Now, from the proposition that mathematical devices are fictional, you must not infer that they are *all* fictional. One must per force conform to reality if you can find out which is the right one. It is only provisionally that all are fictional but ultimately one must be real. After all, the planets do move.

R. HALL: But Ed, just a minute. It's obvious, isn't it, if at this time we know N geometrical constructions to the planet's motion, we can only say that the true motion must be one of this number if we have also established that no other possible system, N + 1, can exist. Now Copernicus didn't say that.

ROSEN: No, he couldn't see that far.

SWERDLOW: He does have reason for rejecting Ptolemy's model; the equant is impossible. You have to have models composed of rotation of spheres. This violates it, so on those grounds he can say that is clearly impossible, but, obviously, there are going to be any number of models to perform uniform spherical rotation.

R. HALL: I am perfectly happy about that. I am merely trying to substantiate something Bob said that I thought was significant, that this issue of fictionalism versus realism becomes more important after Copernicus' system is made known.

SWERDLOW: I don't think one can generalize easily, because you have got to look at each writer in detail. You will discover

---

[*] B. L. van der Waerden, *Das heliozentrische System in der griechischen, persischen und indischen Astronomie* (Zürich, 1970).

that each one holds a different position and within one you will find things that seem contradictory.

Look at a treatise by Viète called *Ad harmonicon coeleste*—it's the most brilliant mathematical astronomical treatise before Kepler or maybe even more brilliant than Kepler in some ways. He has models for everything. He doesn't seem to care whether they are heliocentric or Copernican or Ptolemaic. What he cares about is coming up with rapid and brilliant solutions to mathematical problems in astronomy that no one, not Ptolemy, not Copernicus, had been able to figure out before. He actually has a direct solution to the problem of getting the eccentricity and the direction of the apsidal line from observations of three oppositions.

As to his opinions as to which of these models are true, he is absolutely skeptical about virtually everything. He is skeptical about Tychonic theory, he is skeptical about Copernican theory. Incidentally, he thinks that everyone before him was ignorant and no mathematician.

WESTMAN: I find what you are saying very interesting because I think it is analogous to what I am saying about Reinhold as well. In a sense, the cosmological issue doesn't "grab" him. In fact, he *ignores* the metaphors. What interests him is the beautiful geometry and, particularly, the elimination of the equant.

SWERDLOW: But still there is a problem: if he is so terribly convinced by Copernicus' preservation of uniform circular motion, elimination of the bisection of the eccentricity, why in heavens name does he move back and forth between a bisected eccentricity and a Copernican model indifferently?

GINGERICH: Don't you sense, in looking at the manuscript of Reinhold's commentary, that the great bulk of it is by no means intended for publication? It is a set of working notes ...

SWERDLOW: He says he intends to publish a commentary.

GINGERICH: The early part looks set up for publication; the rest looks like his working notes. I think you have to believe that Reinhold wanted to show the equivalence even if he is convinced about eliminating the equant. Because he shows the alternative, you just have to give him some credit for being curious about mechanism. There is no way he would consider that thing ready for publication.

SWERDLOW: God forbid, but I think he intended to publish a commentary, and I assume this was the draft, which was never finished.

ROSEN: I want to go back to the question how Erasmus Reinhold ever became listed as a Copernican. I think the answer is in Giovanni Battista Riccioli's *Almagestum novum* (Bologna, 1651). You may remember that in that massive work Riccioli was called upon to defend the sentencing of Galileo (that is the fundamental reason for the publication of the book), and in it Riccioli uses a statistical argument. He lists all the astronomers he knows anything about, some of whom *we* know nothing about, especially, a guy called Elephantutius. Some people think that was a little joke slipped in by Riccioli, who was not above such stuff. Riccioli lists the pro- and anti-Copernicans. What he wants to show his reader is that the number of men who accepted Copernicus in whole or in part is smaller than the number of the anti-geokineticists, among whom you find many famous names, from Aristotle to Elephantutius.

The tenth name in the pro-Copernicus list is Reinhold [vol. 2, p. 294]. How did this very astute man, Riccioli, come to this conclusion? I imagine he may have glanced at the preface of the *Prutenic Tables*, where Copernicus' name is prominently mentioned. That, of course, would make Reinhold a Copernican, to a snap judgment.

GINGERICH: There is another good way, too. The only way in which the second edition of *De revolutionibus* differs from the 1543 Nuremberg edition, other than the addition of the *Narratio prima*, is taking the publisher's blurb off the title page and replacing it with the lavish praise by Reinhold as giving sort of a push to Copernicus' book.

ROSEN: At any rate, Riccioli's reputation and prestige meant that the classification of Erasmus Reinhold as a Copernican persisted until the recovery of the lost commentary. There is absolutely no foundation for this classification, particularly if you read the preface to the *Prutenic Tables* where Reinhold disassociates himself from the conceptual basis of Copernicanism quite clearly.

GINGERICH: I think it is almost possible, just from looking carefully at the printed works, to suppose that Reinhold was not a Copernican. You can't conclude he was Tychonian or anything else, but certainly he was not an enthusiast for the new cosmology.

A few years ago in Cambridge, England, I found a manuscript closely related to the material Bob Westman talked about this morning.* It is a set of astronomy notes for lectures at the University of Wittenberg in the late 1560s, when the influence of the late Reinhold was still very strong. It is interesting to see that the name of Copernicus comes up in the notes approximately 15 times. Yet never once does his name come up in connection with the motion of the earth or any cosmological point. He is praised on a couple of occasions for his trigonometric tables, for some things about the motion of the moon, for the apogee of the sun, parameters for Mars (which are cited by book and chapter from *De revolutionibus*), and so on. The lectures are generally based on questions on the sphere of Sacrobosco, and every time it comes to the standard part where the mobility of the earth is discussed, we find only the traditional arguments and not a clue that Copernicus had said anything otherwise. This evidence substantiates what Bob already said about the teaching going on in Wittenberg.

---

* Owen Gingerich, "From Copernicus to Kepler: Heliocentrism as Model and as Reality," *Proceedings of the American Philosophical Society*, vol. 117 (1973), pp. 513-522.

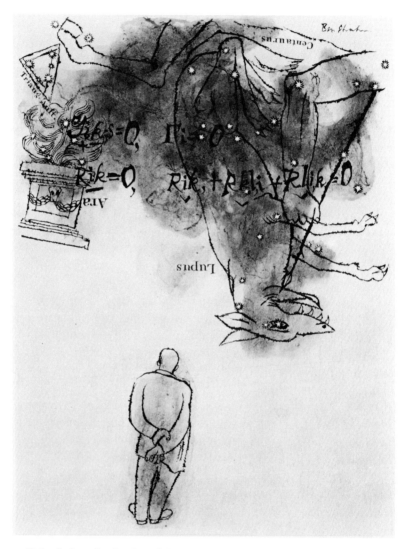

"Man before the Greek and Contemporary Universe," watercolor by Ben Shahn. Einstein's field equations are superimposed on the ancient Hellenistic constellations. (Courtesy of Kennedy Galleries, Inc.)

THE INTERPLAY OF LITERATURE, ART, AND SCIENCE

O. B. Hardison, *Convener*

LIST OF PARTICIPANTS

James S. Ackerman
    Professor of Fine Arts, Harvard University, Cambridge, Mass.
    Art, science, and naturalism in the Renaissance
Herbert L. Anderson
    Professor of Physics, University of Chicago
    Elementary particle physics
Thomas P. Baird
    Associate Professor of Art History, Trinity College, Hartford, Conn.
    Renaissance and Medieval studies
Ronald S. Berman
    Chairman, National Endowment for the Humanities, Washington, D.C.
    Shakespeare
Charles Blitzer
    Assistant Secretary for History and Art, Smithsonian Institution
    Historian and political scientist

459

William L. Bradley
President, The Hazen Foundation, New Haven, Conn.
Philosophy
J. Carter Brown
Director, National Gallery of Art, Washington, D.C.
17th-century northern European painting
Elizabeth Story Donno
Professor of English, Columbia University and Editor,
Renaissance Quarterly, New York
English literature
Philipp Fehl
Professor of the History of Art, University of Illinois, Urbana,
Illinois
Renaissance and the classical tradition
Francis Colin Haber
Professor of History, University of Maryland, College Park,
Maryland
European intellectuals
Pierre Han
Professor of Literature, American University, Washington, D.C.
Comparative literature
O. B. Hardison
Director, Folger Shakespeare Library, Washington, D.C.
English Renaissance and Shakespeare
Edward Haskell
Chairman, Council for Unified Research and Education, Inc.,
New York
Unified science
Richard C. Henry
Assistant Professor of Physics, Johns Hopkins University,
Baltimore, Maryland
Astrophysics
Philip Knachel
Associate Director, Folger Shakespeare Library, Washington,
D.C.
English Renaissance history and Anglo-French relations

Melvin J. Lasky
  Editor, *Encounter* magazine
  Historian and essayist
John U. Nef
  Professor Emeritus, University of Chicago; and Scholar, Washington, D.C.
  History of the Renaissance
Nathan Reingold
  Editor, Joseph Henry Papers, Smithsonian Institution
  History of science
Richard Schoeck
  Director of Research Activities, Folger Shakespeare Library, Washington, D.C.
  Medieval and Renaissance literature and history of law
Merle Severy
  Senior Editorial Staff, National Geographic Society, Washington, D.C.
  Renaissance history
Andrzej Sicinski
  Member, Scientific Institute of Philosophy and Sociology, Polish Academy of Sciences, Warsaw
  Sociology
Yole G. Sills
  Assistant Professor of Sociology, Ramapo College, Mahwah, New Jersey
  Sociology of medicine and interdisciplinary education
Otto von Simson
  Professor of Art History, Free University of Berlin
  Art history
Joshua C. Taylor
  Director, National Collection of Fine Arts, Smithsonian Institution
  19th- and 20th-century paintings and sculpture
Edward N. Waters
  Chief, Music Division, Library of Congress, Washington, D.C.
  Musicology

BACKGROUND PAPER:
THE INTERPLAY OF LITERATURE, ART, AND SCIENCE IN THE TIME OF COPERNICUS

John U. Nef

Of the scientific revolution that several distinguished historians place especially in the time of Galileo, Whitehead said "Since a babe was born in a manger, it may be doubted whether so great a thing has happened with so little stir." He called this rise of modern science "the most intimate change in outlook the human race has yet encountered. . . . The quiet growth of science has practically recolored our mentality so that modes of thought which in former times were exceptional are now broadly spread through the educated world." [1]

While these revolutionary changes in modes of thought should be associated especially with the hundred years that followed Copernicus' death in 1543, we should not slight the interplay of earlier historical forces related to the new human conditions that occurred during his lifetime, the times of Leonardo da Vinci, Columbus, Erasmus, More, Machiavelli, Grünewald, Dürer, Giorgione, Raphael, and Titian.

So we have historically to concentrate not on one period but on two: the lifetime of Copernicus (1473–1543), and the lifetime of Galileo (1564–1642). And the two periods need to be distinguished in all aspects of history—economic and political, artistic, religious, ecclesiastical and constitutional, as well as scientific.

Sir Charles Sherrington contrasted the two periods in science in his little book on the French physician, Jean Fernel. Sherrington applies the term Renaissance to Fernel's period (1497–1558), which is also Copernicus' period, and he says: Renaissance delight "in natural objects and natural occurrences for their own sake" was insufficient to bring about the scientific revolution. There had to be what Sherrington describes as "new-found delight in natural observation for its own sake," in observation for the sake of quantities and precise space relationships, even at the expense of qualities. "More was wanted," wrote Sherrington, "than comment, conciliation and mere systematization. . . . There had to be re-foundation." [2]

That was exemplified by what Brahe and Kepler did with Copernicus' theories of the architecture of the universe. As Weizsäcker has observed, the system of the movement of the heavenly bodies invented by Copernicus was derived from art and theology, from the classical aesthetic concept that "the most perfect curve . . . is the circle," and that in a universe made by God, "heavenly bodies were bound to move in perfect ways." [3] That is why Copernicus believed, erroneously, that the planets move in circles. Kepler (deriving his data from Brahe) replaced Copernicus. Even in its early stages, therefore, the scientific revolution of the century following Copernicus involved a retreat of science not only from religious faith but from art—one of the subjects of our inquiry. It wasn't that art, or even theology, lapsed. Art (and theology too) became more independent of science; science more independent of artistic verification.

The word "interplay" in the title of our collegium leads to the problem of causation. Here, I submit, we are unrealistic if we separate literature, art, and science from the whole movement of thought during the Renaissance and from the new classical and

Baroque civilization that followed after about 1550. If the subject of our inquiry is the European mind, we are bound (if we concern ourselves with causation) to make comparisons with ancient Greece and Rome, and with China. It is a far vaster subject we are entering than the comparatively innocent phrase "interplay of literature, art, and science" suggests.

Art is placed before science in the title and, as we have just suggested, that is more appropriate in dealing with the first of our two periods than in dealing with the second. But notice too that literature is placed before art in the title. This is less appropriate for Copernicus' lifetime than for Galileo's. If literature is distinguished from the visual arts, then it occupies a more compelling place—with Cervantes and Shakespeare, Donne and Milton, Molière, Corneille, and Racine—than in the times of Columbus and Machiavelli.

And with the rise of literature comes the rise of music, which assumes along with the graphic arts and Baroque architecture (Geoffrey Scott's *Architecture of Humanism*) a more intimate place in daily life than had been common in previous history. In fact, the rise of English, Spanish, French, and Dutch literature to a place of great moment for historical development is one of the features of, what are called in England, the Elizabethan and early Stuart eras—a time of comparative peace in Great Britain and of religious and civil wars on the Continent.

The earlier period of Copernicus' lifetime prepared the way for the new developments in science, pure and applied, and in the arts including literature. How?

Looking back into history at the end of the second period, Robert Boyle wrote a significant passage:

> ... what changes in the face of things have been made by two discoveries, trivial enough: the one being but of the inclination of the needle, touched by the lodestone, to point toward the pole; the other being but a casual discovery of the supposed antipathy between saltpetre, and brimstone: for without the knowledge of the former, those vast regions of America ... would have probably continued unde-

tected; and the latter giving an occasional rise to the invention of gunpowder, hath quite altered the condition of martial affairs over the world, both by sea and land. . . .[4]

Boyle would have done well to add, concerned as we are with literature and the visual arts, the discovery of paper and of printing. All these inventions were made before Copernicus' birth, all were exploited as never before during the 70 years of his life. The Chinese had made these discoveries much earlier; they had not exploited them as the Europeans began to do during Copernicus' lifetime and during the century that followed.

Columbus was not the first to discover America, but what happened during and, especially, immediately after Columbus' lifetime was an unprecedented host of exploratory voyages. By the mid-17th century, at the time of Newton's birth and Boyle's maturity, the Europeans had brought an entire new continent with its peoples and scattered islands in the seas into direct and, what proved to be, permanent contact with the other peoples of the earth; to form the one world that besets us now.

The expansive character of the Europeans in the time of Copernicus was manifesting itself in many directions, in novel expeditions into the glaciers and peaks of the Alpine regions, in efforts to despoil the underworld of its treasures in ores and other minerals. So as the history of exploration indicates, it was not the *invention* of the compass, but the *uses* made of the compass and of new kinds of ships that led to the conquest of the globe. It was also true of gunpowder; the Chinese had priority in that invention, but they used it (according to some Chinese sages) as a matter of choice to set off fireworks in the evening. Whereas the European, first and foremost the Spaniards, used it to conquer not only primitive Indians but the highly sophisticated societies of Central America and Peru.

It was so, no less strikingly, of the inventions of paper and printing and movable type. It was multiplication rather than innovation by itself that led to the establishment in Copernicus' lifetime of hundreds of publishing houses with their printing presses in Germany, Switzerland, France, Italy, Spain, and the Low Countries. At the same time the development of woodcuts, etchings, and other kinds of engraving added a new dimension to the arts, also favored by the use of easily transportable canvas for oil paintings.

As a goal in production, almost limitless expansion only began to claim philosophy during the second period, with the works of Francis Bacon and René Descartes. This was a feature of the scientific revolution, the vision of what applied science might eventually accomplish. During the second period—the period of Napier, Harvey, Boyle, and Newton—the novel pressure on the supplies of coal in Great Britain created drainage problems, along with the drainage of the fens in the interest of multiplying agricultural production. That made it seem imperative to replace water power and horse power for driving machinery by more economical steam power.

During my lifetime, we have begun to question as ultimate goals the limitless pursuit of production and transportation, realizing that this very progress threatens the future of the human race. Do not imagine that Napier and Boyle and Newton at the time of the scientific revolution had no concern over the dangers.[5] What was it that reassured their successors and facilitated the ruthless exploitation of the scientific discoveries set in motion in Copernicus' time and immediately thereafter? It was the forces of art, including the advent of printed books and prints, great literature and great painting, the literature of Shakespeare, the canvases of Rembrandt. A cozy life with books and pictures and music extended into the houses of a few, and led the Europeans of the 17th century and of the Enlightenment to imagine that the human race could become civilized. A new place was assumed by these arts—the furniture and the musical instruments sometimes decorated by great painters. Music with Monteverdi and Orlande de Lassus, Heinrich Schütz, Byrd and Dowland, and later Couperin staked out new territory to be harmonized by beauty. To appropriate the title of a recent article in the *Revue des Deux Mondes* by René Huyghe, it was a case of the conquest of the mind by Western art. That conquest had begun in Copernicus' lifetime. European minds, before the Industrial Revolution, were filled with the adventure of art; they felt that humans had the power to assume a new dignity that would humanize applied science.

It is not too much to suggest that the future of the human race depends upon men and women's capacity to show that their ancestors from the times of Callot and Rembrandt and Vermeer to the times of Bach, Schiller, Rameau, and Mozart were not mistaken in the hopes they held for the human adventure. In an age much soiled by dirt and crowding, blood and suffering, they were more realistic optimists than the descendant in the much more affluent societies of our own time. We should welcome the effort, recently reviewed by Hardison, to renew that optimism.[6]

Notes

[1]A. N. Whitehead, *Science and the Modern World* (New York, 1925), p. 2, quoted in John U. Nef, *The Conquest of the Material World* (Chicago, 1964), chapter 7.

[2]Charles Sherrington, *The Endeavour of Jean Fernel* (Cambridge, 1946), pp. 144–145, quoted in Nef, op. cit., p. 284.

[3]C. F. von Weizsäcker, "The Spirit of Natural Science," *Humanitas*, vol. 3, no. 1 (1947), p. 3, quoted in Nef, op. cit., p. 281.

[4]"Of the Usefulness of Natural Philosophy," in Thomas Birch, editor, *The Works of Robert Boyle* (London, 1772), vol. 11, p. 65.

[5]I have touched on their alarm in two of my books, *War and Human Progress*, (Cambridge, Mass., 1950), and in the second edition of *The United States and Civilization* (Chicago, 1967).

[6]O. B. Hardison, *Toward Freedom and Dignity* (Baltimore, 1972), esp. ch. 6: "Demanding the Impossible."

John U. Nef, *Moderator*

DISCUSSION: PARALLELS AND CONTRASTS
OF SCIENCE AND ART

HARDISON: Our moderator this morning is to be John Nef, an expert on the history of technology in the Renaissance, who wrote a classic book on technology in the 17th century, *Industry and Government in France and England, 1540–1640.* Unfortunately he cannot join us until later, so the job of chairing our collegium has devolved upon myself.

Yesterday, we celebrated Shakespeare's birthday at the library. This is an important annual event at the Folger. We have currently arranged a modest exhibition related to Copernicus and the new science of the Renaissance. One exhibition case of "hard science" reaches a climax with the first edition of Newton's *Principia mathematica philosophae naturalis* that we just purchased this year and of which we are inordinately proud. In the other case, we have materials illustrating the influence of science on the Renaissance. John Milton is featured there.

Basically, I suggest that we engage in a civilized and wide-ranging discussion of the interplay of science and culture in the Renaissance. The impact of specific inventions and discoveries on the Renaissance in every field is a very large one. It involves the influence of these materials, particularly in the area of art. I am thinking of an artist like Stradanus who recorded in his series of *Nova reperta* the inventions that he thought were most significant for his age. Another angle might be the interrelationships and conflicts between scientific theory and the theological framework within which Renaissance man tended to view the world. We might examine the way in which certain scientific discoveries affected the Renaissance imagination, the sudden movement from the concept of closed space to open space as it affects Milton's *Paradise Lost*, for example, and as it is studied by Marjorie Nicholson in her classic book, *Newton Demands the Muse*. The question of the unsettling psychological impact of certain scientific discoveries during the Renaissance, the Renaissance equivalent to what Alvin Toffer calls "future shock," might also be an important subject for discussion. I am thinking particularly of the great anxiety that the new science created for society as a whole in the minds of a good many artists and spokesmen, perhaps illustrated by John Donne's famous lines, "The new philosophy calls all in doubt," to which he adds, "Tis all in pieces, all coherence gone, all just supply and all relation." Apparently he was speaking for a good many people in his own age when he wrote those lines.

WATERS: I have been reading several articles about Copernicus. I have read that he studied science, mathematics, theology, law. And the next sentence would talk about his career. Career in what? Is this typical of a man of the period to have a career embracing all four of those disciplines? What was he a careerist in?

SCHOECK: His career was that of a canon lawyer. He did not go to Italy to study Greek. No cleric in the 16th century had been given that luxury. He went to study canon law. Being the brilliant man he was, he could study medicine, astronomy, and Greek while passing his doctoral exams in canon law. Then he went back to the chapter and continued his career as a canon

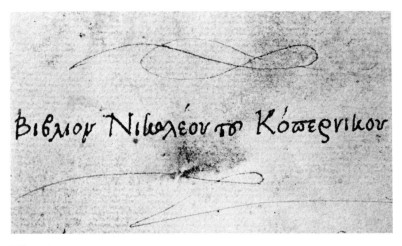

βιβλιον Νικολεου τω Κοπερνικου

When Copernicus went to Italy in the 1490s primarily to study canon law, he also learned Greek. This is his signature, in Greek, in his Greek dictionary, now preserved in the Uppsala University Library.

lawyer, which embraced a wide variety of fields, given the importance of canon law in the 16th century.

REINGOLD: You didn't really have the possibility in that day of having an occupational profession as a scientist except in very unusual circumstances. For quite a number of people whom we think of as scientists, doing science was something that they did for fun. It was something they happened to be interested in. I think it has always been fun for the real scientists. But Copernicus wasn't pursuing science as an occupation.

SCHOECK: There is a parallel, surely, in the history of humanism. In the early days, they were not employed as humanists. They got their jobs as secretaries and what have you. Not until the 16th century, which is late in the history of humanism, are large numbers appointed in universities as humanists.

HARDISON: What about the disciplines of the quadrivium that came along through the medieval curriculum? The quadrivium included arithmetic, which developed into mathematics. Are there instances of university professors who were primarily concerned with the sciences and made major contributions to the development of science in the early Renaissance, or were most of the major contributions made by people who were like Copernicus involved in other disciplines?

SCHOECK: Stillman Drake argues that something like 90 percent of the activity in astronomy in the 16th century was done by people outside the universities, and he allies this with a number of causes, one of which is printing. Copernicus' work, for example, gets printed, and this makes it available to a nonuniversity audience for the first time.

But it does seem clear that humanists in the 14th century, and astronomers in the 16th century, generally operated outside of and in spite of, not because of, the university.

SEVERY: That brings up the question of just how this revolution in the heavens impinged on the lives of the practical people who were going about their business as usual. While we do have the case of "all coherence gone" in Donne, the previous year he had joked at this. It was only in 1611 that it seemed to hit him, and it tied in beautifully with the funereal theme of that poem he wrote for his patron on the loss of a daughter. It tied in with the sense of decay, the pessimism; now, good God, even the heavens are in disarray.

But just how much did this revolution really affect the people—any more than, say, the Einsteinian revolution has gotten to me? I haven't even caught up with Newton yet.

Shakespeare was able to write beautiful plays, talking about the spheres—with Lorenzo discoursing in the *Merchant of Venice* about the music of spheres, which our terrestrial ears are so ill attuned to hear. Milton is able to have a marvelous universe that is all Ptolemaic. In other words, things went on as though Copernicus never existed.

ANDERSON: That is a very interesting point. The fact is that these revolutionary ideas upset the whole world community. Galileo was persecuted for having pursued the Copernican idea that the earth was not the center of the universe. Thus the question arises: If it didn't affect anybody, why did everybody get so upset?

HABER: First of all, Galileo comes somewhat later. In one way, Copernicus took a step backwards by suggesting that his system was true whereas according to the reigning philosophy at that time, you were only making models. And even Sir Henry Savile said when he was asked if he believed in the Copernican system that it mattered not whether his meal was brought to him or he went to the table for it, so long as he was fed. He thought that both Ptolemy and Copernicus were equally effective in explaining what the astronomer needed to know.*

The claim of Copernicus that his system represents reality was a revolutionary step for astronomical philosophy. By the time of Galileo, it was beginning to make its impact and becoming a challenge. But I don't see any evidence that Copernicus upset his own age very much. People quoted him as one of the most brilliant mathematicians and still kept on with Ptolemy.

SEVERY: In other words, society was not ready for Copernicus. Also, he was defused by Osiander's preface, which said, "You don't really have to believe this, it is just another hypothesis."

Things started getting objectionable only when Galileo came along with his optic tube and said, "Look, there are other things up there that are different. The moon has an irregular surface." It is no longer a perfect sphere according to an artist's conception. "And there are four moons gyrating around Jupiter." The new astronomy then started to get a little more real.

---

*Nathanael Carpenter, *Geography delineated forth in two bookes containing the sphaericall and topicall parts thereof* (Oxford, 1625), vol. 1, p. 143: "Here I cannot but remember a merry answer of that great *Atlas* of Arts, Sir *Henry Savile* in the like question. Being once invited unto his Table, and having entered into some familiar discourses concerning *Astronomicall* suppositions: I asked him what he thought of the *Hypothesis of Copernicus*, who held the *Sunne* to stand *fixt*, and the *Earth* to be subject to a *Triple* Motion: His answere was; he cared not which were true, so the Appearances were solved, and the accompt exact: sith each way either the old of *Ptolemy*, or the new of *Copernicus*, would indifferently serve an Astronomer: It is not all one (saith he) sitting at Dinner, whether my Table be brought to me, or I goe to Table, so I eat my meat?"

ANDERSON: That is because Copernicus is conceptual. He realized you could have an alternative system that is simpler when he proposed his theory. In the case of Ptolemy, every planet had its own motion and had to be separately described. Copernicus realized you could have a much simpler scheme once you took another object as the center.

SEVERY: It is an artistic conception that has been simplified.

ANDERSON: This idea has motivated physics throughout the ages, the idea that things should be simple and beautiful. Many of the achievements that we recognized as so important to science came from just that motivation.

SEVERY: Must they be beautiful: Is this an essential part of the scientific thinking?

ANDERSON: I think so.

BROWN: But it doesn't seem to be a two-way street, does it? There seems to be an influence of art on science at that point, this concept of roundness. After all, Copernicus is a contemporary of Raphael, and it is very much the same kind of approach.

On the other hand, Copernicus is so conceptual that he doesn't seem to affect the artist, at least of the period. As far as I can see, the artist was much more influenced a century later by the optical revolution and by the whole concept of things which then are visible and relate to the microcosm that he can see around him. But this macrocosmic conceptualization is something that you really cannot go out and paint.

The whole Renaissance is a narrowing down of the focus and a bringing of man in as the measure. Since the general trend is toward subjectivity, you don't have much influence from Copernicus and from these conceptual revolutions on the visual history of the period.

ANDERSON: My remark concerns a different type of beauty that is connected with simplicity and the idea that you can have a single, simple theme to explain, let's say, the motions of the heavenly bodies. Once you have the conception, once you have reduced it to a simple-enough principle so that you can realize that it has the power to explain a very broad set of circumstances, then you are able and encouraged to go on and find other consequences.

The idea that you could be equipped to have a deeper, broader understanding by having a simpler, more compact understanding is really of the essence of the Copernican Revolution.

He showed that by turning away from the idea that the earth was the center and making the sun the center, you could explain all the motion with a single principle, where before you had to have a separate motion for every planet. There must be some central underlying principle, and Copernicus was the one who found it in that case.

In a similar sense, later on Einstein did the same thing. He found a simpler set of principles, more compact, more beautiful.

HARDISON: We are right in the center of the topic of our collegium—the question of aesthetics as it relates to science and to the arts. Are we really talking about two different aesthetic views—the scientific view seeking the powerful generalization, seeking simplicity? In the Renaissance artists liked *discordia concors*—a unity of divergent elements. They liked a general unifying theme. But they tended to equate beauty with ornament, sometimes exceeding complex ornament, which tended to obscure the basic elements of unity. I am thinking of *Orlando Furioso* and the way it just grew and grew, and Tasso's *Discorsi del Arte Poetica*. Tasso says Homer is the greatest writer because there is more in Homer than any other writer. He compares Homer to an ocean, and the artist to a god in the sense that just as God creates the great world in all of its infinite variety, so the great epic poet creates a little world with almost infinite complexity and variety.

We touched on Milton earlier. Milton was very familiar with the new astronomy. As a matter of fact, he has Raphael come down from Heaven and give Adam a lecture on the heliocentric theory. But obviously, he felt the Ptolemaic tradition with all that gingerbread architecture of spheres, intelligences and music of the spheres and so on, was appropriate for the decor of a great epic poem.

I wonder if we don't have two aesthetic theories operating in the Renaissance—the artistic one tending to emphasize ornate decoration and considering that equivalent to beauty, and the evolving scientific one seeking simplicity and the powerful generalization. I don't know. I raise that as a question.

TAYLOR: I wonder if what you mean by ornate is just multiplicity. And that doesn't necessarily rule out the principle of unitary organization. The complexity that became so popular in the late 18th century was one in which rule was defied. The thing was as interesting as you could not find a rule in it.

HARDISON: Take a poem like the *Faerie Queene*. When Spencer couldn't find unity, he invented it. And he imposed it on the poem through the strategy of a preface in the form of a letter to Sir Walter Raleigh claiming to explain the poem.

TAYLOR: You can talk about the complexity of the *Faerie Queene*, but you know perfectly well how many lines you are going to encounter and how many sounds. It nonetheless is a basic unitary plan. In fact, I think the whole concept was that you could have variety provided you assumed the existence of unity. That is very different from a later kind of variety in which the unity was something you always imagined may be there, but you could never locate. So I think there is a sense of unity and scientific unity that did pervade this Renaissance idea.

ANDERSON: I would only comment that both art and science are done by men. And men have some common elements in their personality and their character that drive them to look for simplicity and beauties in their own particular spheres.

BROWN: We have got to define the terms as to what moment of time we are talking about. If we are talking about late 15th, early 16th century, it seems to me that at least in the visual sphere, there is a tremendous enthusiasm for the simplicity, unity, and finding one overall composition that somehow draws all the forms together. Whereas from there on, having passed through that, people are interested in changing it with more and more complexity. You get from Mannerism into the high Baroque. *Paradise Lost* is a very Baroque conception in that sense.

The early part of the Renaissance breakthrough—and I think Leonardo is the summary of it—was a narrowing down, an impulsion towards a unified scheme, a simplicity. And then I see that being charged with more and more complexity as it goes tearing through the 17th century into rococo.

REINGOLD: I think there are many scientific aesthetics. The definition of simplicity changes in time. It is not a simple rubric you could always apply.

ACKERMAN: May I ask historians of science if they know when the term beauty was first applied to scientific concepts?

REINGOLD: I always assumed the ancient Greeks had that.

SCHOECK: "Euclid alone has looked on beauty bare."

HARDISON: That is Edna St. Vincent Millay. [Laughter]

SCHOECK: But it is an ancient notion.

HARDISON: It is obvious there is not one science in the Renaissance; there are a great many different sciences. There is the tradition of the exact sciences, based on mathematics, but there is also the tradition of a much more empirical kind of science like anatomy. I get the impression that perhaps the exact sciences did not have an immediate impact upon the artistic imagination, but the artist had no difficulty whatsoever assimilating the lessons of Renaissance anatomy and putting them to very good use.

BROWN: I think the artists were very involved with the arithmetic of perspective, for example, right through Dürer, starting with Alberti, and the whole involvement with highly exact orders of space. Leonardo got very much into light, as you know, and did some of the basic empirical research in light. The area where they begin to fade out is in this cosmographic sphere. And it is not true later on.

In the 20th century the effect of the surrounding systems of thought shows up right away in the visual arts. Picasso and

Braque were forging the new language of cubism almost simultaneously with Einstein's original publication of relativity. Cubism has interfaced planes and multiple vantage points and a way of looking at things as a space-time process. I think the whole way of approaching our world has been reflected very sensitively by the artists.

However, it does strike me that in the 16th century, you do not have the conceptual revolutions affecting the artists in that way. Of course, their patronage is often religious, and they are still tied very much to that conceptual scheme.

HARDISON: Let me interrupt our conversation very briefly to introduce our real chairman, John Nef, who has now been able to join us.

John, we are in the midst of a discussion of the rise of science in the Renaissance and various aspects of the scientific method. The general theme of our discussion so far has been the question of whether the point of view of the sciences is also the point of view of the arts in the Renaissance, or whether they conflict and to what degree scientific developments impinge upon the development of art.

NEF: One thing that you may not have stressed is the enormous development of printing which occurs in Copernicus' time. The growth in the output of books of all kinds and of illustrations that went with books from about 1470 to 1540, roughly the years of Copernicus' life, was simply prodigious.

Let me give the case of France. Effectively, there were no printing presses in Paris before about 1470. By 1540, there were roughly 30 publishing houses in Paris, many of which had several presses. The growth of printing in Lyons was almost as rapid as in Paris; and, in addition, there were something like 35 other French towns that developed printing in this period. I want to stress this prodigious development, which is related to the whole subject that concerns us here.

SEVERY: An obvious influence is the dissemination of information. In polemics, you cannot conceive of a Luther pulling off a

Reformation without the printing press and the pamphlet wars that this occasioned.

NEF: Of course, Luther was a great propagandist, but this was not simply a movement of propaganda. It was a movement in relation to beauty. We have only to think of Luther's countrymen, Grünewald and Dürer, to realize the concern with creating beauty as it had never been created before.

Something that intrigues me more and more is that these art objects came into more homes. We have forgotten this.

It used to be a tradition in my youth to say that only the upper classes had things. There is a small Rembrandt, recently sold for millions of dollars in this country. It is said to have been discovered in a farmhouse in Holland near the end of the 18th century by an Englishman who was traveling through Holland, and who asked for board and lodging for the night. He bought this small picture he saw there, by paying for an extra night. It turned out to be a Rembrandt. This suggests that some great pictures got into the hands of people of very modest means, like this farmer.

BROWN: We have some sociologists in our group we haven't heard from.

SILLS: A basic premise of the sociology of knowledge is that the production of ideas reflects the class and generational interests of the people who produce them. Both the sociology of science and the sociology of art and literature are subsets in this field. The sociology of science is concerned specifically with how scientific ideas originate, are produced, and disseminated—and how they are accepted or rejected. The sociology of art views art as one of the products of culture and examines the artist's status in society. We should remember that until the 15th century artists considered themselves—and were considered by society—to be superior craftsmen who created decorative objects of beauty through technical skill. It was only later that the status of the artist became more exalted. In view of this, a possible explanation of the phenomenon of finding the small picture in the house of a farmer is that it was considered decorative art.

BROWN: The fact is well documented that in 17th-century Holland there was a new ownership of works of art. There was a study done by F. Loerke in 1905 on this and there is Arnold Hauser's book, *The Social History of Art*, which takes the sociological approach. It is very well documented, but it is to me more relevant as economic history, the question of the bourgeois culture of the time. Art became a commodity. They speculated in paintings, and it became a status symbol to have one around your house.

What is more interesting in terms of intellectual dissemination was the history of print-making. There ideas began to flow. Marc Antonio Raimondi did the Raphaels in prints, and this is what got them around.

Dürer, coming back from the south, brought a whole perception of the Renaissance. Printing and the printed book provided the means for dissemination of ideas. Dürer made a print about stars—he did that beautiful one of the signs of the zodiac—but it didn't have much to do with Copernicus.

Because music was taught as part of the old curriculum and as a way of training the mind, it relates to a mathematical approach to life much more closely than the visual arts. It would be fascinating if there were breakthroughs in music that might somehow be analogous to the kind of intellectual approach that we saw in astronomy.

WATERS: Earlier this morning Mr. Brown himself stressed the contemporaneity of Picasso, Braque, and Einstein. We can add Stravinsky and Schoenberg, and today we have the ultra *avant garde* such as Stockhausen. Schoenberg certainly made a conscious attempt to build a hypothesis on his 12-tone system of music. I presume he felt that he was approaching closer to the truth of beauty (if you can use such a phrase) than the older system, which people like Wagner and Strauss seemed to have exhausted.

In music and in beauty generally, I find everything much more dependent upon psychological factors than I do upon scientific or demonstrable factors.

In terms of beauty and simplicity—and someone almost equated those two terms this morning—no matter how complicated Schoenberg's complications, I don't think they have increased the content of beauty per se. Beautiful as some people think Schoenberg may be, they all admit—and maybe this can be related to Copernicus—that the supreme artist as far as musical beauty is concerned is Mozart.

NEF: Didn't printing have some influence on music in the Renaissance?

WATERS: Yes, great. Very early in the 16th century, 1501 or 1502, the first printed music from movable type was published by Petrucci in Italy.

LASKY: What kind of musical instruments were used? How strong was conservatism and the notion of what a proper musical instrument was? How audacious was the sense of the new? After, all, that is what links Copernicus to modern audacity.

SEVERY: The Renaissance had a multiplicity of instruments. All the deuces were wild in their musical deck. They played almost anything that was on hand. The music was not written for any specific instrument; it was played by an instrument within that range. But then later, in the Newtonian period, the classical orchestra was formed by winnowing out instruments, getting rid of those rough-edged ones. Then we got our standard orchestra with its pairs of flutes, oboes, clarinets, horns. Instead of a variety of strings we have just violins, violas, cellos, and basses. So we lost a lot of colorful instruments; we lost serpents and bombards.

HARDISON: I would like to ask the musical historians whether there is any relation between the rise of polyphony and the Renaissance and the scientific point of view?

WATERS: That goes back considerably before the time of Copernicus. I think it connected with the development of notation, when musicians learned to write more than one line on a staff. That goes back to about 1000 or so, to Guido d'Arezzo

Music, together with astronomy, arithmetic, and geometry formed the medieval quadrivium, here symbolized in Gregor Reisch's *Margarita philosophica* (Freiburg, 1503). (Permission of Harvard College Library.)

where notes are delineated on the fingers of each hand. As one voice was added to another, composers began to write entirely in a horizontal manner rather than vertical, and they did it with consummate skill.

By the time they got down to Copernicus' time, the contrapuntalists contemporary with him had a skill that has never been equalled or surpassed.

ACKERMAN: Concerning the relationship between music and intellectual and scientific life, I think it is a case in point that Galileo's father, a great and revolutionary musician, was a member of the Camerata in Florence. Its effort was to restore monody, that is to say, the single line, which the members thought to be Greek. It seems to me this fits in with the description of simplification in the approach of Copernicus. At least, it is a kind of effort at naturalism because the Camerata, too, is trying to get the voice to express the emotions, not in symbolic terms, but in direct terms.

I don't see a clear relationship between this kind of approach and Galileo, but I do see both art and science attempting to approach nature more directly.

HARDISON: If I remember correctly, E. A. Burtt argues that Osiander's preface should be taken seriously and that Copernicus offered his theory as a kind of mathematical model, which was important for computation and so on, but not as the ultimate truth unlocking the secrets of nature. Only later, with Galileo and Bacon, was there this great urge to concretize science and to say that science has to unlock the secrets of nature and show us the ultimate truth or reality of nature.

Burtt argues that with Galileo and Bacon, science took a giant step backwards; that the point of view of Copernicus is really much closer to modern scientific methods; and that, in fact, the influence of Galileo and Bacon and the popularizers of science retarded the development of science rather than advancing it.

LASKY: Just the opposite of Dr. Oberman's thesis in his lecture yesterday. It was a very dialectical lecture with which I disagree, but which I admire tremendously. He argues that Copernicus was not the man of conjecture, but thought he had found the truth of things. Oberman argued that the semi-scientific churchmen who kept on saying, "This is probable conjecture" (meaning it is no damn good), were closer to the future scientific tradition than Copernicus who said "This is the truth and you have just got to believe it."

HARDISON: Let me add to that, then, a point made by Professor Randall in his study of the history of science. He suggests that the tradition of the exact science is basically a medieval and scholastic tradition, that it migrates with the expulsion of the Averroists from the University of Paris down to Italy, and particularly to the University of Padua. The humanists considered it obfuscatory and impractical. The ultimate product of this Averrostic tradition was the development of modern astronomical theory, particularly Galileo. So I offer these as questions for the historians of science: Was the tradition of the exact sciences medieval rather than humanistic and was there really a loss of sensitivity to scientific method between Copernicus and Galileo?

SEVERY: Doesn't this involve the rift between technology and science? I think that an artist in studying perspective, Alberti, for example, or Piero della Francesca or Leonardo, is a scientist in working out his artistic means. Also I would consider Paracelsus and Vesalius to be scientific. Yet I read that the pure historians of science will say it is only after Descartes mathematicized everything into numbers and separated the technology that you really get science.

ANDERSON: If you think of the names Copernicus and Galileo and Newton, you can usually characterize each one by a very simple statement, almost a simple word. You can say that Copernicus is heliocentric and immediately you get the conception of what he really introduced into the world of thought. Falling bodies, let's say, would be a simple one for Galileo. And for Newton, $f = ma$, a simple mathematical relationship that generalizes all of the motion and dynamics of bodies.

This kind of statement emphasizes the interest of these great men in finding universal principles. I think it is that search for unity and universal principle that characterizes the scientific enterprise. In our effort to relate the scientific things to artistic and literary effort, is there some principle like this that we can find when we look at the works of artists and musicians and literary people? I think it becomes very clear when you talk about the scientific work, but it is a little more difficult to pin down when you talk about the effort of the artist or musician.

LASKY: On this point of art and science, I would tend after the discussion this morning to hold to the rather extreme opposition between the two. Although I think I was the first publisher of C. P. Snow's *Two Cultures*, I have always disagreed with him intensely on that. I think there have always been two cultures and always should be. And even if we thought it undesirable, there would still continue to be two.

Consider the notion of beauty. We expressed some doubt about it this morning: whether it is real beauty; whether a scientist doesn't convert that which seems ugly to every other scientist into a new, beautiful form; whether he doesn't, by the very truthfulness or effectiveness or experimental improvement of his ideas, create the audience which then appreciates this beauty. But leaving that aside, let's just call it beauty and say art does share that quality with science. But I would remind you, the logic of science by its very nature, when confronted with two theories, one complex, one simple (and both effectively deal with data and problems), will ultimately choose the simpler one. Yet this is not at all true in art. In art, one might choose the complex one, or the simple one, or any combination of the two.

Secondly, a scientist would have to base any theory on a whole tradition of true propositions. Whatever Copernicus used, they would have to be, in his mind, a true and accurate computation of some figures. Most artists have tended almost willy nilly to go to the false as far more interesting, whether it is Picasso (a little palmistry in the outstretched hands of his "Guernica") or John

Milton who knows a good deal about Galileo, but still goes on with the "music of the spheres" because that is a harmonic element in poetry that for literary reasons he doesn't want to give up.

In addition to these two points—the complexity versus the simplicity, the falseness as against the truth—I would add one about incompleteness. A work of art on canvas or a sculpture can be incomplete. There is a profound incompleteness in art and literature and poetry which can make for its greatness. Whereas if a scientist were offering new work in astronomy and physics and left a big hole in the middle and went right around it and said, "It is more beautiful if it has got a hole in the middle, more beautiful if it has a few illogicalities about it," it is simply read out of scientific court.

Fourthly, I would consider the matter of coherence and incoherence. There is something about the human effectiveness, the human purposefulness, of art which transcends its creative incoherence. Some magic happens which raises it to a higher level of coherence, whatever the aesthetic theory. Whereas in physics or in astronomy the incoherent elements, where the things don't add up, get immediately pointed out, and are corrected and changed.

And, finally, I would say something about the two differing personalities. The element of responsibility in a scientist is central. Responsibility-Irresponsibility, I think, would be an extreme way of summarizing the personal aspects of Science versus Art.

FEHL: I would say it all depends on what subject the artist is dealing with. In certain areas, the responsibility of the artist is very great. He can make us a better or a worse man, improve our actions and thoughts, or seduce us to approve of, or even enact, something contemptible. If the task is high, his obligations are high. Michelangelo gives an example of great moral integrity in his art as well as in his private life. Most artists, of course, did not live up to that but many, it seems, honored it in the orientation of their hopes. The dignity of art is based on this.

LASKY: I wonder if we are not talking about two different types of responsibility. I am talking about the internal "responsibility" of the artist; you are thinking of his public responsibility.

TAYLOR: I think the whole concept that art or artists can put aside integrity is sheer nonsense. You couldn't exist as an artist without integrity. The idea of completeness as opposed to not complete is to miss the whole sense of completeness. Nor can I imagine an artist who would suppose what he was doing was capricious for the sake of being capricious.

In another sense, of course, the scientists are also capricious; I am unconvinced about a logic of science—almost every breakthrough has been made by denying what has already been accepted as logic. I think it is particularly true in contemporary science. There is much more dependence on what is called an educated hunch than on the simple matter of logic, because our structure of logic is a tradition and has no necessary relationship with the truth.

To suppose science is logic and has responsibility and consequences whereas art doesn't is to suppose that a creative, imaginative mind is not important, but our physical existence is. I don't think the two can be separated, certainly not in an idea system.

LASKY: I didn't deny the artist's integrity. We are doubtless dealing with fine, great, serious, heroes of history, whether in science or in art. I would say the expression of the artist's integrity, certainly from the post-Copernican time, is a form through errant individualism. He follows in the truest possible way his own expression of his vision of what is, take him where it may.

SEVERY: That is what happens in a scientific breakthrough. It is errant individualism upsetting the apple cart. So we have one revolution after another.

NEF: Perhaps we can go into that tomorrow because the time has run out. And I think it behooves us to pursue this responsibility of scientists and artists and compare it with the responsibility of the statesman and the warrior.

John U. Nef, *Moderator*

DISCUSSION: DISCOVERY IN ART AND SCIENCE

NEF: Welcome, ladies and gentlemen. It is a great joy for me to be here again. I have the advantage today of the perfect chauffeur, namely, my wife, so I am not late. I am particularly happy that my old friend and collaborator, Otto von Simson, is here today.

SIMSON: Yesterday in the Widener Library I jotted down the following passage from a man born almost a century after Copernicus, Pierre de Bérulle, the founder of the Oratoire.

NEF: The great figure in the French Catholic Church, early in the 17th century.

SIMSON: Bérulle writes that "the novel opinion, little regarded in the science of the stars, has its uses and should be followed in the science of salvation." On the margin of this sentence Bérulle has added the name of Copernicus, "an excellent mind of our century." The passage seems to me extraordinary. I would like to make two observations on it.

In the first place here is a great theological and mystical thinker, one of the great figures, I think, of religious thought in

Tintoretto's *Last Supper* in the San Giorgio Maggiore in Venice was an innovative study in perspective. (Permission of Alinari/Scala.)

France. He is described as such by Henri Brémond in his classical history of religious thought in France. And the interesting thing is that here, far from opposing Copernicus, far from seeing anything irreconcilable between Christian theology and the discovery of the great Polish scientist, according to Bérulle a real stimulus for Christian theology was provided when the heliocentric system replaced the geocentric system. Now, Brémond describes Bérulle as a revolutionary in terms quite similar to the way in which I would generally describe Copernicus. He says Bérulle has brought a revolution in theology by replacing the anthropocentric system by theocentric thought.

For Brémond the greatest exponent of "anthropocentric" theology is St. Augustine whose importance, of course, Brémond does not deny; but he feels that in the theology of St. Augustine in a sense man came first, whereas the theology of Bérulle is centered in God. Whether or not we admit this, it seems very interesting, this paralleling of the religious thought of Bérulle with the scientific thought of Copernicus.

May I, as an art historian, add a second observation, which perhaps may steer us a little further: There is something very interesting to our concern here, the interplay of literature, art, and science. In 1594, Tintoretto painted as one of his very last works the great "Last Supper" in the San Giorgio Maggiore in Venice. This is a most extraordinary painting because it is arranged in such a way that the central figure, Christ, appears far away, luminously, while entirely accessory figures, such as the servants serving the supper and so forth, appear much closer. Christ is far away, yet drawing all our glances toward him. He is, as it were, the Sun in Tintoretto's "planetary system." This is a revolutionary picture even in terms of composition. And here again, I would submit for your discussion whether there is not this kind of structural similarity or affinity, perhaps, between Copernican thought and the language of art.

SEVERY: I don't think I should leave the statement from yesterday lying on the table, about art being contrary to organization. My understanding of Renaissance art is that it was a very deliberate attempt to organize experience: all the studies on perspective, ways to organize what you see and to present this in a formal manner. You see the difference developing from the Gothic art style where there will be a number of different scenes that take place at different times—a linear form that can be related to the linear aspects of Gothic architecture, and to polyphony in music. You see art being drawn closer into a frame, into a single presentation of an experience. And finally in Raphael, for example, you get the pyramid, or the circle that again brings in the concept of circular movement in the heavens. It is a way of organizing your experience. So I must say that *form* is crucially important in the arts.

ANDERSON: I wouldn't disagree with that. I would like to emphasize two points. There has to be in all of these disciplines some comprehensibility. In order to have man appreciate it, he had to understand it in some fashion.

Second, in science the objective is always very clear. You look for the underlying laws and the simplest possible expression of them. I don't think that is the objective of art.

NEF: But there is a conflict only in appearance I think. The artist aims at universality.

HASKELL: The artist expresses it in a different way. He epitomizes. He will say, "Thou canst not stir a flower Without troubling of a star." But Newton writes an equation, the equation for gravity: $F=GmM/r^2$. These are two statements of the same thing. The one is *epitomizing*, talking about the effect of a flower on a star; the other is *nomothetizing*, that is, law-formulating. Both are expounding the same universality.*

It seems to me that these many different artists, epitomizing these many great truths, are in the end saying, in their way, what the scientists are saying in a different, more abstract and orderable way; the way that gives rise to technology.

SIMSON: Wouldn't it be correct to say that Copernicus, among other things, was concerned with the relationship between the observing subject and the natural phenomena observed and, of course, the relative significance and possible faultiness of such subjective observation? Now, this kind of question had been posed in the arts with the advent of the central or one-point perspective even before Copernicus established his new paradigm in relation to the planetary system.

I think at least for an artist it is difficult not to understand the immense transformation going on. And I may add, transformation from the viewpoint of form, as well as subject matter. Suddenly the observer appears in the picture. Suddenly the entire picture is ordered according to the perspective of the observer.

Probably a good many of you have read the very stimulating book by Thomas Kuhn, *The Structure of Scientific Revolutions.* Kuhn, whose field is the history of science, has also been intensely interested in the possible affinities between art and

---

*From Harold G. Cassidy, *The Sciences and the Arts—A New Alliance* (New York, 1962), p. 6.

science. What he says there is that the history of scientific discoveries is not a cumulative process, but suddenly through a "discovery," something is being created or seen or defined that makes more sense; I would say the same thing is true for art. The artist, if he is worth his mettle, makes a statement like the one the scientist does with entirely different means, but he does make a statement about the universe.

WATERS: How does an artist make a discovery? This perplexes me.

HASKELL: The sonata form or perspective. These are great artistic discoveries. Perspective is both scientific and artistic.

SIMSON: You could quite rightly object, I think, that perspective is a scientific and geometric discovery, but not an artistic discovery. The question is, how perspective is being applied. It is the "how" that matters in art.

Let me give you one example: Giotto. There is no question about three things.

First, since Giotto, art never has been the same.

Second—let me put it in the words of a humanist at the end of the 14th century, Vergerio of Padua, who said—"Usually artists would consult different sources to produce a good work. But in this century of ours, there is no artist who would not direct his work after the example of Giotto."

Third, and most important, this was also felt to be the case by Giotto's contemporaries. I remind you only of the unforgettable sentence in Dante's *Purgatory* where he speaks of the nature of fame and where he says that Cimabue was believed to lead the field in painting, but now the name of Giotto is in the mouth of everyone, so that the fame of Cimabue is obscured.

We have here the example of a very great artist, Cimabue, who lived to see his work abandoned, forgotten, obscured, because there had appeared a younger man (believed even to be his pupil) who had eclipsed what he had done. Why this should have been so is perhaps beyond anybody's explanation, but the fact is quite indisputable and equally true for a number of other artists whose stature can be compared to Giotto's, such as Raphael, or the master of the Cathedral of Chartres, or Cezanne. But the nature of these artistic "discoveries" remains impalpable. That is what distinguishes them from the discoveries of science.

SCHOECK: Doesn't that suggest that in the history of art, there can be a turning away? You quoted Dante. What could be more dismal than the fortunes of Dante in the 18th century when Voltaire and everyone else called him a barbarian? It took a century of rediscovery to find out what Dante was all about.

It seems to me, to borrow from the mathematicians, the history of art is not a river or single line. It is a topographical curve, and tradition recircles and regroups whenever a major artist comes along either to make discoveries or to subsume what has been going on around him.

BRADLEY: Is it possible to say that an artist can discover something as a scientist discovers it? Perhaps it might be in this sense that the artist discovers a new way of asking a question, which is not a verbal question. It is not a logical question, let's say, but a new way of looking at experience or the universe. Once that has been given to mankind, then it is part of the heritage, and you can never pretend that it was not made known.

I think there is a parallel to this when R. G. Collingwood points out in his *An Essay on Metaphysics* that in philosophy thought builds upon thought. The problems remain the same, but once you have had a Plato, then you have a new insight into the possibility of thought. You can't go back beyond that.

HENRY: In scientific discovery, even though we stumble and go back, there does seem to be a path. I don't have the impression that is true in art. In the direction art has gone, has it been influenced by this inevitable direction that science has gone in? Has that interplay occurred?

BRADLEY: I should think that the principle of inevitability could not apply in art. There is creativity here that is less explainable even than in science.

HENRY: If that is so, has science sort of dragged art along

The landlocked Indian Ocean in Ptolemy's *Cosmographia*, the great woodcut atlas printed in Ulm in the 1480s. Copernicus used and annotated the Frombork Cathedral's copy of this work. (Permission of Harvard College Library.)

with it?

SEVERY: Or has art dragged science along with it? I could give an example from cartography.

Why do you have north at the top of the map? East was a perfectly fine direction in the Middle Ages. Go to Hereford Cathedral, and on the wall there you will see a perfectly logical ordering of what they thought the world was about in terms of what was important to them. What is at the center? Jerusalem. And the continents, what they knew of them, were clustered around.

Let's take our Mercator maps. These are all conventions and artistic representations. This is shorthand for the way we think the world is. Earlier artistic conventions determined what men thought their world was.

Ptolemaic maps ringed the earth with ocean. But the Indian Ocean was enclosed by a land mass; this was thought to be not only logical but aesthetically pleasing. We don't like to leave blank spaces. So we will fill them up with land.

There always seems to be an artistic conception at work, molding scientific ideas, even in something as tangible as geography. When explorers were coming back with reports that certain places were not there, these still appeared on maps because artistically it seemed as though they should be there.

HASKELL: Man is a complicated organism whose mind has conscious, subconscious, and preconscious aspects. Flashes of insight come first, and it may take decades to work them out. As a matter of fact, there are several books on the flash of insight, the sudden seeing: for instance Hadamard's *An Essay on the Psychology of Invention in the Mathematical Field*. For example, before F. A. Kekulé discovered the benzene ring, he first had a dream of a snake holding its tail in its mouth. And then, later, as he was about to step on a bus in Paris, he saw the benzene ring, the hexagon. That was the basis of the whole development of organic chemistry. The point is, that the first flash may be art.

KNACHEL: But isn't there also involvement of what the textbooks so often refer to as the "scientific spirit" of the 16th and 17th centuries? This seems to imply a skepticism, not only in science, but a carry-over of skepticism or of the "scientific method" into other fields as well.

I wonder, really, to what extent that is true—not only in areas like religion or politics, but indeed in science itself. Isn't it Thomas Kuhn who says that once you set up a new paradigm, you operate within it until finally the paradigm no longer offers adequate explanation—at which point a new insight is needed, incorporated in a new model.

SEVERY: Science is based on great faith. You have to have faith in the hypothesis. The heliocentric theory is an act of faith. Copernicus is asking you to believe that such and such is so in terms of his vision.

LASKY: It is almost a commitment. But if somebody gives me a better explanation of the things I am explaining, then I would revise and alter my opinion. A better explanation is either more identifiable in terms of evidence, or it is more coherent, or it is more logical. From the point of view of the psychology of the intuitive-leap mechanism, there is only irrationality at the other end. In science where there is a statement, the scientist abides by the principle that revisionism always rules. And in that sense, not faith but heresy is scientific.

ANDERSON: Those of us who went to hear the symposium speeches by Heisenberg and Holton yesterday were impressed that they emphasized the idea of intense motivations. One of the things that drove Heisenberg was, you may say, a religious feeling that there was a simplicity in the universe that he as a man could find. It drove him to develop some of his great ideas.

We were impressed when the talks yesterday indicated a kind of irrationality that was invoked when a scientist developed his ideas. There is no *logical* way for Einstein, for example, to develop his theory of relativity. He had to take up the idea just from a sort of feeling that there was a unifying principle involved. His belief was that a simple statement would describe more accurately the way in which the universe was organized.

HASKELL: In his book *Physics and Beyond*, Heisenberg said: "In the final analysis, our compass must be our relationship with a central order." I used that statement as the motto of our book, *Full Circle—The Moral Force of Unified Science*. Of this book Heisenberg wrote me saying that he heartily agrees with its direction, its effort.

In an article in the *Scientific Monthly* [vol. 54 (1942), pp. 545–551], I called this central order "The Religious Force of Unified Science." I said that the reason why I don't speak to scientists about a divine aspect of Man is simply because that is not the scientific idiom and vocabulary. But personally, privately, I certainly agree with Dr. Anderson.

ANDERSON: What are the motivations that drive scientists to do their work? One of the points brought out yesterday that I thought might be pursued somewhat further is the idea that the scientist is motivated to seek unifying principles and to find the underlying laws. The artist doesn't necessarily seek out unifying principles, but may move in the opposite direction. He may try to make it a more complicated structure.

HENRY: Concerning your remark on Einstein, I think it is interesting that the same feeling Einstein had, that nature had to be a certain way, the feeling that led him to relativity, caused Einstein to reject the probabilistic quantum theory. Everything since then has tended to indicate that the quantum theory and probabilistic view was quite right, and Einstein was wrong.

So it seems to me that it is not that there really is this sort of "god" underlying nature such that if you follow "his" path, then you will succeed in finding the underlying laws of the universe. If a person has a certain kind of spirit or way of thinking and the spirit happens to fit the particular spirit of the time, he is going to be successful insofar as the philosophy of the time is capable of success. It doesn't mean his idea or approach is right in any larger context.

ANDERSON: I don't quite agree with that. I think it is quite true that Einstein did not find the kind of unifying principle that he sought all his life. But I think there is quite a significant body of scientists today who are looking for that. They are trying to unify relativity and quantum mechanics, for example.

NEF: I am told that Heisenberg is a very good player on the piano, by the way. He was in residence at Cambridge not too long ago and they asked him if he would play.

He sat down at the piano and played from beginning to end Opus 111, the last sonata of Beethoven, which is an absolutely unique work. All the dons were more and more overwhelmed by this music, and there wasn't a sound when he finished.

Heisenberg is reported in this connection to have discussed the difference between science and art. "If I had never lived, someone else would probably have formulated the principle of determinancy. If Beethoven had never lived, no one would have written Opus 111."

DISCUSSION: RAMIFICATIONS OF THE WORD
"REVOLUTION"

NEF: Now I want to call on the editor of *Encounter*, Mr. Lasky.

LASKY: I want to return to Copernicus and the interplay of Copernicanism with a field that really hasn't been touched upon. That is politics, society, political philosophy. I would like to pay a curious tribute to this man who with the title of his book, *De revolutionibus*, began the popularization of word and concept "revolution." I am always amazed to see Copernicus talked of as a revolutionary and to hear talk about the Copernican Revolution. It is a double entendre all the way through. It is an anachronism.

It is as if Nero used a box of matches, or even better, a Ronson lighter, to put the fires to Rome. Because revolution was, astronomically speaking, the turn of a full circle. In that sense to call it the "Copernican Revolution" is a tautology because he surely didn't return to any full circle. What we mean by it is something profoundly different and something new, which Copernicanism imposed on the modern world. Today we are incapable of thinking of politics or of society without saying the Industrial Revolution, the American Revolution, the French Revolution, the Russian Revolution, the Communications Revolution. None of this means, of course, what Copernicus meant by it.

In this Copernican year, I think it would be a very good time to become self-conscious and to observe this marvelous semantic development which, in the English language, happened one century later. I have done a little work on this in the British Museum. I discovered that the word "revolution" was used politically to mean a "rebellion," to mean an "overturning," to mean "a change of masters," as early as the 13th and 14th century in Bologna and Florence and Milan.

It comes into the English language from the Italian, but its general popularization is a result of the Copernican "revolution." By the 1650s, it is for the first time in some dictionaries in England as a hard word, very rarely used, referring to the full cosmic circle.

When you observe it in politics it suggests, with very rare exceptions, this circular notion. For Hobbes, the English Revolution was not the Cromwellian insurrection and overturning (as we think of it), but it was the restoration of the King in 1661! And there, he said, "Now, we have our revolution, the true revolution, because events have turned round to their point of origin."

We find certain other anachronisms if we look at the translations of some of Milton's Latin works. If you look to see where modern translators use the word "revolution," you will discover nine times out of ten he didn't use *revolution* at all. Machiavelli never uses the word "revolution," although in our English translations, his writings are full of revolution.

Why is it impossible for us to avoid the word "revolution?" Why is it that this great celestial drama in the sky became the profoundest drama of modern man? And what was there about it that makes it impossible for us to say the Russian insurrection of 1917 or the American revolt of 1776? We have to say "revolution," and we have to say that this inventor or artist made a very important contribution because he was a true revolutionary. In my view all these matters are related to the profound shock that Copernicus and Galileo delivered to the modern world. It is the beginnings of the modern world at its most sensitive semantic point. The word and the image shocked by changing the drama of the world as it had been known.

Copernicus not only became the towering figure in the history of science, but the man who accidentally and unconsciously imposed the one central concept of modern history: the idea of revolution.

SCHOECK: In the 1530s in England, because there was no word for revolution in the sense that you have been adumbrating,

Thomas More had to reach back into the Church Fathers for the word "debellation," which means the overthrow or overturn. Fortunately, that word has dropped out of our vocabulary.

The topic of politics suggests another, very probable impact of Copernicus, one which connects Copernicus with Louis XVI— namely, the ceremonies and rituals of kingship necessarily project the king to the world order. After 1600, the ceremonials and rituals had to change. While the idea of sun as a metaphor for king is one that goes back into antiquity, nonetheless, in the 17th century and largely because of the reordering of the sun, the sun as an image for the king suddenly became far more potent.

The deliberate effort of Louis XVI to avail himself with all of the overtones of sun as a metaphor for kingship, I think is ultimately traceable back to Copernicus. This idea was posited in a seminar by Ralph Giese here at the Folger Library last fall.

WATERS: It is interesting how this word revolution has become part of our everyday language, though, even with phonograph records: 78 rpm, 33 rpm.

LASKY: We still use it in the old way, "revolutions per minute." Yet it is a quality of the human mind to be able to keep two contradictory ideas in play at the same time.

HABER: There is also an idealogical contradiction. You have not just the heavens revolving, but a whole system revolving. In the Reformation period there was the expectation of certain conjunctions spelling a whole upheaval in the affairs of men. For instance, Stoeffler predicted that with the great conjunctions of Jupiter and Saturn in 1524 the world would probably be completely, catastrophically overthrown. So you have this strong identification between the conjunctions of the heavenly bodies in their revolutions catastrophically affecting the affairs of men.

SEVERY: How did this change, though, because of Copernican theory?

HABER: All I am suggesting is there is a very definite link between the revolution in the heavens and the idea of revolution in human affairs.

SIMSON: I don't think we should lose sight of the fact that Marx, so far as I know, does not use the word "revolution" in the sense I am going to use it. But behind his entire teaching is very much the idea, it seems to me, of the return to a phase of society, to a pristine form of human existence that has been vitiated by modern development.

HABER: Isn't there a very strong millennial overtone in every one of these great revolutions, the idea that essentially you have that capacity in human affairs to purify all the corruptions away? It is the old image of the phoenix returning and purifying itself. After all, this is basic in the Reformation. You must find the original text to purify away the corruption.

Is not the Copernican system ultimately an attempt to go back to fundamentals, to purify, to get rid of the excrescence? Even Einstein in his one autobiography describes his own work in somewhat the same terms. I wonder if your conception of the French Revolution, the American Revolution, Russian Revolution, doesn't also carry that idea of the radical going back to roots, and the upturn, the upheaval, the restoration, of the original order.

LASKY: There has, also, always been the image of Fire and Water. Fire is the only way in which you can burn out that which is old and diseased, and Water is the only way you can wash anything clean. So Fire and Water, according to my very simple anthropological reading, become those two elements which will change and purge.

HARDISON: There is a very charming metaphor applied to revolution—I believe it is Lenin's image: The revolution is an old sow that eats its children.

But in the midst of all these very psychological comments on revolution, particularly in the Renaissance, I think it is worth remembering there was a very modern attitude towards many of the scientific discoveries, a pride in them, a feeling they had improved life without causing adverse side effects. I think the great example of that for the Renaissance would be the series of engravings by Stradanus called *Nova reperta (New Discoveries)*. They were very popular throughout Europe. They are a series of engravings of a printshop, a silk-weaving factory, paper-making, various other discoveries, which around 1610 seemed to Europeans to have transformed the quality of life, to be fascinating for their own sake, and to be entirely positive in their effects.

ANDERSON: I think the whole idea of having a Copernican celebration after 500 years somehow implies that we cherish our past, we don't try to throw it over all the time, and the values of those ancient works that we preserve today continue onward. I wouldn't want to go home closing on the note of revolution.

NEF: I am so glad you brought this up; it is a good word to end on. I can't thank Wilton Dillon enough for having had the idea of this admirable setting for a discussion, and O. B. Hardison for acting as our host here at the Folger Library.

Years ago, Gertrude Stein came to the University of Chicago. According to a report—I wasn't there—she went to dinner with the president of the university, Robert Hutchins, and there was a long discussion. As she was going down the stairs leaving with Alice B. Toklas behind her, Alice B. Toklas turned back and loudly said for Hutchins and others to hear, "Gertrude has said things tonight that will take her years to understand."

The theologian and astronomer, symbolic of this collegium, converse in
Cardinal d'Ailly's *Concordatia astronomie cum theologia* (Venice, Ratdolt,
1490). (Photograph by Charles Eames, from the Honeyman Collection.)

SCIENCE, PHILOSOPHY, AND RELIGION
IN HISTORICAL PERSPECTIVE

Walter Shropshire, Jr., *Convener*

LIST OF PARTICIPANTS

Ruth Nanda Anshen
  Editor, *World Perspectives*, New York
  Philosopher, specialist on Alfred North Whitehead
H. Vashen Aposhian
  Professor of Microbiology, University of Maryland, Baltimore
  Biomedical ethics
Eulalio R. Baltazar
  Professor of Philosophy, The Federal City College,
    Washington, D.C.
  Evolutionary philosophy, process theology
Daniel J. Callahan
  Director, Institute of Society, Ethics and the Life Sciences,
    Hastings-on-Hudson, New York
  Philosopher and former editor of Commonweal
Robert Cope
  Pastor, The Unitarian Church, Princeton
  Artist and writer

John Crocker, Jr.
    Episcopal Chaplain, Massachusetts Institute of Technology, Cambridge
    Ethics of technology in society
Bernard Davis
    Professor of Bacterial Physiology, Harvard University Medical School, Boston
    Medical ethics and the new genetics
John Dillenberger
    Professor of Historical Theology, Graduate Theological Union, Berkeley
    Interaction of Protestantism with scientific thought
Frederick Ferré
    Professor of Philosophy, Dickinson College, Carlisle, Penn.
    Scientific empiricism in science and religion
Langdon Gilkey
    Professor of Systematic Theology, University of Chicago Divinity School
    Interaction of science and theology
Michael Hamilton
    Canon, The Washington Cathedral, Washington, D.C.
    Theological implications of genetics and ecology
Werner Heisenberg
    Director Emeritus, Max Planck Institute for Physics and Astrophysics, Munich
    Physicist
Samuel K. Klausner
    Professor of Sociology and Director, Center for Research on the Acts of Man, University of Pennsylvania, Philadelphia
    Psychiatry and religion
John Maguire
    President, State University of New York College at Old Westbury
    Philosophy of religion

D. Williams McClurken
Director of Broadcasting and Film, National Council of
Churches, New York
Ethical questions of religion, technology, and science
James McCullough
Life-Science Specialist, Library of Congress,
Washington, D.C.
Biomedical ethics
Ernan McMullin
Professor of Philosophy, University of Notre Dame,
South Bend, Ind.
Interaction of science and religion—Galileo and Darwin
Uta Merzbach
Curator of Mathematics, National Museum of History and
Technology, Smithsonian Institution
History of mathematics
Heiko A. Oberman
Director, Institute of Reformation Studies, University of
Tübingen
Nominalism, history of the Reformation
William Pollard
Executive Director, Oak Ridge Associated Universities,
Oak Ridge, Tenn.
Physicist; interaction of science and religion
Larry Rasmussen
Assistant Professor of Christian Social Ethics, Wesley
Theological Seminary, Washington, D.C.
Christian social ethics
Roger Shinn
Professor of Social Ethics, Union Theological Seminary,
New York
Ethics and science
Walter Shropshire, Jr.
Assistant Director, Smithsonian Radiation Biology

Laboratory, Rockville, Maryland
Biophysicist; interaction of science and religion
William J. Thompson
Professor of Humanities, York University, Toronto
Historian, philosopher, futurist
Joseph Walsh
Assistant Professor of Science and Society, Stockton State
College, Pomona, N.J.
Impact of science on non-Christian religions
John Archibald Wheeler
Professor of Physics, Princeton University
Physicist and cosmologist

Tuesday, 24 April 1973

Canon Michael Hamilton, *Moderator*

DISCUSSION: CLAIMS TO TRUTH OF FAITH
AND SCIENCE

HAMILTON: We are here at St. John's Episcopal Church, where every President has had cause to pray at some point in his life. Perhaps there may be a similar visitation one of these days! Our topic this morning is the historical relation of science and religion. I have long been concerned to bring together various disciplines—secular as well as Christian—to discuss these matters together.

Our first speaker, Ruth Nanda Anshen, has taught all over the world and is the Founder and Chairman of "The Nature of Man Seminar" at Columbia University. She is also the editor of the *World Perspectives*, 50 volumes; the *Science of Culture Series*, 10 volumes; *Religious Perspectives*, 28 volumes; *Credo Perspectives*, 30 volumes; *Perspectives in Humanism*, 10 volumes; and she is the author of a new book, *The Reality of the Devil: Evil in Man*.

ANSHEN: I submit that there would be no disputing the fact that science became scientific for the first time in the 16th century in Europe. It was then that the disruption of Western

Christianity and the rise of modern science took place. And it was Copernicus himself who may be chosen as the most representative figure responsible for changing man's conception of the nature of the universe.

Although Copernicus (in his *De revolutionibus orbium coelestium*) resorts fully to the mathematical techniques elaborated by Ptolemy and by all those centuries during which Aristotelian cosmology and Ptolemic astronomy dominated Western thought, the inspiration of Copernicus goes beyond Aristotle to the age of Pythagoras and Plato, philosophers who have dominated Western ideas from the beginning.

The *scientific* importance of Copernicus is being elaborated with great skill and wisdom by my peers in physics and astronomy. Yet the extraordinary *philosophical* importance of Copernican astronomy was that it removed the earth from the center of the universe and placed it among the planets. It undermined the very foundations of the traditional cosmic world-order with its hierarchial structure. It undercut the qualitative opposition of the celestial realm of immutable being and brought into prominence the terrestrial or sublunar region of change and decay. For the immediate effect of the Copernican Revolution was to spread skepticism and confusion. Its philosophy cast doubt over everything: God, nature, man, the universe itself, in a certain sense.

This conception of the history and destiny of mankind, deriving from the Christian story in religion and a mutilated classical tradition, dominated the social and political thinking of Europe for a millennium. I believe that the names of Columbus and Copernicus may be taken as symbolic of the time and space frame of reference, a notable expansion effected by geographical exploration and discovery, on the one hand, and the growth of scientific and historical knowledge, on the other. For if Columbus and his successors enlarged the earth as a habitat for man's practical activities, Copernicus and his successors deprived it of its traditional primacy in the cosmic scheme.

The crisis in European consciousness, as the Renaissance may be said to describe it, is the direct result of the development of the new cosmology, which replaced the geo- or even anthropocentric world of Greek and Medieval astronomy by the heliocentric, and later by the centerless universe of modern astronomy, and profoundly influenced the social, philosophical, religious, and spiritual changes.

The Church—and men—of the Copernican age could never recover from the shock that Copernicus had created by opening the cosmos from a closed to an expanding universe. And it was Copernicus who inspired Galileo to declare that "Scripture may teach man how to go to Heaven but not how the heavens go."

To leap now from Copernicus himself to his influence on modern man we must recognize that the positive triumphs of the last quarter of a century are beyond question. The scientific vision has been followed wherever it led. The search for an understanding of the interrelationships and the interactions of knowledge and reality is at the heart of all scientific inquiry, of biology as well as of physics.

In science, inquiry is not inhibited but rather stimulated by the conviction that generalizations now held to enjoy a high degree of probability may in the light of new evidence and new experiences have to submit to radical revision. Men of science take it for granted that finality is a conception alien to their undertakings. It is often argued that finality is foreign to the spirit of science merely because the inductive method, supposedly the scientific method of ultimate authority, precludes any generalization of having more than probable validity. While this is in part true, it does not express the most important aspect of the situation. The relinquishing of finality on the part of science is not merely a matter of method. For particular generalizations, those, for example, that have to do with the revolutions of the planets or the circulation of the blood, claims may be made that are tantamount to indubitable certainty. Such is the competence and seduction of the inductive method. The foe of finality is the spirit of science as such, irrespective of the method it may employ and regardless of the established verity of some of its conclusions. For nothing in science is ever so conclusive as to fail to entice the scientific spirit to further exploration, of which the result is always a crop of fresh problems and a harvest of unexplored insights.

The spirit of science is an incessant search and endless struggle, of scrupulous criticism. The temper of science is a temper aglow with defiance of all that limits and arrests inquiry, analysis, experiment. Therefore, in this sense, the question of finality is utterly incongruous with the critical and mobile spirit of science.

Our *aim* in philosophy and religion is to show that science ought to embrace not only purpose, or theology; it ought also to show that life processes transcend physics and chemistry, statistics and equations; the need to point to the error in Newton's celestial mechanics, in which he stated that man does not influence the objects of his observation. On the contrary, he is constantly influencing such objects since he brings his perception to them and indeed the most creative and seminal minds in science do act in just this way. They become metaphysical.

Nature operates out of necessity; there is no alternative in nature, no will, no freedom, no choice as there is for man. Man must have convictions and values to live for and this also those scientists who are at the same time philosophers recognize and accept. For they then realize that duty and devotion to our task, be it a task of acting or of understanding, will become weaker and rarer unless guidance is sought in metaphysics that transcends our historical and scientific views or in a religion that transcends and yet pervades the work we are carrying on in the light of day.

HAMILTON: John Dillenberger is Professor of Historical Theology at the Graduate Theological Union at Berkeley. He has taught in a number of universities and has edited works on Calvin and Luther.

DILLENBERGER: I have always been fascinated about how Copernicus got to be so important. I am interested by the fact that Bacon and Pascal did not accept his views. And I put down as an axiom, which we can debate later, that a thinking, cultured, learned person, in the year 1650, would have had more problems in accepting than rejecting Copernicus.

Now, what does that mean? Copernicus published his book in 1543; we are talking about 1650. That's a long period of time. I find that scientists have a selective way of dealing with their own past, that they look back and extract from everything that was said whatever happens to have been right. Unfortunately, in the history of theology, we have had to live with our mistakes. Everyone has reminded us of them. Nobody reminds scientists of their past mistakes. It's a very odd reading of history.

There is the question of Luther's role. I am inclined to take Luther's statement on Copernicus, "the fool would upset the whole art of astronomy," as the mildest thing that Luther ever said about an enemy. The statement is based on hearsay evidence, but I do not want to make a case that therefore it should not be taken seriously.

In terms of statements, Calvin comes off better. I have never found that alleged passage in which Calvin was against Copernicus. The statement probably comes from A. D. White, *History of the Warfare of Science with Theology in Christendom*, which has been repeated over and over by reputable historians. Individuals who are reputable historians in another area occasionally will quote A. D. White, whose graduate students wrote the book for him. (That is not to their discredit.) But it is a book that had a powerful influence, even though more junk hasn't been assembled between covers in a long period of time.

Now one can read Copernicus as if he were a modern person, when as a matter of fact, if I understand Heiko Oberman's address yesterday, Copernicus was rather backward looking compared to potential predecessors—if not predecessors in point of influence, at least in point of time. Professor Oberman gave a picture of nominalism, which is at least 120 years prior to Copernicus. In a very fascinating analysis of the scholastic nominalist distinction between *potentia absoluta* and *potentia ordinata*, he suggested that nominalism provided a transcendance of God, which at the same time freed a world for exploration, including the exploration of observation and the exploration of imagination. I have no way of refuting this kind of analysis, and find it very suggestive and imaginative; I began to think that nominalism had had an interesting history.

Nominalism appears at the junctures of the world where change apparently occurs. One can think of Luther in that nominalistic context. That reminded me of a statment by Gerald Holton, "The best scientists are people who are poor metaphysicians." Nominalism, of course, is a type of philosophy; but Holton's statement is a way of saying that certain types of affirmations about the world do get in the way of seeing new emergences.

This is exactly the problem that Copernicus represents. People differ on how to assess his discoveries. Insofar as Copernicus was looking back, as the humanists were, to previous documents, he could claim that he was not an innovator; he said he was building on the ancients. Older possibilities and the re-emergence of the classic world were considered to be viable possibilities for serious thought.

My reading of Copernicus is that he happened to have been right about the matter of the earth. But if we are dealing with mathematical tables, and if one were a scientist at the time, one could do just as good a job with non-Copernican alternatives. So the question of accepting Copernicus is an open question for a long period of time.

The other side of the problem emerges when one recalls that in this period science is natural philosophy, and not what we know today as natural science. It is, if you want, as metaphysical as any theology of the period; and that is where the real problem emerges.

Today scientists bracket out the natural philosophy from the science they want, forgetting that natural philosophers at the time had no interest in such a separation. If theology was all or nothing, in that sense one could say Copernicus was all or nothing, or Galileo was all or nothing. I think the Church was right in challenging Galileo, but for all the wrong reasons.

Galileo essentially was a rationalist. He believed that more adequate and significant knowledge of God could be found in the Book of Nature than in the Book of Scripture. As far as the Church in any period is concerned, that is about as heretical as one can get; but it had nothing to do with science.

So often in history the right war is full of wrong battles. In this period all the wrong battles were fought to the confusion of the real issue. The scientific warfare was with a theology that by 1600 had developed a total alliance of thought with Aristotle and which, therefore, was unbending in a way foreign to Calvin and Luther.

Galileo was a Christian rationalist, and his rationalism pushed nature to greater prominence than Scripture for a knowledge of God. Now one can say, all right, on scientific matters, it should. But the issue was not that simple, because the form of rationalism that Galileo was pushing had implications for the way in which redemption was understood. The proof is that the subsequent rationalist development understood redemption in a totally rationalist framework. One could say that John Locke's *The Reasonableness of Christianity* represents this tradition.

If the Church was totally wrong in the defense it took, the problem was that the natural philosophers had an equally inflexible conception of the nature of reality. So the battle began between the Book of Nature and the Book of Scripture. Galileo was right in not wanting Scripture to tell him anything about nature. But one must recall that throughout the early history of the Church, it was felt that Scripture helped clarify nature. By the time of Bishop Butler, natural philosophy was wanting to return the courtesy. After all, nature could help to clarify

Scripture.

This matter of nature clarifying Scripture forms the new problem. This is part of Galileo's platform, and if you want to see where this goes, reflect on Newton. We usually think Newton was in his dotage when he wrote his theological works. But in point of fact, it interested him more than his scientific work. Having finished the first, scientific revolution, his great ambition was to do the second one, namely, to straighten out the matter of religion. And how does one do that? By finding the clues and secrets to interpret Scripture. Having found the secret of nature, he now wanted to find the secret of Scripture. Ergo, the second revolution would be accomplished.

The idea is that, on the basis of analogy, we can use the Book of Nature to gain clarity about Scripture. Clarity in that sense meant, not getting rid of the spurious mysteries, but getting rid of the mystery itself. But that is a religious-philosophical issue. The problem until and even during Newton's time is thus one of competing philosophies, not a matter of science and religion. The possibility of an adaptation between religion and science was thus made impossible.

So what I am suggesting is that we are dealing with a very mixed period of history, and that at least until Newton, a learned person did not have to accept the germ of truth that Copernicus had accidentally discovered, a truth which Galileo clarified, and Newton made inevitable. But that represents a long, ambiguous history, that is, from 1543 to 1687. Therefore, there are no saints and sinners here, but a battle over philosophical orientations that finally did have to be sorted out but in the meantime confounded both theology and science.

HAMILTON: Thank you very much.

The last person I'm going to call on to make an introductory statement is a visitor, Dr. Heiko Oberman. He is Director of the Institute of Reformation Studies at the University of Tübingen, and a Professor of Church History there. He was born in Utrecht, Holland, studied at Oxford, received his doctorate at Utrecht and he has taught at Harvard. Welcome back to our country.

OBERMAN: I will be as brief as I can this morning in making my own comments, because I am rather interested now to get reactions and critique on the paper I read yesterday. But there are a few points I should like to make that have become of major importance for me while working on this paper in the last three months.

What I find of lasting interest is the debate around Osiander's anonymous preface to *De revolutionibus*, on the nature of hypotheses. One of the first enemies of Osiander, and one who has influenced the further debates about it, is Kepler. Kepler took great offense in Osiander's effort to show that the Copernican presentation and cosmology were "hypotheses" and that they could not make a claim at final truth. I think that that issue might, for our argument, prove to be especially significant, even when we want to leave the 16th century behind us and use it rather as a point of departure to think for ourselves about the claims to truth of faith and science. In this debate around Osiander, we have an historical document from which to start.

This having been said, I would like to point to two books that have not become as prominent in Great Britain and in this country as they are at the moment in Germany. These are the two books by Hans Blumenberg: one is called *Die kopernikanische Wende* (Frankfurt a.M., 1965), and the other *Die Legitimität der Neuzeit* (Frankfurt a.M., 1966). Because in Germany these play such a great part in the discussions about the history of ideas and are so generally accepted, I thought that it would be useful to say a few words about the views of this most learned scholar.

Parallel to Nietzsche's attack on the Christian emphasis on humility—you may recall Nietzsche's bitter comment about the "worm existence" that Christianity inculcates in many by stressing the ideal of *humilitas*—so Blumenberg in an exact parallel, without ever mentioning Nietzsche's name, claims that the campaign against curiosity is the main sin for which Christianity has to stand. In the Patristic period and throughout the Middle Ages, curiosity was a sin, and curiosity is translated by him with the word *"Neugier."* Is there not an English word with an Anglo-Saxon root for curiosity? Inquisitiveness—it still has a Latin origin.

In the Middle Ages this curiosity was sheer pride, and therefore no one could enter Heaven who stood guilty of curiosity. According to Blumenberg, you find the beginning of modern times when all those issues re-emerge that had been repressed by the campaign against vain curiosity. "Modern times" have started, thus Blumenberg claims, precisely at the very moment the heavens are freely investigated, whereas before it was not allowed because that would be the sin of curiosity.

For Blumenberg, the main culprit in this whole tradition is nominalism, which goes back to Saint Augustine. His argument runs somewhat like this: Because Augustine did not allow questions about the relation between our reality and the reality of God, and because Augustine said God made this world because he *wanted* to create a world in this way, therefore one has to accept this world as it is. And that is the end of all possibility of all curiosity, of all investigation, of all that makes man man.

Thus nominalism, which garners out the tradition of Augustine, is the end of the road, the end of thinking man. Therefore the real importance of Copernicus is that he broke away from that nominalistic attitude that would say, "The way the heavens are, we can't know. We can only accept that God wanted it. But you can't investigate it. We are in a world that does not allow any further investigation."

Blumenberg calls that Christian resignation. And against the backdrop of this Christian resignation Copernicus emerges, the first free and thinking man again, the symbol of humanism, and therefore the symbol of the beginning of the new modern times. That is the function and stature of Copernicus for Blumenberg and for a wide layer of the German intelligentsia at the moment. This book—*Die Legitimität der Neuzeit*—has now reached considerable sales, is read by students in all universities, and is on the desk of many of my colleagues, because it is both well written and well documented with its many footnotes—which in Europe, at least, is still an indication that it may be worthwhile to ponder it.

Here we have a real solid opponent to fight with and—though politely—fight with him I did, yesterday. Partly because it seems to me that here is a hidden attack on Christianity as such, just like you find with Nietzsche. Now, we are used to that; as long as Christianity is attacked with that much vigor, it is still taken seriously as an opponent. But in our context far more crucial and to the point: It seems to me that there are distorting elements in Blumenberg's efforts to retell the story of the history of science. Hence my emphasis on the importance of nominalism as freeing us from prescriptive metaphysics.

I do not know whether the German distinction—*Weltbild* and *Weltanschauung*—is translatable into English. *Weltbild* is what you can chart, what you can debate, what you can quantify, what you can prove. *Weltanschauung* is the metaphysical webbing. Now, I have nothing against *Weltanschauung*, if it is developed by induction, drawn out of your *Weltbild*. But when it is an *a priori* statement, claiming to describe how the world really looks, and when *a posteriori* you have to fit in your *Weltbild*, then it takes on that directive and coercive force out of which nominalism has potentially freed us.

Perhaps you will allow me to close with a personal comment. At the time of the space shot—with its statement "That's one small step for a man, one giant leap for mankind"—I was in this country on a lecture tour. Everywhere I went in academic communities, I heard the same kind of criticism: We have the inner city problems, we have our mess in the United States itself, and yet we are spending millions and millions to get up there—that is just nonsense and just pride. Pride! It is the Icarus story again. In academic circles, time-honored words are being used: "*Quae supra vos nihil ad vos*," a warning not to be concerned with the

The Apollo 11 astronauts lifting off on the first manned lunar landing mission, 16 July 1969. Neil Armstrong's words "That's one small step for a man, one giant leap for mankind," have become enshrined in the folklore of the space effort. (Courtesy of National Aeronautics and Space Administration.)

heavens, not to penetrate them. It is not realized that this attitude is the product of a very dangerous alliance of earth-parochialism and *space-angst*. It is in the name of this alliance that we are called upon to stay on this earth, to mind our own business, and not to be "waylaid," to think in terms of the place where we "really are." Yet—and this is to be said with all possible clarity—our only chance for survival is in space itself.

FERRÉ: I am interested in how a contemporary nominalist like yourself, with an enriched notion of nominalism, deals with the question of reality. I would like to come right out at your distinction between the *Weltbild* and the *Weltanschauung*. As I heard you, you were talking about the legitimacy of the *Weltbild*, the picture of the world that is verifiable, quantifiable, capable of confirmation, and measurement. You were even willing to tolerate the *Weltanschauung* that was inductively reached from the *Weltbild*. This seems to me to be unfortunately one-sided. Don't you, as an enriched nominalist, want to recognize that the treatment of language is creative and it involves critical expectation?

Nominalism, as I have usually used the term, has been related closely to the conventional school of interpreting what it is that we do with our language, rather than the realist interpretation of science, in which you are trying to elicit the laws and the universals that are somehow what is there. And so I am wondering to what extent you are interested in reality? How do you stand with regard to these fundamental questions of reference?

DILLENBERGER: I take it you thought it was an advance when one went from *Weltanschauung* to *Weltbild*. But *Weltbild* becomes another *Weltanschauung;* then one has a new problem.

OBERMAN: To answer first John Dillenberger's thesis about the relation of *Weltbild* and *Weltanschauung*: A thoughtful nominalist would answer here by pointing to the great advance that has been made at the moment that astrology became astronomy. In this transition nominalism was a major factor. The content of this transition is exactly the move away from *Weltan-*

schauung and the move towards *Weltbild*.

Every individual nominalist will have a lot of residual a priori *Weltanschauung* in the expectations he brings to his experiments. But his method is such that he can be critically asked to become aware of the bases of his *Weltanschauung*, and to unmask them as such. Here it is that the seeds lie for a rather optimistic view of progress: Every next scientist will be able to unmask the bases of the *Weltanschauung* operative in the conclusions or expectations of his predecessor and to discard them in the scientific results that have been made.

One of the very interesting reasons why I have become a nominalist is that I am living on the critical edge—and as I hope a "fertile crescent"—between an Anglo-American culture and a German culture. In the German culture, the lectures and the books that are written reveal a basic, all-pervasive confidence that language will be able to solve problems. Allow me a short illustration: You know the saying that we in Europe like to quote to ridicule American pragmatism: "If you build a better *mouse*-trap the world will come to your doorstep." But you can also put it the other way: "Build a better *mouth*-trap and the world will come to your doorstep." This is an equally valid caricature of the German confidence in *-isms*.

ANSHEN: I understand this collegium to be not exclusively a kind of historical exegetical study, learned and profound though it may be. We are at the same time confronting the tremendously serious problem concerning the generation following us, and the difference between the second law of thermodynamics, the law of entropy, which leads to chaos, and the law of evolution, or anti-entropy, which leads to organic life.

Our own era, at least the last two generations, represents a fundamental break with all past history in which the collective social processes appear as inexorable events ruled by pseudo-natural laws. I want to point to our rapidly growing ability to control the forces that throughout history hemmed in individual decision making. We have eaten from this new tree of knowledge, and what fifty years ago seemed to be *fate*, has become the subject of our deliberate choices. What is decisive is that the trend is irreversible, I believe.

Now, if choices are to be made, the first question is, on what foundation of knowledge? The second question, what are the sources, the criteria for our choices and judgment?

OBERMAN: Mrs. Anshen has alerted us to the fact that there is a way of dealing with history that does not allow us to deal with the problems of the next generation. For me, the excitement of my paper grew while working at it, because I saw how those two are related to each other. We cannot deal with Copernicus without taking a stand about where we find ourselves now, what our coordinates are.

It is on this borderline that nominalism becomes crucially important. In order to understand this, one has first to demythologize the concept of nominalism. Look for example at the works of Paul Tillich, eminent philosopher of religion. Yet he is the man for whom everything that was wrong and fallacious in the whole history of the thinking of mankind could be designated with the one short word "nominalism." In the hands of the philosophers, the word nominalism has become a war slogan rather than the designation of a historical movement that has been culled out of the sources.

I think that it is important to realize that nominalism "grew" with the times. Its historical roots lie in the realm of logic. But in the later Middle Ages it takes on a larger significance. At that time it starts to become a movement of considerable radiation, basically because nominalism became a factor in the Great Awakening, the awakening process of man. There were debates and rifts that split up the universities in the 14th and 15th centuries far more deeply than any split related to the student unrest today.

Father McMullin is already shaking his head at someone who can speak with that much fervor about nominalism. I do so with such glee, because I am a nominalist myself.

McMULLIN: I was fascinated by Dr. Oberman's paper yesterday. But I must admit that I find the role of nominalism in early modern science much more problematic than he does. First, a point of agreement. I fully agree that Copernicus was in no sense a revolutionary, except possibly in the *consequences* of his thinking, consequences of which, however, he could have had no conception. I don't see him in any interesting way as the initiator of the scientific revolution. One can locate that revolution more plausibly in the 17th century, if one does not decide (in the light of recent controversies) to avoid the notion of "scientific revolution" altogether.

Even if one were to confine oneself to the "revolution" in astronomy, one would immediately have to link with Copernicus four other people on whose achievements the completion of the "Copernican" system depended. He needed Brahe for his observations. Copernicus made virtually no observations on his own; indeed, he based his astronomy to a surprising extent on ancient results, some of them going back as far as the Babylonians. Second, instead of circles he needed ellipses, which Kepler provided. Third, he needed some way of uniting the physics of earth and sky; this Galileo made possible. And finally, he needed a dynamics to complete his work as *science*, and that, of course, Newton provided. So that Copernicus was no more than the beginning of a beginning, and to try to make of him a "revolutionary" thinker, of someone who quite altered the direction of scientific thought on his own, is quite wrong. That is the point of agreement.

But a major point of disagreement between us has to do with the significance of nominalism as a theory of science. The inheritors of nominalism in our own day were, of course, the logical positivists, the Vienna Circle of the 1920s and 1930s. But as a theory of science, logical positivism is dead. One catches a whiff of it still in law schools and psychology departments. But among philosophers, it has been abandoned, even by its former proponents. It died in the fifties; the reasons for its demise were many, but one of them (some would say, the most important one) was the failure of its nominalistic assumption about the relation of language and the world. The positivists assumed that this relation was an unproblematic one akin to naming, and that correspondence rules could be specified for even the most complex theoretical concepts.

Much of the effort in recent philosophy of science has gone to showing that this simply does not work. We do *not* know the "names" of things, nor do we assign "names" in a once-for-all way. Language is not a "given"; it has to be laboriously and tentatively constructed. And the growth of science is largely a matter of language modification, of controlled conceptual shift, something that nominalist accounts of science have proved poorly equipped to handle. If the important innovations in science are in most cases innovations in networks of theoretical concepts like energy or force, a theory of science that makes meaning a relation between name and particular and rejects any hint of a universal is faced from the beginning with intractable problems. Despite the most energetic attempts of Goodman, Quine, and the other admirers of the spare economies of a nominalist ontology, nominalism has had little success with the language-structures of modern theoretical science.

But getting back to the historical issue, it seems to me that if there is one antecedent that Copernicus could *not* have had, it is surely nominalism. There are scientists in the 17th century who can plausibly be linked with some aspects of nominalism, Descartes, for example, or Boyle. But Copernicus, no. As several speakers pointed out yesterday, Copernicus harkened back to the Pythagoreans. He thought of his work as a return to older sources, sources much older than, and indeed quite antithetical to, the *via moderna* debated in the schools of his day.

Copernicus was a powerful mathematician rather than an observational astronomer. Not more than about 30 contemporary observations are cited in the entire *De revolutionibus*. There is no suggestion in that work that he saw induction or observation to be at the heart of some sort of new scientific enterprise. If, therefore, one wants (with Dr. Oberman) to propose nominalism as the antecedent for a new science characterized mainly by inductive method, then the break between the old and new would not come at Copernicus. Conventionally, one would think of it coming at Bacon or Galileo.

But even there I would want to enter a *caveat*. I am not sure whether the scientific "revolution" of the 17th century ought basically be regarded as a revolutionary change in *method*. Even if it is, it certainly cannot in any simple way be characterized as a changeover to inductive method. Galileo himself, in the *Dialogue on the Great World Systems*, significantly picked as his *opponent* an empiricist, someone who insisted on the ultimacy of observation as the means of warranting scientific statements. Recall the famous passage where Galileo's spokesman puts forward his claim as to where a body falling from the mast of a moving ship would land. When his opponent asks, "Have you ever performed that experiment?" Salviati replies, "Without experiment, I am sure that the effect will be as I say, because it *must* happen that way." Replies of that kind played an important part in 17th-century science. It is true that the greatest figure of that science, Newton, often characterized the methods with which he arrived at his results as "inductive." Yet as we look at them today, we cannot altogether agree; they are diverse and complex, and their antecedents are equally diverse.

I want to suggest, then, that the role of nominalism is an equivocal one in our discussion. It is equivocal first because nominalism had virtually no impact on Copernicus. It is equivocal also because links between the nominalism of the 14th century and the physical science of the 17th are quite difficult to establish. Where would one look in Galileo's work, for example, to find such traces? It is true that Bacon, Boyle, Newton, and the others stressed the role of experience just as the nominalists had done, and that theology was influential in their thinking generally. But to what extent was their emphasis on experience a consequence of voluntarist theology, to what extent can their departures from Aristotelian science be construed as a rejection of a theory of science as necessary truth, in which God's freedom to create the world He chooses is compromised? We cannot, in the present state of research at least, give the confident reply that Dr. Oberman does.

One last point: Dr. Oberman quite rightly stresses the great hostility that the nominalists displayed toward metaphysics. Yet it would be important, I think, to recall here the point made so often by historians of science from Burtt onwards, that a grasp of the metaphysical background of 16th- and especially 17th-century science is absolutely essential to our understanding of it. When Copernicus asserts that the sun *must* be the unmoving center, he is not just making an inference from astronomical data. The influence of neo-Platonic "light metaphysics" is clearly pervasive in his thinking. To speak of the metaphysical *foundations* of modern science may be too strong a metaphor, but it can scarcely be denied that metaphysics of the sort rejected by the nominalists played a powerful role in the shaping of the new science. This would once again suggest that we should be wary of claims that nominalism itself was a determining influence in this process. Dr. Oberman's capacious "nominalism" needs dissecting into many different strands. The central strand of historical nominalism (a specific theory of language) will, I think, turn out to have had very little effect on early science. The other strands (notably the empiricist ones) will be found to derive, in important and complex ways, from other philosophic traditions beside the nominalist one.

OBERMAN: First a word to your thesis that nominalism has found today its heirs in the logical positivists and that, therefore, it is a dead horse. As a historian, I would be inclined to say that to the same extent as you are right in claiming that there has been no scientific revolution, it is just as true that there are no dead horses in history.

Secondly, you have declared that the horse had to die because, as we now know, there is no language of concepts available to us. I regard this as one of Ockham's main points; his point was exactly that we cannot presuppose a language of concepts that refers to and jibes with reality, that therefore we have to analyze critically the language we use. Language is not "given," it is not a supernatural, or metaphysical pre-existent entity.

Your third point: nominalism is not the antecedent of Copernicus. There I completely agree with you. That also marks the relativity of Copernicus' importance. Here I associate myself directly with what John Dillenberger said. To a certain extent Copernicus was falling back to a pre-scientific level of thinking, exactly to the extent that he did not relate and agree with nominalistic achievements before him.

Fourth, the inductive method is not the whole story about nominalism; you do not do justice to the whole enterprise of the nominalists. Another aspect is the emphasis upon imagination and to the importance of making mental experiments, forecasts or "expectations" of experiments. Particularly in astronomy, because the heavens are so far away, imagination has to fill the gap between us and the reality that we want to test. The nominalists discovered that sheer experience will fool you. When Copernicus used to say that the sun only *seemed* to rise, he used results that the nominalists had worked out in warning that sheer empiricism will mislead you—and there Galileo stands in the same tradition.

Finally, look at the conclusions of Anneliese Maier. Though in the school of Étienne Gilson, and hence inclined to be rather critical of nominalism, she has to agree on the sheer weight of the evidence that in nominalism you find the first stirrings of what we call "modern science." I think all of us would have to acknowledge that Anneliese Maier has worked with the sources with an intensity and precision that no one can easily equal. She has no ax to grind when she sees in nominalism what one may call dramatically the birth of modern science.

MERZBACH: I suppose I really must be, if not a devil's advocate, at least a mathematician's advocate. When we talk about Copernicus, we are talking about that which we know on the basis of his book and a very few other writings, and I wonder why we don't take him by his word?

Now I do not think taking him by his word means putting him into a pre-scientific position, whatever that means. It does mean putting him into the position of a mathematician. When he cites Pythagoras, he does not go into an involved philosophical business. He refers to a tradition, presumably the one to which he was exposed in Italy. But what is it he says basically? He says, "Everybody is talking about the Ptolemaic system, and yet nobody really seems to have looked into the technical details of it. I have looked into it and I don't like what I see. So therefore I am going to go back and see what the alternatives are." Then he cites the ancient authors, but he states definitely that he feels he has a better mathematical system, a better explanation of the world machine. I really think all of this is rather far removed from the philosophical and theological traditions that we have been discussing.

I have always thought, in support of this argument, that it was rather significant that for purely reasons of technical convenience, Copernicus does not put the sun exactly at the center. He can account for the phenomenon more neatly if he just moves the center over a bit. Now that seems to me a terribly mathematical, technical, engineering type of way of handling the thing.

OBERMAN: We find the invocation of the Pythagorean tradition in the dedicatory letter where Copernicus says, "Shouldn't I have the freedom of research that earlier generations have had in disagreeing with Ptolemaeus. At that time it was possible to disagree on the movement of the earth. Why not today?"

In the first place, I think this is an argument to show that he is not an innovator. In the second place, I think that this appeal is pre-scientific. Rheticus would later write to Ramus, "I don't want to look any more at the classical authors and authorities. I want henceforth to look at reality." I think that is a step further and a significant advance. It is an earlier stance to look at your sources instead of looking at the heavens themselves.

Finally, all the achievements of the great nominalists of the 14th-century Parisian school were being published and were available to form the climate in which Copernicus was soon to be introduced.

MERZBACH: But I am suggesting that for the practicing mathematician, the discussion of method may be secondary to doing something with the system. Isn't this precisely what Copernicus is saying? He is tired of people talking about the thing, without getting into that hard work of those circles . . . . As a result of everyone's being upset with Osiander, we tend just to play on that.

MAGUIRE: Dr. Oberman, I want to put two questions that follow the exact form of your presentation yesterday, which started with a fascinating litany of contemporary problems, and then plunged back to the past and finally came forward to the present again in discussions based on student unrest and space.

I wonder if I properly sense the presence of Karl Barth in your remarks yesterday, particularly the Barth who first appeared and was called a Christian nominalist. You so stressed the absolute transcendance of God, that it does give you the antidote to *space-angst*—everything is ultimately all right, God is really in the heavens, totally uninspectable. That then relativizes and renders everything else fragmentary, but it invites you into the inspection of everything, including those instances where God makes himself known in the world. I am just interested to the extent to which those early Barthian modes of theological thinking are reflected in your own current thinking, and whether it would be fair to call you a Christian nominalist in that sense, that is, the absolute transcendence of God, the utterly relativized character of everything else and, therefore, the invitation into the study of everything else. If so, that would certainly be sympathetic with the spirit of modern science. That is question one.

In your talk yesterday I became acutely aware of this enriched or expanded use of nominalism when you discussed the imagination and a lot of other things that we have not typically associated with that nominalist tradition. I began to wonder whether there was a telescoping going on in which a concept of experience that got rounded out in much more complete forms 150 to 200 years later is symbolically being represented by Copernicus. It is the use of experience in that rich sense of imagination, feeling states, all of what is later going to be associated with extreme forms in the Romantic tradition, which is somehow getting read back to this period at the beginning of the 16th century.

And so my other question is whether the key to this enriched concept that permits you to call yourself a nominalist may really be an appropriation of the experiential tradition and *calling* it the tradition of nominalism?

OBERMAN: You are asking that wonderfully "private" question that would not be so readily asked in academic circles in Germany: Where do you stand yourself? How do you tick? Does that explain what you have been trying to convey, or perhaps how you have been distorting reality?

I would be inclined to say that I am first and foremost a historian. Yet, I am able to reconstruct those particular phases in history best where I feel empathetic. Here it is that the Barthian tradition may come in. There are some points where I find myself in great sympathy with Barth, for example where he emphasizes the transcendance of God, provided this idea is applied in the sense in which the nominalists did it, by putting "the Heavens," as part of the whole of creation within the same mathematical world that can be penetrated by our scientific thought. Confusion results—and mythology starts!—when God is understood "metaphysically," instead of seeing in God the absolute transcendance, leaving us no other ladder to his "Heavens" than his revelation.

I think you are entitled to know where a man stands, to be warned against his biases. But do not discard, therefore, his historical analyses, his research results, because he may be in an advantageous position, exactly because of particular points of view that others have not been quite able to see.

Secondly, this "imagination" concept is not a reconstruction, a telescoping of history or a "plugging in." I got it directly out of the sources. If you had asked me this a year before, I had not noticed this, though I had been working for years with nominalism.

This imagination theme is more on the level of Jules Verne—the science fiction writer, the man who conceived of the possibility of a submarine and of space flight. There is a level where science fiction becomes scientifically relevant. It is there where you start to think in another dimension and start to play around with mental experiments that cannot yet be carried out. They are suspended in judgment until they can be tested, but they allow me to direct my research to other questions than the ones that are prevalent in the society I am now living in. In that way there is not only the transcendance of God, but also the transcendance of the *status quo* in scholarship and its prevailing research axioms. I think that is the importance of "imagination" in the realm of science.

MAGUIRE: I would say, Heiko, that's a scholarly project that might very well issue into a fairly thorough-going reinterpretation of nominalism. Your talk suffused the nominalist tradition, if I may put it that way, with this power of imagination, with a strong experiential component, which then also issues into forms of modern scientific experimentation.

Now, that, I must confess, is something quite beyond what I had associated with the nominalist tradition. If all that is in those sources, then I would think the next move in the scholarly project would be to exhibit them for us and tell us that story in ways that force us to reconceive and stretch our conception of what the tradition is.

OBERMAN: May I ask you to take the footnotes of my paper very seriously, as much as the text. I really tried to document everything that I brought in the text.

DILLENBERGER: I have had the advantage of looking at the footnotes. But the question is not, was this in nominalism? Oberman's evidence convinced me. He resurrected the material, but now my question is, what happened to it? Was it influential? Did it not die out as an influence?

MAGUIRE: Ernan McMullin gave a pedigree of nominalism, which petered out in forms of 20th-century empiricism. But if it comes to expression later in the Romantic movement, and not into the dead-end that Ernan described, that is important for us to know.

OBERMAN: I think nominalism has for a time gone underground. The force of the nominalist vision is that it is able to keep theology and science together in two powerful hands, but two hands of *one* body. Then along the lines of the Reformation, there has been an effort to make the Reformation discoveries acceptable by reintroducing Aristotelianism, and on the other hand, a scientific era emerges that discards theology altogether and makes the *Weltbild* into its new religion.

Only when we can bring these two together again in a consciously enriched nominalism, will we be able to relate and solve the problems that we are confronted with here today in the relation of faith and science.

GILKEY: I am very impressed with your defense, Heiko. I am interested now in turning it over and asking about the radical distinction you implied between theology and this very rich form of inquiry that you developed out of nominalism. I have the feeling that both you and John Dillenberger have enriched nominalism to include what we would call the scientific methodology, and even a good deal of phenomenological use of understanding of language, symbols, and their dialectical interaction with experience.

And then I began thinking: that is what I mean by theology. But you have got theology up there somewhere, dealing with truth, whereas this is only dealing with accuracy. When you accept the radical influence of historical relativity with regard to theological language, the relation of symbol to conceptuality and experience, I wonder if you can hold this very strong distinction between Heaven and earth, or between theology and your new enriched nominalist feeling? If one says the distinction cannot be held, that in their different ways theology is much like what you just described the nominalist methodology to be in science, then can it remain as antimetaphysical, anticonceptual?

Don't you really get back, it seems to me, not so much to Barth as to Tillich? I think you have subverted that distinction between truth and accuracy by the ability with which you defended yourself against the attack on nominalism. I do not understand what you mean by truth versus accuracy, once accuracy has almost disappeared into the rich search for symbolic language that will be an interaction with experience. That begins to sound like theology to me, and even metaphysics.

OBERMAN: Let me try to answer that in my halting fashion, because I am pushed a little bit to the limits of where I feel historically safe, where I have footnotes and predecessors who have answered this exact point.

Accuracy is the symbol for the truth that I can find by my own investigation, by following my own curiosity. Truth, as I used it yesterday, in the context of theology, is the verity that I can only find where God has been willing to expose himself. I cannot penetrate beyond what God has made known about himself. That is what I tried to say about vain curiosity that works itself out so differently in the sciences—when it is appropriate—and in theology, where you try to penetrate the mysteries of God instead of accepting the limits of what we know about God, what He has made known about Himself.

I think there are, therefore, two very different avenues of approaching research in theology and in the sciences, a difference often not respected. In theology, we have on the one hand turned pietism into a dirty word, and therefore put a cloak of suspicion over all the experiences of the inner life. On the other hand, we have felt that our theological appeal to God's own revelation is so fragile that we have to base it on all kinds of natural theology. There we have metaphysics back again. Then we try to use the results of our accuracy to support the truth of theology, and that is, I think, what has been so confusing for us.

GILKEY: Well, I agree with you fundamentally that in both areas, the scientific and the theological, one is dealing with a given reality, and that one does go to experience in revelation, common experience, and so forth. But the demands of contemporary theology are such that this imaginative attempt to restate the symbol, puts us in somewhat the same position vis-à-vis the attempts of the scientist to restate what the new richer interpretation of experience is.

I have a feeling that what you imply in the theological realm is the a priori language you do not want in the scientific realm, whereas I am thinking of theological language as having that imaginative and groping character.

McMULLIN: I am troubled by a suggestion that seems to underlie a lot of what has been said here, that one can accept what was originally a nominalist way of separating, and holding apart, the realm of God and the realm of man (or nature). It makes me uneasy to hear metaphysics and theology and the humanities described as the domain of "truth," with science characterized, by contrast, as the domain of "accuracy." That sort of compartmentalization, with its unmistakably positivist overtones, I cannot accept.

Let me go back to Augustine, whom you mentioned yesterday many times. Augustine is one of the sources for this whole discussion of *curiositas*. The entire thrust of his writings on creation is to see the world as a sign of God. He constructs an extremely elaborate semantics, in which the quest into the structures of nature is seen as a quest into the structure of creation itself. Augustine construed nature as the outpouring of God's omnipotence, an outpouring which is most amenable to man's power of inquiry. He is quite explicitly defending the propriety of human inquiry; it becomes *vana curiositas* only if the "sign" character of its object be lost from view. Natural science thus becomes a pursuit of God's truth in the real to which our senses give us the most immediate access.

In the nominalist tradition as Dr. Oberman interprets it, the goal of natural science is not truth; it is something diminished, something to be contrasted with the deeper truth of revelation, the deeper truth of metaphysics. There is more than a suggestion that man by his own effort cannot arrive at an insight into reality, but only at a kind of peripheral account. The Divine "illumination" that Augustine in his theory of knowledge saw as aiding man in his attempt to grasp the structures of God's world no longer plays any part. This world is now a collection of opaque particulars, admitting no more than instrumentalist generalizations.

If this is a fair interpretation of nominalism, I would want to stress how remote from this the realism of 17th-century science was. And moving up to the present, I would also want to recall the arguments brought by contemporary philosophers of science in favor of a broadly realistic understanding of science. Take the constructs of the contemporary scientist, a concept like the gene or the chromosome or the DNA spiral molecule, for example. Ought one regard these as authentic insights into the real? Or are they simply to be described in some sort of conventionalist categories, as fictional or as accurately predictive devices, products of the nominalist imagination?

OBERMAN: What can be reached is accuracy in describing *quae supra et circa nos,* our cosmos. It only becomes truth when it also can speak about *qui supra nos,* the origin and the "originator."

How did this whole cosmos come about? Are you satisfied with the big-bang theory? Soon enough that can be accurately described. At the moment it still needs a lot of imagination; it is waiting for verification, and I may still be alive when that verification is found. But all this is still on the level of accuracy. It describes, as Augustine would say, the creation *(creata)*, not the Creator. That demarcation line makes it impossible for me to have any confidence in natural theology.

McMULLIN: Can there be any truth about *creata* at all, then, other than revealed truth?

OBERMAN: Yes, "this truth I hold"! I assume with my whole existence and vocation that it can be. As you know, my words "accuracy" and "truth" are symbols to distinguish different aspects of reality. It was a "Renaissance fallacy" that we came to believe that man can unravel all aspects of the universe. Here humility has its scholarly place, *humilitas* as used by Thomas Aquinas for the man who stays within his own limits. This ideal is in the process of being transferred now to the modern university, but our students fight against it. They are disappointed when they come into our classes and we do not offer all aspects of reality. Whereas they are, understandably, hungry for truth, we can give them—at best—"only" accuracy.

WALSH: How then are you not using truth equivocally, if it applies to both, the symbol for the Creator and the Increator? What is the commonality of the terms?

OBERMAN: That's why I distinguished between accuracy and truth. That's why I tried to give distinctive references by saying the one (accuracy) refers to the *quae supra nos* and the other (truth) to *qui supra nos*. It is very important to see that this creation is a result of a person, rather than the outflow of a being. I can't put it more philosophically than that.

"*Creavit quod voluit. Creat quod vult*": God created and creates according to his will. That is not the tired end of the road, as Blumenberg saw it. That is not the expression of a philosophical "know-nothing" movement. It is the acknowledgment of the boundaries of our quest for accuracy. There is a decisional aspect to our creation that does not allow us to penetrate beyond the original decision, to enter into the mysteries of God himself.

ANSHEN: May I just quote from an ancient, hieroglyphic tablet from the 17th-century dynasty in Egypt?

"When man will have learned the secret of what moves the stars, the Sphinx will laugh and life will be destroyed."

HAMILTON: That's an interesting note to close on.

Wednesday, 25 April 1973

William Pollard, *Moderator*

POLLARD: Our assigned topic this morning is "Methodology of Science and Religion," a question of scientific knowledge and theological knowledge. We will ask how can we have knowledge in either, though most particularly, how can we have knowledge in religion?

We have for the first discussion of this topic Langdon Gilkey, one of our leading American theologians, Professor of Theology at Chicago. He has written extensively and, to me, very persuasively on a variety of theological topics. He was teaching in China during World War II and got interned by the Japanese for two and a half years when they took over China.

Then, before we start our discussion we will hear from Frederick Ferré. He specializes in both the philosophy and history of science and religion, is Professor of Philosophy at Dickinson College, and travels around the country in his own plane. Later Professor Heisenberg will join our discussion.

GILKEY: As you will discover, Fred Ferré, the other commentator, has nerve, and I don't. That is to say, Fred waited to hear what Professor Heisenberg was going to say, and then worked out his remarks. I didn't have that kind of courage, so I worked out my remarks totally ignorant of what Professor Heisenberg said yesterday afternoon. However, I have addressed myself to the question of the tradition of science, which was his theme. My remarks could be subtitled "Reflections on the Identity Crisis of Science from the Perspective of a Venerable Ex-Queen."

THE STRUCTURE OF ACADEMIC REVOLUTIONS

GILKEY: The history of academia presents us with many surprising aspects. One particularly relevant to our subject, the tradition of science, is that of palace revolutions—to borrow a phrase from Professor Kuhn—queens that rise to dominance and then are unseated and replaced by usurpers who in turn become queens. For each discipline within the court intrinsically and apparently inevitably feels itself capable—and wishes to be—queen: that supreme discipline which provides the basic principles for the self-understanding and the working out of all the other disciplines. For me as a theologian, this is a somewhat nostalgic account hardly relevant to the present status of theology in the world of the intellect! For science such an account may possibly illumine some of the sources of her current anxieties and the crisis of identity that she seems now to feel—the crisis, incidentally, that was set before us very vividly yesterday afternoon in the various addresses.

A glance at the history of these court revolutions shows that two essential characteristics are required for a discipline to succeed in grasping and holding this regal role of queen: (1) The first is what we might call *autonomy;* independence from other intellectual disciplines in her fundamental theorems or postulates, in her methods of procedure, and in her data, and thus, by implication, freedom from dependence on other aspects of culture. To be a queen, the supreme principle by which all else is to be understood, a discipline cannot be a variable in relation to other modes of inquiry, a function of the work of other disciplines; for then each of the others would be equally primary and none would be queen. Rather a discipline must be "a se," founded on herself and her own work alone, so that only what *she* produces is determinative for her conclusions, and through that of the work of other disciplines. (2) The second requirement is that the discipline in question produces the kind of knowledge that is, in some way, of *ultimate value* for the cultural life over which she reigns; she must possess a recognized power to serve culture uniquely, as nothing else can, and with regard to that culture's "ultimate concern."

In the past, especially in the Medieval period, theology was queen because of these two factors. She was autonomous because her own fundamental postulates and principles, her methods, and her data were not derived from either nature or culture, or the various disciplines that studied them, but from divine revelation and divine tradition, which she alone could interpret, understand, and formulate. Other disciplines: physics, astronomy, metaphysics, ethics, and law could help her, to be sure, but only in the role of "handmaidens" since they did not control either these sources of primary knowledge or the ways of getting at them. Correspondingly, she provided the culture with the knowledge considered of ultimate value—knowledge concerning ultimate salvation. The supremacy of theology was dependent on the fact that both formally and materially her truth was sacred, divine and eternal truth, and so independent of and uninfluenced by other forms of truth, which thus had a lower logical status and a vastly lower value.

Since the 16th and 17th centuries, natural science gradually has usurped this role, upset the queen mother and herself taken the throne. Again the same two factors, albeit in instructively different forms, explain her success. Increasingly she insisted on the *autonomous* character of her work, separating herself from metaphysical foundations, confining herself to the given, "objective" data alone, and understanding her thinking as purely logical, that is, mathematical and inferential alone, as the work of a *neutral* mind on *objective* experience guided only by recognized canons of logic; and consequently as productive of perfectly clear, exhaustively definable, and essentially verifiable concepts. This familiar description of scientific understanding may not have been accurate; but the *role* of this description is clear enough—as is the reason for its present *passionate* defense: namely to preserve the absolute autonomy of scientific inquiry as based on herself and her work alone and so as independent of other disciplines and the factors *they* study. The autonomy that theology achieved by means of divine revelation and so by a truth transcendent to history and culture, science has achieved by the claim to a truth gained solely through the neutral—in that sense transcendant—intellect working on objective facts, and so a truth separate from the political, economic, personal, psychological, social, and historical pressures of culture. As the transcendence of theology was expressed in the eternal and changeless status of her dogmas—while other truths were historically relative—the separation of scientific truth from cultural ups and downs was expressed in the *objective* claim to embody a purely *cumulative* tradition, likewise unrelated to history since nothing but the data and *logical* inference affected the development, and since only the *last* stage need be understood in order to understand all the rest.

Secondly, natural science achieved this role through offering the kind of knowledge that increasingly the culture found to be of ultimate value: the knowledge that leads to control over nature and over nature's forces—technological knowledge. It might also be added that both queens have in their time succeeded in living well: The Church ended the Medieval period owning one third of Europe, and science's hold on corporate and governmental grants and funding have been the envy of us all!

In both cases another tendency manifests itself, namely the urge of *other* disciplines to accept the tutelage of the new queen after the coup d'état. In the Medieval period all other disciplines —natural philosophy, social theory, metaphysics, law, the humanities, and the arts—understood themselves according to fundamental principles generated in theology. Correspondingly in the modern period, all disciplines that could do so fell over themselves in their eagerness to model themselves on the new queen, as "science," that is as empirically based on objective data, and as guided by mathematics and the logic of inference alone. Those that could not convincingly so interpret themselves were hardly allowed in the new court! All have, however, valiantly tried: even psychology, literary studies, history, philosophy, and, *mirabile dictu,* theology and religious studies! Not unsurprisingly, the revolutionary forces—again in both cases—have appeared from those disciplines—or that discipline—which could not be comfortably fitted into either model: history.

For in a strange way it has been history, and its challenge to the absolute autonomy and unambiguous value each queen claimed to represent, that has unmasked both queens and revealed their common, ordinary, *human* status—and *that* has been very hard for both of them. The dawning realization through historical inquiry that the divine sources of theology, in Scripture and tradition, and so its own theoretical expressions in dogma and doctrine, were *historically relative,* compounds of a variety of cultural forces—economic, political, social and religious—has stripped theology of its claim to absolute autonomy and so to transcendent authority. Since the 18th century, theology has been regarded, and seen itself, as a cultural discipline among others, subject to influences from other aspects of culture, and thus a variable of forces outside her own preserve. Also, awareness of the vast ambiguity of religion's legacy in history— in greed, intolerance, confessional wars, immoral morals, and oppressive politics—has eroded its claim solely to provide an ultimate and unambiguous salvation to men—and men have turned to other saviors.

This dethronement of theology through history may shed light on the present identity crisis in the tradition of science. To take the second point first, historical awareness: The awareness of the present results and future prospect of technology has eroded the claim of the new queen to provide saving or healing knowledge. For in controlling nature, we have desecrated it and our world; and an expanded technological control threatens—or seems to threaten—to dehumanize man. Thus have those other disciplines not subsumable under science appeared recently as again significant in academia: speculative and normative social theory, ethics, religion, literature, and the arts.

And it is thus, I would argue, that we should interpret both the value and the threat of a *historical* interpretation of science, an interpretation that sees science as *not* autonomous or independent in its historical development, and so not as pure or as simply "cumulative" as is a logical progression. Rather for the *historian,* if not for the logician, science appears as influenced from top to toe by religious, philosophical, and social ideas, as generated by strange, often sudden and even irrational intuitions, and so as resulting in a proximate system of symbols more metaphorical and poetic than directly logical, empirical, and verifiable. Whether David Strauss in the 19th century was ultimately right or wrong in his *historical* interpretation of the New Testament is not important; what is clear is the vast significance of his role in showing the historical relativity of the foundations of theology—and thus is it also clear why the dogmatic theologians detested him!

In like manner whether Thomas Kuhn was right or wrong in his theory of scientific revolutions in the history of science is not so important as that more than anyone else he has exploded the myth of the absolutely autonomous, and so nonrelative, nonhuman, character of scientific truth and its development. Much as Michael Polanyi and Feyerabend have also done, he has shown science to be one human discipline among others, influenced by other nonscientific factors in personal and social life, in philosophy and in the arts, and determined at crucial points by nonlogical aspects of human creativity. Thus has he also been resisted so *passionately* by the scientific community; he threatens to unmask a queen, to topple an idol, to render relative that to which we had given an ultimate devotion.

One may note that in revealing the humanity, i.e., the historicity and relativity of a discipline of inquiry and of its results, one does *not* thereby establish the nonexistence of its referent. As both Professors Heisenberg and Bohr make clear, the approximate, metaphoric, and partial character of scientific symbolism does not imply the denial of the reality or objectivity of nature; nor does the relativity of theological symbolism imply the denial of its ultimate referent.

These anxieties about usurpation are well taken. A dethroned queen, as theology has found, has a precarious life at best at the academic court. Her autonomy now gone, she seems merely a quack and charlatan, to be interpreted solely by other disciplinary viewpoints; and no one now knows what *good* her sort of knowledge does to the academy or to the race. In recent days we have seen science, too, scorned by many as a delusive abstraction from reality, creating a mechanistic and objective world in which man loses himself. And we have also seen science subordinated to the viewpoints and goals of other disciplines, especially the economic and the political. The frantic fears of many scientists and advocates of science that if the humanity, relativity, and ambiguity of science are admitted, all will be lost, are thus *not* idle fears. But theology has had to learn the hard way that she does not have to be queen in order to exist and to be herself as one discipline among others. So possibly the dethronement of science as queen, if it should happen, need not be the signal for her demise but the signal for a growth in her creative and her human wealth.

Allow me then in conclusion to suggest a more democratic, nonautocratic interpretation of academia and so of the scientific tradition. Academia, the world of the mind, is a community (not a court) of *interdependent* but *relatively autonomous* disciplines. Their interdependence or relative dependence on one another—so that none is absolutely autonomous or queen—springs from their common *human* character, as activities and creations of man. And man is in some sense a *whole* of interdependent faculties. How he knows is determined not only by the data and the logical rules of inference, but also by his prior view of the world as a whole and of his relation to it, and by the ways imaginatively he can conceive what he finds there. Hence even in the most methodical cognition historical presuppositions, ontological symbols, imaginative models and paradigms function importantly in his most precise knowledge. His cognitive work is thus always "theory-laden," suffused with and structured by attitudes and symbols drawn from the rest of his life. The actual influence of such presuppositions and symbols on science—the influence of *extra*-scientific traditions—has been historically documented over and over: of Platonism, Stoicism, and Aristotelianism; of Jewish and Christian doctrines and attitudes; of Enlightment and Romantic ideas of development in history, in morals and in social institutions—and now of technology with the model of "programming." It is always *man* who knows in science, not pure, logical mind; and man is *historical,* subject to tradition and the language and symbols given him by tradition, a *total* historical and social tradition—not just that of his white-coated predecessors in his area of research! He knows what he knows within the terms of that total tradition, within its fundamental horizon. This interdependence of the disciplines in academia, that none can be fully understood without the others, that science *itself* cannot be understood without a close look also at history, philosophy, art, religion, and social theory, expresses the unity of academia, founded on the mysterious unity of man as knower, intuiter, believer, imaginer, doer, and hoper.

The relative *independence* of the traditions of academia from one another is based on the variety of dimensions within human existence, and the variety of aspects or facets of man's own being. There is a world to which man is in intrinsic relation, and he wishes to know and understand that world—hence science. But man is also a knowing subject, and that subject is in turn formed by his social world: hence the humanities, psychology, social theory, and phenomenological and transcendental philosophy. He lives and knows within an ultimate horizon shaped by his most fundamental intuitions and expressed through his most basic mythic and speculative symbols. Out of this variety of dimensions in his world and in himself the various disciplines of academia spring. All of these interact as he does anything he does: knowing, loving, believing, acting, or hoping. Hence each discipline is inescapably involved in understanding any of the others.

The tradition of science is thus as complex, as multidimensional, as is man the knower and the world he seeks to know. For this tradition is a *human* activity, and it is borne by a *community of persons* with a common commitment to inquiry and to the truth. On the most basic level this tradition and the community that embodies and bears it are sustained by their shared and so relatively invariant goals, standards of excellence, and criteria of validity, by, that is, their common "mythos" of symbols. Secondly, this tradition is characterized by changing or variable fundamental notions, models and paradigms through which the community of inquiry expresses its understanding and orders its specific modes of inquiry and testing. Here there is variety and change, even possibly "leaps" and "revolutions," in ways of understanding, modes of inquiry and methods of verification; and here there is clear dependence on other aspects of cultural life. Thirdly, there is the invariant constancy of the referent of inquiry: the real world which we seek to know in our varying ways. There is as well the constancy of man's perceptual and rational faculties however he may chose to utilize them.

The mythos of the scientific community, and their commitment to it, form one major aspect of the relative autonomy and independence of this tradition—and yet interestingly these are precisely the most personal and inward elements there. For commitment to these shared symbols holds each member spiritually or inwardly fast to the *objectivity*, which is their common goal. Out of this personal factor arises the fidelity to the data and to the rules of valid thought that is the major ground of the autonomy and independence and *glory* of the community. But the forms of that commitment, the ways the data are interpreted, and the symbols used for understanding and communication are *historical:* variable, relative, and cultural—as is all of human thinking. Ironically, the spiritual, inward, and personal, is what is invariant to the scientific tradition; the intellectual and logical is what leaps and jumps in this tradition. There thus is no absolute autonomy here, no simple accumulation, no unambiguous service or salvation for man; for science is human not divine, a citizen and not a queen. But no tradition and no community in history expresses more clearly the wonder of human integrity, responsibility, and courage; the magnificence of human imagination, insight, and clarity of vision; nor so well incarnates whatever hope we may legitimately place in human powers to understand our world and to deal creatively with it.

DISCUSSION: THE NATURE OF SCIENTIFIC KNOWLEDGE

Ferré: I would like to give some footnotes to Professor Heisenberg's address yesterday. They are organized along the lines of his remarks; that is, in terms of how traditions relate to the *problems* of science and religion, to the *methods* of science and religion, and to the *concepts* of science and religion.

His position on the relationship between tradition and the problems of science was at first hearing very attractive, it seems to me; but I would like to raise one or two questions. In what Thomas Kuhn calls normal science, what counts as a problem for any given generation of science is normally defined in terms of some overarching understanding of the nature of things. In the vision of science that Kuhn gives us, we see an important distinction between the degree to which the problem faced by a given scientist is just the derivative of a paradigm defining this status in normal science, or to what extent there may be discontinuities, what Langdon Gilkey has just now referred to as leaps and jumps.

Discontinuities can occur when the scientific community itself redefines what counts as a problem, through the shift of the fundamental substructure of the community's self-understanding. To give just one example, notice the shifting back and forth as to whether or not the attempt to explain the gravitational phenomenon counts as a scientific problem. In the Cartesian model of physics, the vortices try to give some concrete mechanism whereby we can understand the gravitational phenomenon, but it is displaced by admiration for the Newtonian scheme, which explicitly rejects the attempt to give hypotheses that would account for gravitational phenomenon. After the Newtonian scheme becomes the paradigm of what counts as the problem, the issue of pursuing what it is that causes gravitation is no longer a scientific activity, but rather something having the odor of the occult about it. The *hypotheses non fingo* remark of Newton served therefore as a kind of barrier against solving what had been a problem not long before. Today the shift back through Einstein's general theory of relativity makes it once more a scientific problem to try to understand what is behind gravitational phenomena. More recently, the University of Maryland and other experiments in gravitons and gravity waves are introducing, once more, the notion of this as a problem.

I simply raise this as a question, or a comment, on the possibility of discontinuity, so that we do not look exclusively at the problems of the scientist as being somehow completely determined in a particular moment in history. There was a note of a sort of fate in the remarks of Professor Heisenberg yesterday, that we simply find ourselves thrown into history and there are our problems, something about which we can do very little. There is truth to that, but there is also the recognition that fate operates not always in continuity. Sometimes there seem to be creative disjunctions in what counts as formulating problems themselves.

The religious situation strikes me as having certain analogies. What counts as a theological problem is also normally tradition-dependent. I think that there are some dis-analogies as well, because in theology, it is notoriously the case that there is no field-defining paradigm. There are schools and movements in theology, however. Within these schools there is a kind of fatedness about what counts as a theological problem; but again, traditions are not themselves finally without some possibility of creative disjunction. We all remember the revolutionary problems that were created—not merely inherited by tradition—by the enterprise that St. Thomas is remembered for, of trying to understand Christian imagery, the central symbols of faith, through Aristotle's categories of understanding. Or other revolutionary traditions, such as Rauschenbush's attempt to understand central concepts or symbols of faith through their social applicability.

There's one further footnote I want to make about the question of tradition and problems. Professor Heisenberg mentioned yesterday that sometimes people may find certain problems are exhausted and move on. It seems to me there are two very different notions of exhaustion at work here. In one notion, to be exhausted means to be emptied of promise. A problem is no longer perceived as giving possibilities for interesting new discoveries. The other sense of exhaustion, having to do with the young people and their attitudes toward science these days, has the notion not merely of being worked out, but rather, of tired out. This notion of exhaustion raises squarely the problems of faddism in both science and theology, as over against the question of more fundamental kinds of inadequacies that may exist in these two areas.

Why do people turn away from normal science and why do people turn away from normal theology? Is it because the big issues have been exhausted so that somehow there is no more ground for fruitful work? Or are they disproven? Are these views somehow no longer capable of being intellectually defended? It seems to me unlikely that they are disproven and unlikely that they are played out. I would say it comes down to a third issue, a crisis of confidence. This is the old Kuhnian theme again: A crisis of confidence that is, however, based on more than a single notion. It is based on evidential anomalies, things that simply don't seem to fit. There may be also aesthetic criteria, and a growing sense of triviality, but I would submit that in both cases of science and of theology there are important moral considerations as well.

Now more briefly, I'd like to touch on the other two points Heisenberg made, the relation between tradition and *method* and the relation between tradition and *concept*.

It has long been recognized (and Whitehead's famous statement about methodologies in *The Function of Reason* puts it well) that ways of doing things can become obstacles for free speculation, that methodological dogmatism can be a real difficulty in trying to break free into some new way of handling materials that we wish to understand as human beings. Whitehead coined the definition of obscurantism as the refusal to speculate freely on the limitations of traditional methods. (We get the notion there of method and tradition together.) In fact, he goes further; obscurantism is also the downgrading of the importance of such speculation. As he points out, the obscurantists of any generation are those who are entrusted with the dominant methodology of the age. In the old days, when theology was queen, the clerics were the obscurantists, and now that science is queen, it is the scientists who have been obscurantists by refusing to speculate freely on the limitations of their traditional methods. Therefore, the extent to which tradition is a matter of continuity is a danger as well as a help.

The objective methodology, which Heinsenberg states as a seamless fabric of his work, is, of course, deeply under attack these days. In the final chapters of Roszak's *The Making of the Counter Culture,* for example, he urges us to rid ourselves of the objective consciousness and to look at the world with eyes of fire and to see the magical aspects of this world. So we have before us the need to speculate about new methodologies, to free ourselves from traditional methods, but, if I may add a cautionary word, without abandoning the self-critical elements in man's "organ of novelty," as Whitehead puts it. Reason is the inherent power of mind to control itself and keep itself from radical chaos. If we can somehow speculate about new ways of allowing our organ of novelty, of mentality, to exercise itself without turning into a radically uncontrolled chaos, at that point we may find that the future will beckon with a kind of science that will not easily confirm Goethe's fear—the fear that Heisenberg mentioned yesterday—of reducing man to the bondage of a traditional methodology rather than enhancing the human qualities.

The final point with regard to tradition and *concept* in science and religion brings us back to yesterday's question, nominalism. Because yesterday we were talking about concepts most of the time.

Heiko Oberman recognizes that tradition is both inescapable and also sometimes a hindrance, that the language that we employ comes out of a larger context of tradition, but also it is we who creatively employ it.

How much constraint does the rule of words place on the scientist? If Kuhn is to be trusted, he insisted, at least in the original edition, that we see the world differently depending upon our expectations; there is no mutual observation in language. But still intractable areas of experience must be admitted, or otherwise we would never be able to recognize anomalies when they arise. Otherwise everything would fit perfectly into our expectations, if expectations rule totally. Since there are anomalies and surprises in science, there is in science at some point the intractable that is not relative to our tradition and our expectations, and that is where the objective control of science comes in.

Now, I ask you, where in religion is the analogue to the intractable that we find in science? Where are the possibilities that are analogues of the anomalies, of the surprise, of that which somehow forces us to restructure, to stretch and change our language in theology? I am not asking the old question about conclusive falsifiability. I am asking the question of how we avoid the image of the quack or the charlatan, as Langdon Gilkey put it this morning. It is in this domain that some would appeal to religious experience and others to authoritative tradition or Scripture: I am just raising this as a place for some further discussion.

With regard to the scientific side, there has been a remarkable return in the last few years to an understanding that science's motivation is to tell us truth about reality. There was, of course, in the 19th century and the early 20th century a great deal of sheer conventionalism, that is, the old understanding of nominalism in which you were free to call anything by any name you wished as long as somehow the words went together. It was self consistent, but not necessarily attempting to claim that there was a realistic reference behind it. Now it seems to me that sheer conventionalism does not reflect the prime motivation of the scientists who give their lives to doing their theoretical work; they are interested in reality, not simply in what kinds of poetry they can create that will—as Toulmin said yesterday—be pleasing to the mind.

It seems to me that this move toward realism in science comes out of an understanding of science's motivation, but it is still *possible* for the scientist to shrug off the question of truth, in the sense of our models and imagery actually being isomorphic with the structures of reality beyond. The scientist can say, "That's not my bag; what I am trying to do is to provide you with pictures that will be satisfying to the mind and, also, on the basis of which you will be able to anticipate experience."

I think it's a different logical situation for the theologian. For at least the theistic religion, where another center of subjectivity is the reference—that is, where we are dealing with whether or not there is really a God who really is aware of us and who really has providential concern—I think it is impossible for a theologian to shrug and say, "All I am responsible for is giving you a picture that hangs together, that is satisfying to the mind, that is fulfilling to life, that is capable of grounding a community." It seems to me to be impossible for the theologian to say "I don't care whether there is another 'Thou' out there, as long as somehow this works for me."

So these are my remarks, my footnotes on Heisenberg.

KLAUSNER: One source of the alleged identity crisis within science is that members of our several disciplines in science are recruited from different social positions and so each brings a peculiar social and cultural baggage to the analysis of scientific problems. Our problem is not so much that of a shared mythos as of the series of myths we bring to, and which seem to clash in, the common arena of science. The scientific community is a cluster of communities.

While I feel it important to account for culture in terms of social structural variables, when Fred Ferré finally refers to such variables, he does so in the context of a negative judgment of science. He seems like a neo-Rousseauan treating the social as a form of constraint. From within the scientific community, tradition does not appear as constraining, but as a necessary basis for taking the next step. Factors within the social structure of science contribute to the very advance and development of science. Consider, for instance, the discussions on the structural requirements of innovation, the effect of peer group interaction and peer review of articles as a means to progress or the requirement that, within the rules, one say something new.

GILKEY: To respond, I think our remarks differ in what I mean by a historical view of science as opposed to a scientific view of science. I do see science as a community of persons devoted to inquiry, a tradition of people bound together by certain commitments and standards; that's what I meant by mythos.

When you discussed constraint in your statement, what interested me was your witness that for the scientist, the tradition is freeing. Now I think this is true of every community from the inside. This is what is meant by the difference between external and internal history. Within the history of theology one finds very clear statements to people outside the community, that grace is freeing, that the mind is free now under the authority of revelation. What would seem an incredible constraint in the negative sense of the word to someone outside of that tradition may seem strangely freeing to people within it.

I think Mr. Roszak would feel the scientific disciplines are incredibly constraining. The scientist quite rightly should react and say they are not, that they are the position from which to see things clearly. I would agree; the tradition of science would be freeing to the scientist.

MAGUIRE: One of the things I find most striking about the conversation this morning is that less than a decade ago, in such a gathering as this, participants would be stressing the incommensability of science and religion; that is, science and theology aren't comparable structures. Now today you have as a kind of transformation equation "human-beings-in-community." You look at how human beings behave in these two communities of inquiry and begin to look for similarities and dissimilarities.

Having made that observation, I want to ask a question: In the communities of faith right now, there is very strikingly what I would describe to be a very dead period in theologizing, at a time when you have this wild, almost ersatz, resurgence of faith expression; you have young people and others expressing religious vibrancy at a time when the academic traditions of reflection are really pretty desiccated. If you compare these two enterprises of inquiry, is there anything corresponding to that in the community of people who do science? Could you have a time when the theorizing was going dead, and yet there was a kind of grass roots excitement and vibrancy that would roughly correspond to these experimental surges?

THOMPSON: Your point is very basic. We see now a collapse in the seminaries at the time that registration in religious courses is highest, and religious studies in the broadest cultural sense have the primacy now. This I think is also going on in science. The thing that most fascinated me in conversations this year first with von Weizsäcker and then with Heisenberg, and then it came up again in Gunther Stent's *The Coming of the Golden Age: A View of the End of Progress,* is that scientists themselves who have achieved historical impact in their disciplines are willing to say we have reached the end of the age of natural science. Heisenberg's image was that people are now cutting stones and putting them next to one another with meticulous attention to the work of the stone cutter next to them, but no one has the sense of the architecture of the whole, or the sense of why one should build the cathedral in the first place.

Thus with a maximum expansion of science, there seems to be a routinization of excitement and charisma, and even the leading scientists themselves are becoming demoralized. When von Weizsäcker presented this to me in conversation for the first time, I thought it was incredible, and I said, "Isn't this just like what Rutherford said at the turn of the century, that atomic science is all over?" We were actually in the Max Planck Institute, so there was an historical irony in the fact that it was Max Planck who had proved Rutherford wrong. I said, "Isn't this just a case that we might say that physics is over, but that things are going on in biology?"

I had great hope for biology or the interface between, say, solid state physics and biology. And then, a month or so ago, I read Gunther Stent's book in which he said biology has gone from the period of charismatic innovation and is now in the dogmatic and academic period. I can accept the argument in the arts, but I was surprised to see it in biology, and even in what was the most charismatic field of science with the Crick-Watson discoveries. Then Stent makes the point that gets to your point. He shows how Crick was able to take the genetic code and put it into a table of 64 units and he goes on to say how incredibly isomorphic it is to the Chinese I Ching and then talks about binary mathematics and things of this sort. He makes the point that the ability of these people to intuit cosmic structures by some imaginative, miraculous, or religious aspect 4000 years in advance of their scientific corroboration is utterly staggering.

Now if we look at the kind of science described in the isomorphic Chinese I Ching and the genetic code, we don't have a kind of empirical science, but we have an a priori *logos*. This is like Pythagorean science, like Chinese science, like the mathematics of the ancient Maya. This presents the way to go beyond the limit. Rather than trying to manipulate nature to control it, we learn that we can't dominate nature, and that the contemplation of nature in all its radiant forms becomes the most proper and fitting relationship of consciousness in nature and in that sense man doesn't have to dominate nature to do it.

My remarks are kind of a gloss on a gloss—there is a kind of isomorphism to what you were saying. You start out first with the age of religion as the queen, which is the age of gods; the crown then moved to the age of science, which is the age of heroes; and you then end up with history, the way in which we, through a sociology of knowledge, analyze the scientific process and the religious process. We see a common structure of imaginative innovation, which is the *logos*, the way the mind knows, and see that now we can see all around us, that there is a way that the mind is interacting with pure structure, with pure form, in which there is a convergence of disciplines into a new mythos, and, back to stage one, into a new cosmic mythologizing—a sociological re-enchantment of space, time, and self. We get the sense in which Heisenberg is a perfect Pythagorean, working in music, mathematics, and poetry, and seeing a kind of correspondence in all these.

This cannot be administered by the academicians, because they express *ratio* rather than *logos*. The charismatic aspect of the *logos* is a totally different thing than anything we know about consciousness or nature. In this sense, we seem to be moving into chaos. There is a distinct relationship between information theory, that an increase in information generates an increase in entropy, so the more we measure the more we take apart.

It seems to me we've got this fantastic split in the culture and that, in fact, we are the ruling elite suffering an utter failure in confidence. The forces of charisma may have been taken outside of the professional structures. This is true of science, theology, and art, as well.

POLLARD: Perhaps at that point we should take a break, so that when Professor Heisenberg comes, we will be ready.

DISCUSSION WITH PROFESSOR HEISENBERG

POLLARD: Professor Heisenberg, we are delighted that you would consent to join our discussion. We've been discussing, for the first hour and a half, your talk yesterday.

HEISENBERG: That's kind of you. Are there questions which I could answer?

McMULLIN: In some of your work, you have suggested that we are moving into a sort of "Pythagorean" phase in science. Your argument has two parts. First, you argue that analogies derived from physical experience, which have guided mechanics so fruitfully over the centuries, are exhausted, incapable of further development. The metaphors of matter, force, energy, not only do not suggest how the anomalies presently facing mechanics should be overcome, they can be positively misleading in their associations. Even, the concept of a physical individual, so central to the atomist tradition, no longer works for us. Your second point is that in these circumstances, there is only one place where the physicist can find alternative sources of conceptual creativity, and that is in mathematics. He must turn to group theory, to symmetries, to formal properties. I see this as an extremely disciplined sort of approach. But the question I'd like to ask is this: Have we any real reason to suppose that mathematics will be able to bear the strain of such a demand? Is there any reason for believing that a turn to purely formal symmetries is likely to provide the sort of resources in the face of experimental anomaly that the older physical analogies did?

HEISENBERG: I would definitely believe that this is so. First of all, I should emphasize that already at the time of Copernicus, the decisive change was not towards empiricism, but rather the opposite. Before that time we have a kind of descriptive science, which was Aristotelian. Then Copernicus, and especially Galileo, introduced mathematics into the picture; they asked not what are the phenomena in nature, but how can we prepare the phenomena of nature so that they show their mathematical structure. That I think was a very important step, and actually, the enormous success of this method was due to this mixture of empiricism and mathematical structure. Without the mathematical structure, I think one would have been stuck very soon again in that kind of descriptive science that we had before.

In spite of these clear Platonic features in the development of science since Galileo or Copernicus, the mechanical views after Newton have been rather Democritan, especially atomic physics in the last century. I cannot doubt that at present we have reached a new stage where again we must realize that the Platonic view is really the deepest of the views.

Once I gave a lecture in Athens and I said that now we have learned from recent developments that the whole struggle is between Democritus and Plato—Plato was right and Democritus was wrong. It is a nice way of describing the situation.

Unless we get still more deeply into this mathematical side of nature, we will have difficulty in understanding it. After all, we are going farther and farther from our primitive, daily life experience, and we are going to a realm of very remote phenomena. Either we go to the distant stars, or to very small atomic particles. In these new fields, our language ceases to act as a reasonable tool. We will have to rely on mathematics as the only language that remains. I really feel it is better not to say that the elementary particles are small bits of matter; it is better to say that they are just representations of symmetries. The farther we go down to smaller particles, the more we get into a mathematical world, rather than into a mechanical world.

This, as I see it, is the successful way of proceeding. We should not look for the smallest particles, as Democritus did. Of course, I will not object against such experiments. Every experiment can lead to useful information. But still, I feel that, for example, this ever-continued search for the quark particles is a consequence of the wrong philosophy.

McMULLIN: May I carry my question one stage further? Your answer raises for us the crucial ontological question as to what the reference of science is. Since your formal symmetries are a product of the *mathematical* imagination, are you not making some sort of prior assumption about the nature of the world itself, that it is in some sense a product of the mathematical imagination? The physicist (unlike the mathematician) has always had to cope with anomalies, with a physical reality that continually escapes his own conceptualization. The object he tries to understand is an intractable entitly opposed to him experimentally, rather than something that is freely constructed as a formalism obeying only the broad demands of consistency. I wonder if the analogy of physical tradition ought not lead one to be very uneasy about the step that you are taking. That is why I am raising it in the context of your discussion of tradition. If I am not mistaken, you are proposing a quite violent break with tradition.

HEISENBERG: Yes, I'm sure that you are just close to an essential point in philosophy, namely, the ontological question of whether matematical structures are only forms in our mind, or whether they are there before the human mind ever was created.

There is a very great difference between this kind of objective idealism of Plato and, let us say, the more subjective idealism of the 19th century. I would definitely be in favor of the objective idealism of Plato.

I would simply say that these mathematical forms are what, if I can express it in a theological manner, are the forms according to which God created the world. Or you may also leave out the word God and say the forms according to which the world has been made. These forms are always present in matter, and in the human mind, and they are responsible for both.

I would not say that mind is something entirely different from the material world. Therefore, I am quite happy about modern medicine, where we learn that whenever we think, there are some chemical processes in the brain belonging to thought. That doesn't disturb me at all. On the contrary, I think that is quite satisfactory.

But I would like to say that the mathematical structures are something behind the whole thing, or beyond the thing, not only in our mind. All mathematical laws would hold also on the distant stars. If there are some beings which, for instance, would develop the concept of number, then they would have the same theory of numbers as we have. Thus the mathematical structures are actually deeper than the existence of mind or matter. Mind or matter is a *consequence* of mathematical structure. That, of course, is a very Platonic idea. But I would always feel that is a reality.

DAVIS: The discussion in the first hour and a half brought out a rich variety of comments on Professor Heisenberg's speech yesterday, concerned not so much with the question of the nature of reality but with the question of the identity crisis in science today. It was pointed out that ten years ago, any discussion involving scientists and theologians would have emphasized the incommensurability of these two queens, these two aspects of intellectual activity, and that, today, we find very much more in common between them.

As a scientist I'd like to suggest that perhaps there is a weakness in this whole discussion, in that the word "science" is being used to mean more than one thing.

There is certainly a crisis in science today, and some people are claiming that because of this the whole validity of science is questionable, and its objectivity may not be as reliable or as valuable as it once seemed to be. But I would suggest that, in fact, science is not only a body of knowledge, but also a process, a methodology creating that body of knowledge. I believe that the crisis is primarily sociological, a situation in which people are reacting against the social consequences of some of the products of science and therefore against the social process that created those products. But the further extension of the protest, to the validity and objectivity of science as a coherent conceptual scheme, is superficial. It might therefore be a mistake to consider that we are facing today a real crisis with respect to the nature of science in any deep sense.

HEISENBERG: I would certainly agree with you that what we face is a sociological and not a scientific crisis. I think science always has more or less the same structure. Science changes considerably in its philosophical aspects, but this I would not consider as a crisis of science. It is not a revolution, it is evolution of science, and a very interesting one.

I am very interested in this sociological crisis as well, because I tried to talk to many young people, and I must confess that I don't know any good answers because I cannot really quite understand its roots.

But I would never consider it as a crisis in science. As I tried to emphasize yesterday, the problems of science are still given to us by the historical forces, and so long as this is the case, then there is no reason to worry.

THOMPSON: In a very interesting essay you have written on

art, called "The Idea of Nature in Contemporary Physics," you use a phrase that is perhaps better than crisis of science. You talk about limits.

In a very poetic metaphor you say that humanity finds itself in the position of a captain whose ship has been built so strongly of steel and iron that the magnetic needle of its compass no longer responds to anything but the iron structures of the ship; it no longer points north. In this situation we have to reorient ourselves by looking at the stars again.

You make a very interesting point that I think is coming up in many areas of contemporary culture, the idea of limits; we are hitting our heads on the limits to growth of industrial society, the limits to the avante garde, the limits to scientific innovation. And in that statement you say that perhaps now the limit itself can become the means by which we reorient ourselves.

Where do we stand and where are we going, in this moment?

HEISENBERG: This comparison means that as soon as you feel you have no way of orienting the ship, when you don't know how to steer it because you have no compass, then you start to think, and you try to find new sources of orientation.

The stars are a symbol for religion, and I would put it that what the young people actually desire in this sociological crisis is a new kind of religion. Of course, they don't say it that way. The names never are called. But I am sure that what they hate is this feeling of everything turning around. Whenever somebody speaks about ethical values, the next one comes and says that's all nonsense, forget about it.

I feel that any healthy society has always had certain values and principles that you should not discuss, not prove, not disprove, never doubt. You should just accept them. That is the basis for the life of this society. Nowadays, a funny thing is that everybody thinks that freedom means that you can doubt everything. But I think that is a fundamental mistake. Our liberal societies only have been in good condition so long as they never doubted the value of freedom. As long as we really believe in mental freedom, and we feel that everybody has to ask his own conscience, and finally we don't doubt that you shouldn't kill your neighbor, then it's all right.

THOMPSON: Do you feel with the movement in this new direction of religion with the young it will be necessary to lose the scientific emphasis from the previous generation? Is there a way we can be scientific and religious and Pythagorean?

HEISENBERG: Well, it may be that they consider this over-emphasis of rationalism in science as a danger. They feel that this over-rationalizing everything really makes it impossible to get to any kind of stability and, therefore, they tend towards irrational.

Now everybody, of course, knows how dangerous political movements can be when they become irrational. I am thinking of my country. But of course you can look at other countries nowadays and it's almost the same thing. This entering of irrational motives in political life is a great danger, but at the same time it may be necessary in order to get the new stability, because rationalism alone, so far as I can see, is not sufficient to establish a sound basis of a social community. This is simply because, when you go to the rational arguments, they always lead from assumptions to results, but you can never prove whether the assumption is right. This is the basis of my logic, and that's how religion always has been.

Would you agree with these statements about the sociological crisis?

DAVIS: I would agree. But I would go a little bit farther in implicating science: namely, it seems to me that the success of science has depended upon profound skepticism with respect to beliefs received from authority, and it has also depended upon separating those areas where you can get value-free judgments. Men can come to an agreement in science much more completely than in any other area, so that we have a consistent body of scientific knowledge. This leaves a vacuum in other areas: science cannot solve moral problems, but it has made us skeptical about traditional solutions.

More specifically, science has taken over, with great success, many areas where our previous beliefs invoked the supernatural to explain natural phenomena. It has therefore shaken our belief in the supernatural as a source of guidance. It seems to me that perhaps much of our current crisis arises from this loss of a set of stars to guide oneself by. The world hasn't found a substitute for the supernatural as a basis for a generally accepted code of behavior.

HEISENBERG: You say that science is based on extreme skepticism, but I doubt that, and therefore I have put in my talk yesterday something about the theological basis of the new science of Galileo and Copernicus. Actually, I am really convinced that these people did not want to doubt things, to be skeptical. Rather, they would say, "Now we have found a new way towards God. We have seen that we can recognize the laws which God has given to the world in these mathematical structures." An extreme skepticism came about during the 18th century, but I am sure that science is not necessarily connected with that kind of skepticism.

I would not agree with what you say about the supernatural, because there was not such a clear distinction between natural and supernatural. Kepler and Galileo just found the actions of God or the will of God in these phenomena that they observed. So for them, the distinction between the supernatural and the natural was not there; they simply felt it was the same thing.

I would say the distinction now is only that we claim that we restrict our science to the natural things, that we leave the supernatural to the priests or to somebody else. Science is not forced to make this distinction between the supernatural and the natural. Therefore, I would hope that science would find out that what it actually does is to discover things that also go well together with the other side of the world, which has been called the supernatural, but which is also the natural.

DAVIS: Your examples come from a long time ago. In terms of

modern science, couldn't I defend the statement that scientific activity does involve a rather profound skepticism? The graduate student who questions the voice of authority and does it successfully is the one who breaks through and makes an important discovery. But he has to question authority.

HEISENBERG: That is perfectly all right. You should question authority. I doubt that you would take Einstein as a special example of a very religious person. But anyway, Einstein said of himself that he was very religious, in the sense that behind the structure of those things he observed as phenomena, he always had a strong feeling of some general connection which he could only admire, but could not rationalize.

Well, I can't quote by heart now what Einstein actually wrote, but I know several of his statements which pointed exactly in this direction, that in science there is no clear distinction between the natural and the supernatural, and that he, being in a very critical and skeptical time, was perfectly willing to see this structure behind the natural.

I will say it in just the same way that my friend Pauli argued: Really good scientists always knew that this distinction between the supernatural and the natural is an illusion. I will put it as strong as that.

MAGUIRE: Dr. Heisenberg, in historical interpretation, a pair of terms are often contrasted, reformation in contrast to revolution; and another pair of terms, continuities and discontinuities. Prior to your arrival this morning we discussed whether we may be in a cultural moment in which the new themes and the new approaches that you described yesterday may be so strikingly different from what has gone immediately before that a radical discontinuity approximating a revolution may be occuring. Would you talk about where you see that critical moment, such that if we went beyond it, you would have to say that genuine discontinuity has occurred? Can you forsee this sociological crisis reaching such proportions that the human responses to it will constrain us to acknowledge and perhaps celebrate discontinuity and revolution, in contrast to continuity and reformation?

HEISENBERG: On the sociological side I always am against revolution and in favor of evolution, simply because I see that normally revolutions are constructed so that at first everything is destroyed; then, of course, one later has to reconstruct it, and something comes out that very often is not so different from the earlier shape of things.

In physics—or in the sciences generally—the really great revolutions are not brought about by destroying everything that has been and then inventing everything new. Actually, it saw a more gradual change which afterwards could be called a revolution. For instance, when Planck discovered the quantum of action he did not intend to make a revolution in physics. Then gradually people like Einstein and others saw that these new ideas were spread into all physics, so that at the end, physics was really in a complete reconstruction. But there was no time at which anybody, even Einstein, would have said, let's just tear the whole house down and start anew.

FERRÉ: It seemed to me yesterday, when you were talking about the futility of looking for the particles on and on, that you yourself were showing an interesting skepticism that could lead, in retrospect, to quite a revolution if your recommendations were adopted. For example, where you warn against putting too much stock in the search for the quark, and point out that if the quark is found, then there will be anti-quarks, and so on and so forth, and the whole process continues; rather, you said, one should look for symmetries instead of entities.

I wonder to what extent your recommendation is based on a sense of the exhaustion of a line of reasoning, something that is based more on a sense of the hopefulness of an alternative, or just being tired of what you might perceive as triviality or aesthetic messiness of the former line of looking at things. That is, what is your ground for making a recommendation that would, if accepted, turn us away from looking for entities in the Democritean way at all?

HEISENBERG: My points would be that I am always hopeful for new developments to come, but I think one should not insist upon looking for the new developments just at the very point where one believes that they might be coming. One should be prepared for many changes. Unfortunately this problem is part politics and part science. Since I had to act sometimes as an adviser to our government, I had the disagreeable task of being confronted with problems in which I did not always agree with my colleagues.

You can either argue we must build bigger and better accelerators, spend many billions of dollars in order to get still bigger ones, or we should, rather, stay with those which we have and wait for ten years and see how things develop. I have actually taken the stand that we should not expand too much. That was, for instance, eight years ago, there was a discussion in Europe whether we should build the so-called storage rings, which were relatively cheap, or whether we should build such an accelerator like the Batavian machine.

In fact, the storage ring was built in Europe, and the big machine was postponed. This decision was really a great success because the storage-ring machine is at present the most interesting; it produces collisions of particles of the highest energy and has given very new and interesting information.

In ten or fifteen years we might find new and extremely interesting phenomena in astrophysics and then we should put our money into astrophysics, or maybe molecular biology. I am rather convinced that such a field as elementary particle physics can come to a close like thermodynamics, or like optical atomic spectra, which has come to a close.

FERRÉ: This must come deeply out of your sense of the Pythagorean structures of reality. To this extent, then, science is creatively directed at some points—not always—but at some crucial points, by some basic metaphysical or religious or other transscientific consideration.

HEISENBERG: Yes, I would say so.

MCMULLIN: Because you yourself were one of the initiators of one of the most important revolutions of the century, we have an unsual opportunity today to ask you a question that has been fiercely debated by philosophers of science ever since the appearance of Thomas Kuhn's work on the nature of scientific revolutions. It has to do with the nature of the rationality that scientists display during such revolutions. Is there one kind of rationality that governs "normal" science, the sort of cumulative systematic inquiry that goes on at the non-exciting periods, and a different kind of rationality that governs at times of fundamental change of paradigm? Some, indeed, have suggested that during these latter changes the criteria of rationality, governing proof, and understanding break down entirely. My question can be put very concretely. In those crucial years between 1924 and 1927, when radical conceptual changes were occurring among the Copenhagen group, what sort of factors proved to be decisive in bringing about the changes? Kuhn, in a controversial analogy, suggested, for example, that there was something akin to religious conversion in such a change. There would be, he argued, an incommensurability between the old and the new, making it impossible to produce cogent arguments in favor of the new, without begging the question. Would it be fair to say that some minds proved totally immune to conviction, as Kuhn suggests?

HEISENBERG: That's a very interesting question. I would say, first of all, we were for a number of years, say, between 1921 or 1922 until 1927, in a state of continuous discussion, and we always saw that we got into trouble, because we got into contradictions and into difficulties. And we just could not resolve these difficulties by rational means. One would argue in favor of the waves, and the other in favor of the quanta. I have heard many discussions between Bohr and Einstein, and Einstein would argue for the light quantum, and Bohr would say, even if you had a definite proof that you had found the light quantum, if you would send me a radio cable telling me of the new research, this cable couldn't arrive here without the waves. So we actually reached a state of despair, even when we had the mathematical scheme—which was to every one of us, to begin with, a kind of miracle. Then we saw that mathematics did things we couldn't do ourselves. That, of course, was a very strange experience.

I do remember that when I first studied this matrix calculus, I saw this was a mathematical scheme which worked, but we did not know what kind of language to use, nor how to talk about it. Out of this state of despair finally came this change of mind. All of a sudden we said, well, we simply have to remember that our usual language does not work anymore, that we are in the realm of physics where our words don't mean much.

I remember that I once had a discussion with Niels Bohr where he doubted whether we would ever find such a mathematical scheme. He felt nature might be so irrational that we could never get any kind of good mathematical description. So

At a seminar in Copenhagen in the 1920s. In the front row are Niels Bohr (with hands folded on railing), Werner Heisenberg, Wolfgang Pauli, and George Gamow. (Courtesy of the Margrethe Bohr Collection, AIP Niels Bohr Library.)

he was quite surprised when it turned out that there was a mathematical description; we couldn't doubt that this was the correct scheme, but even then we didn't know how to talk about it.

After these infinite discussions, well, they first threw us into a state of almost complete despair, and only afterwards we felt this kind of liberation that things really were consistent. Those other physicists who never really went through this ordeal of discussions, they never could quite follow. I had a long discussion with Einstein in 1954, a few months before his death. A very nice discussion. But we could never agree on quantum theory, and I would say we never agreed because Einstein had not participated in those four years of discussions.

DAVIS: I wonder whether Kuhn doesn't tend to divide science too excessively into cut-and-dried categories. It's straightforward filling-in if you have evidence for a novel conclusion and everybody quickly accepts it. In the revolutions that are the interesting parts of science, the novel paradigms meet great resistance. But actually, it seems to me, there are all sorts of degrees of skepticism that get gradually overcome. Even a minor step forward, with straightforward, conventional evidence and logic, can encounter strong resistance if it contradicts an alternative that somebody else has staked his reputation on.

POLLARD: Concerning what Kuhn is talking about in terms of conversion: I started an undergraduate major in physics in 1930, when quantum mechanics was very new. The older members of the department, and the graduate school, too, were uneasy with it, or not attracted to it. It wasn't physics as they had known physics in their experience. It was almost a new generation that came in for whom quantum mechanics was natural, and the older generation had by now died off. I am sure there are situations in science like that, even in molecular biology, where a lot of older traditional biologists have been very strongly resistant to changes.

DAVIS: Now we are talking neither about philosophy or sociology, but about psychology.

POLLARD: Yes.

GILKEY: No, I think that's a mistake. I think Kuhn is talking about the logic; I don't think he's talking about conversion. Now he may be wrong, but it isn't a psychological category, though that's relevant, namely, as to whether one is converted or not.

Kuhn's point, which stands or falls, is the logic, that is, the methods of testing, the fundamental paradigms, and so forth. Now here in your discussion, you have indicated a continuum of mathematics, which I think Kuhn doesn't take account of in his book. What he is interested in is the almost total revolution of methods of testing, of what you are looking for, and of conceptuality.

Now you have given us, it seems to me, a most interesting picture of a continuum of mathematics and an almost total revolution of conceptuality. The continuum of mathematics means it wasn't the total revolution at all. Is that what you were saying?

HEISENBERG: It's almost correct. But the mathematical scheme of course, also underwent discontinuous changes. That is, matrix mechanics were discontinuously different from the Bohr-Sommerfeld theory. The difference just came in by the difference of concepts that you emphasized, so you cannot quite separate the concepts from the methods.

In Newtonian physics, the state of the object was described by giving its position and velocity. It was, mathematically speaking, a point in an infinite phase space. But if you put this system in connection with the measuring equipment, the whole definition has changed completely, and, therefore, the new mathematical scheme is very different from the old one. The surprise was that such a scheme existed at all. Bohr had the impression before that time, that we know that Newtonian mechanics doesn't work, and that may mean that nature is so irrational that we can never get any consistent mathematical scheme for description. Then it turned out that we did get the new and extremely mathematical scheme.

McMullin: But it wasn't just a mathematical scheme, was it? The criteria for its acceptance were distinctively physical. Its symbols were mathematical in their syntax, but physical in their semantics. Their meaning was thus derived, in part, from their application to physical contexts. Besides the observable quantities of your matrix mechanics, there were the more theoretical part-physical notions (like probability-of-finding, spin . . .) whose warrant was their fertility in guiding the development of the theory. In a period of revolutionary theory-change, such as that between 1922 and 1927 in mechanics, if in fact the mathematical formulas proposed had not been embedded within a theoretical account leading to a new level of physical understanding, then conviction would not have come about. And it was gradually accepted as a new level of understanding in part because it opened up new empirical data of an unexpected kind, and suggested ramfications that had not been thought of before.

Kuhn and others following his lead have not perhaps sufficiently recognized the persuasive character this sort of "fertility" possesses for experienced scientists. Even in the most revolutionary of changes, even where there has been a drastic shift of meaning between the old terminology and the new, there can in practice be a high degree of skill-learned agreement about the indices of a theory's fertility. The fact that it ordinarily takes time for this agreement to be reached testifies not so much to the subjective character of the considerations involved as to the fact that it takes time for the fruitfulness of a theory in the organizing of data and in the directing of research to manifest itself and be properly appraised. If it were only a matter of estimating the support a particular theory derives from the data available at the moment at which it is put forward, the time element could be disregarded. But what really serves to convince (it seems to me), when the new theory is as paradoxical and disturbing as the new quantum theory of matter was, is not a static assessment of that sort. Rather, one must live with the theory, so to speak, over an extended period, and follow it closely as it responds to successive problem-situations.

Would you agree that what brought about a resolution of the debates of 1924-27 was a progressive assessment of this sort on the part of those who were most intimately concerned, on a day-to-day basis, with the vicissitudes of the new hypothesis?

HEISENBERG: It was certainly the mathematical element in the phenomena which after Pythagoras has again and again demonstrated its fertility, especially in the time of Galileo, and in our century again in relativity and quantum theory.

POLLARD: There is still a miraculous quality about it. There is the extraordinary thing that we still really don't understand, say, that Riemannian geometry was worked out as a pure product of the human mind at the end of the last century. And then, in a kind of miraculous way, in general relativity, the curvature tensor in Riemannian geometry and the energy momentum tensor in physics joined.

HEISENBERG: The mathematicians, of course, look for many different forms, and then, in many cases, they see possible forms—but they are not very interesting. However, there are a few that turn out to be extremely interesting for physicists, and of course, they are developed.

For instance, matrix mechanics. Mathematicians were not especially interested, but then when physics came it became useful.

GILKEY: Might I return to the question of skepticism. I am not a scientist, but having heard scientists speaking in the last days, I can't really believe that skepticism is the basic character of science. I think this is a myth in the minds of scientists, in relation to other kinds of authority. But as I have heard you, and Toulmin's and Holton's discussion of this yesterday, there is actually a deep faith in the order of things.

I would take the word skepticism with a grain of salt. To describe this sense of the deep order of the congruence of the human mind to what is being studied, this confidence that even though I don't see the order it is there and, therefore, I look for another kind of mathematics or another conceptuality, this—I don't want to say faith, because that sounds pejorative—this confidence. . . .

DAVIS: Well, it is faith.

GILKEY: Yes, all right, it is faith that there is a reason, and the real questioning of authority is whether they are in touch with that reasoning, that is, that deep fundamental *logos* of existence and of mind. This can hardly be described as skepticism.

Science really could not exist in a culture or a human mood in which there was a really deep skepticism about the order of reality, about the utility of thinking, and I think, also, about the use of this in being a human being, and in its social utilty.

HEISENBERG: I would like again to quote a sentence by Pauli. He always said: "I am so skeptical that I am even skeptical against skepticism."

Extreme skepticism is not really in disagreement with a belief in some kind of order.

POLLARD: I'm sure I speak for the whole group in expressing our profound gratitude that you would agree to meet with us.

The great spiral galaxy in Andromeda, M31, about 2,000,000 light-years away, with two elliptical satellite galaxies. From the Hale Observatories' 48-inch Schmidt telescope.

Thursday, 26 April 1973

William I. Thompson, *Moderator*

DISCUSSION ON
"THE UNIVERSE AS HOME FOR MAN"

THOMPSON: Professor Wheeler has graciously accepted our invitation, but he has only an hour, so we thought we would begin with general discussion and reactions to his address yesterday.

WHEELER: Three points might serve to start the discussion. One is a question and two are comments additional to what I said yesterday.

How does it come about that we are able to set up more or less workable moral standards in a world whose ultimate purpose we still are far from knowing? What a miracle! Where does the legitimacy come from? Does it come from the law of life, from the fact that in this evolving world only those societies that do have a moral standard are able to survive? That's my question.

The two comments that I would add to my report of yesterday have to do with life on the one hand, and logic on the other.

Life I did not mention in my talk [as originally given], because it is difficult to know what weight to give it in the structure of the universe. But in brief, one is led to ask whether the universe has the physical constants and the size it does, and whether it lives for the length of time it does, because only in this way will life be possible; what is the good of a universe where there is nobody around to look at it?

But then the question comes up, why not a bigger universe with more particles and more suns? Is there at work some law of economy? Is any universe "too extravagant to be allowed" that gives birth to consciousness in more than one locale? Do we live the "cheapest universe" that will allow life and mind to flower?

If so, then one's attitude toward the presence of life on other planets and in other parts of space gets quite drastically revised. The present thinking of many of our most distinguished colleagues is well known: Man is an unimportant consideration in the economy of the heavens. The universe is a vast machine that just happens to be there. We are an accidental bit of dust in a remote corner of the universe. We live in one of the hundred million solar systems in this Milky Way; and there are a hundred million Milky Ways. To put it in the arithmetic of exponents, there are 10^{22} solar systems altogether. According to these numbers, the reasoning goes, the chances are overwhelming that there is life elsewhere in the universe. But this reasoning presupposes that the universe is here a priori, that we are only an accident.

What if the reverse is the case? What if the central consideration in the structure of the universe is the requirement that there should exist somewhere, for some time, the consciousness of that universe? What is the "cheapest way" to meet that requirement? The "cheapest universe" with which one can get by and still have mind somewhere is a universe barely large enough, and long-lived enough, to bring mind into being in one place. Cut the content of the universe from 10^{22} to 10^{11} solar masses and find that it lives only one year from big bang to collapse, not long enough by far to form a star, let alone heavy elements and life.

Of course, we have no good means today to detect life elsewhere. Imagine that each time somebody is born a flash of powerful radiation is developed, and we have our telescope and watch the sky. Then the one judgment says there will be flashes all the time, here and there in the universe. The other judgment says we can look and look and look through this magic telescope and never see a flash nor any sign of life. This absolute difference in assessments indicates how great our ignorance is about the universe we live in. The range of the presently conceivable is so great that it is difficult to see how any thinking man can dogmatically defend one or the other view as obvious: (1) life is an accident in the plan of the universe, or (2) the requirement for the development of consciousness is central to the structure of the universe.

The second comment has to do with logic. We would like to look forward to the day when some kind of mathematical magic would let us understand the whole machinery of existence. I suggested that the most natural place to look for that kind of idea is in the realm of logic, which is central to the whole structure of mathematics, and where the most revolutionary developments have taken place in the last few decades. Logic is the only branch of mathematics that knows how to "think about itself." This circumstance suggests a natural reason to think of logic as a foundation for any mathematical description of what goes on.

GILKEY: This may encourage you to answer your first question; I'd be very curious for you to enlarge on the real meaning of your very important phrase, "participatory universe," that ended your very interesting talk.

I can understand the meaning in terms of the influence of the observer, an epistemological reason, so to speak—the influence of the observer on what he is observing. But you seem to give it much more meaning than that, what one might call an ontological meaning.

I assume you didn't mean man creates his universe as he knows it. But somewhere between the merely epistemological effects on the object he is knowing, and the concept that man is creating the universe so that it really is a child of man—somewhere between tells us what you meant.

If I could draw you out on that, it might have some relevance to your question about morality. In fact, the participatory universe of man, and man and universe, is one way of starting to talk about the religious question.

WHEELER: Very much so. In this connection, let me cite two remarks. One is from the wonderful address that Thomas Mann gave on the 80th birthday of Freud. Mann recalls that each of us is in some sense the stage manager of his own dreams. He goes on to ask if we ourselves also through the work of our own minds create and stage-manage existence itself; are we "actually bringing about what seems to be happening?"

The other quotation is from Leibniz, where he reassures us that, "although the whole of this life were said to be nothing but a dream and the physical world nothing but a phantasm, I should call this dream or phantasm real enough if, using a reason well, we were never deceived by it."

Is there any evidence from physics that the universe is "participatory" in character? Tantalizing evidence! When we investigate the world of the very small, we find as central point the quantum principle, or the "Merlin principle," as I called it yesterday. One way to state the mathematical and physical content of this principle is known as the Everett interpretation, which has awakened so much interest since its first publication in 1957 that it's being republished this year in much more extensive form by Princeton University Press as *The Many-Worlds Interpretation of Quantum Mechanics*. When we observe a system, a certain irreducible quantum of interaction occurs, which renders uncertain the outcome of the observation. In this sense our intervention is not merely permitted; it is compulsory. One cannot even think, one cannot even be conscious, without participating. The consequences of this intervention, however, are not deterministic; they are probabilistic. "God plays dice." This long known and most remarkable feature of nature Everett describes in a new way, as a "branching of history." In one branch of history, in consequence of the observation, the electron is revealed to have its spin axis pointing up, *and* the observer recognizes that the electron spin is up. In the other branch of history, the electron spin axis points down, and the observer recognizes that its spin is down. But neither branch of history communicates with the other. The one universe, containing the one "replica of observer-plus-electron," is unaware of the other replica of observer-plus-electron. This is the sense in which Everett provides a "many-universes" interpretation of existence itself. Or as William James said far earlier, "Actualities . . . float in a wider sea of possibilities from out of which they were chosen; and somewhere, indeterminism says, such possibilities exist, and form part of the truth."

Where is the "participation" in all this? The observer is inescapably a participator. He chooses what observation he will make. In so doing he decides whether the history of the universe will split into "electron-spin-up" and "electron-spin-down" branches *or* into "electron-spin-axis-pointing-to-the-left" and "electron-spin-to-the-right" branches. Having decided this much, however, having so to speak "picked the game," he has no choice in the outcome of the game. He cannot say on which of the two branches of history he is going to find himself. That is decided, according to quantum mechanics, by "throwing dice." In other words, the future is irretrievably affected by the decisions of the observer, as participator; but this participator does not make all the decisions!

There is a very lively sense, therefore, in which this universe is participatory. So far this participation has only been manifest to us at the atomic level. We have not seen how to reach out and influence the stars and distant galaxies. We can only influence things near at hand. What we have to date is very far from a full-dress participation of life and mind in the constitution of the world. I am as far as anybody can be from a proper understanding of what it would mean to say that the whole universe is brought into being by our participation.

At first even the words sound absolutely ridiculous. After all, we know that the scale of man's existence is extraordinarily short compared with the whole time scale of the universe. So it is preposterous to think that life and mind in these few hundreds of thousands of years of self-conscious existence will have any control on or influence over the development of the universe in the ten billions of years past and in the several tens of billions of years yet to come.

But then one has to return for a second and more cautious look at these deep questions. I had referred to the big-bang beginning of the universe and the collapse at the end of the universe. Regarding this predicted scenario the question is always raised: What about other cycles of the universe? Suppose we compare one cycle of the universe with another. Which of the conceivable cycles of the universe permit life and mind and which do not? Which, in the words of David Hume, are the "botched and bungled" universes; and which world has *meaning* because it allows *mind?* Actually Einstein's general relativity does not give us cycle after cycle. It predicts only one cycle. Our fascination with the idea of cycle after cycle is, I suspect, only a way station towards thinking of something much deeper, (1) a chaos prior to the moment of commencement of the universe, and (2) a chaos after the moment of collapse of the universe, both of which are completely disorganized, and neither of which even deserves the word "before" or "after," much less the word "time," to tell "where" it is, plus (3) a guiding principle of "wiring together" past, present, and future that does not even let the universe come into being unless and until it is guaranteed to produce consciousness of consciousness, at some point in its history-to-be.

DAVIS: In your talk yesterday I think you stated that of all the the scientific generalizations that have been made since Copernicus, Darwin's theory of evolution was the most important. Since you raise the question of where morality comes from, it seems to me that we must really consider the consequences of

Darwinism in some greater depth.

Ernst Mayr in a beautiful essay in *Science** last year pointed out that the implications of Darwin for our understanding of nature of man have hardly begun to be appreciated by our society at large. Most people don't have any understanding of it. Thoughtful people must really consider this problem very seriously.

You raise the question of where altruistic behavior can come from. Darwinism originally had an unfortunate history, in that the "social Darwinists" extrapolated the principle of natural selection to society at a time when they could see the selective value only of competition but not of cooperation. The result was an apparent biological justification for ruthless competition in human affairs. Today we understand how genes that promote sacrifice of individuals can have evolutionary survival, if the sacrifice promotes survival of others in the group, carrying the same genes. The obvious example is the social insects, such as bees and ants, which have built into them an instinctive behavior that doesn't raise any question of what they do for themselves. They function only as parts of the large society.

Man has the problem of being intermediate between animals whose instincts are highly social and the animals whose instincts are feral. Therein lies part of our problem. But I don't think that it is hard to see that man can have developed altruistic drives through the necessity of having these traits for survival as a species. If we accept that view, then do we really have to raise the question of whether there is anything outside the accident of evolution to account for all the things that make us unique, including morals and aesthetics? If one can become convinced that evolution can account for altruistic drives, what would be the reason for seriously considering such alternatives as, say, a whole universe that evolved in such a way as to produce life with this marvelous product?

---

*"The Nature of the Darwinian Revolution," *Science*, vol. 176 (1972), pp. 981-989.

If I am not mistaken, it has been calculated that if the entire time scale of the Earth itself is set at 24 hours, *Homo sapiens* has been here for about one second. And in the cosmos, with its 10^{22} solar systems and with man here for one second of the 24 hours of our Earth, isn't it a little bit contrary to the Copernican tradition now to try to go back and say we must still find something outside the natural evolution of the universe and natural evolution of the living world?

WHEELER: "Natural selection of the universe" sounds preposterous and yet, on the other hand, one thinks of a flower evolving. The plant may be enormous compared to the flower itself. You have the growth of the plant, and then the flower is exposed for a brief period: that is the whole purpose of the plant, to make this flower, and then it fades away. The evolution that gave the flora and fauna that we know proceeded by generation after generation of accidental mutation plus natural selection. For the universe, too, we are talking here of natural selection, but without benefit of generation after generation in which to operate. There is, according to Einstein's standard dynamics of the universe, one cycle only; so no principle of selection can pick out its structure from other conceivable structures, which does not encompass in a single gaze past, present, and future—something in this particular respect very different indeed from Darwinian natural selection, but absolutely dependent upon it. The two kinds of natural selection join together, on this view, into a closed loop: The universe allows mutation and Darwinian natural selection to proceed; this Darwinian evolution leads to consciousness, and consciousness of consciousness; mind at this level gives meaning to the universe; and without mind at this level (somewhere in the history of the universe), and "participation" (as participation is embodied in the quantum principle), the universe with which the discussion started could not even have come into being in the first place ("natural selection of the universe," as spelled out here simply to try to assess the idea).

It is hard to know how to judge this concept that the universe just wouldn't exist if it were not for the consciousness around to be aware of it. It reminds me of that limerick—I wish I could remember the words—about the tree in the quad.

DAVIS: "There was a young man who said, God
Must find it exceedingly odd
To think that this tree
Would continue to be
When there's no one about in the quad."

"Dear Sir, your astonishment's odd.
I am always about in the quad.
And that's why this tree
Continues to be
Since observed by, yours faithfully, God."

WHEELER: Perfect. That Balliol rhyme makes so pressing the physics issue how (on the view we are trying to assess) are past, present, and future to be understood as wired up to the participator. Only so can the existence of life and mind in a limited part of the history of the universe be a precondition for that universe ever having come into being. The purpose of the gigantic plant (the universe) is, on this view, to produce a single little flower (mind). The idea is incredible. Anybody who looks at the size of the yucca flower and compares it with the size of the yucca plant is convinced that the flower has nothing whatever to do with the existence of the plant, except that biological friends come along and tell him that is the whole purpose. Is it the same for mind and universe?

THOMPSON: I am struck by your point about man's proportion, that we are the flower and the universe is the stem, and that consciousness is very basic to the universe, that a universe without organisms conscious of the laws of nature is dead nature. Put a man in the universe who is conscious of the laws of nature through science, and you have the feedback of the consciousness on nature, which creates culture. You are creating a model of a dynamic culture, which is itself an event in the universe.

Now it seems to me that the way we conceive this culture is most often negatively in terms of disorder. Man disorders the ecological system of the planet; that is the impact of culture.

WHEELER: Your mention of "feedback" is a happy simile indeed. However, I would think of man's role as more than merely cultural. But you give me courage to try to restate the central point of the idea that we are talking about. Actually it is not even an idea; it's an idea for an idea. It is the idea that the universe simply could not exist without mind being somewhere present.

It is not that there is a universe and then one has his choice whether he will or will not put man or mind in it. It is that in some sense the universe absolutely could not exist without mind in it somewhere, at some time. This is the concept, right or wrong.

CALLAHAN: Are you saying there could be something without man, namely, a random activity? Or are you saying that there could be no universe without man, but then implying that by "a universe" we mean something which has been in some way ordered, or has had an order imposed upon it?

WHEELER: I would think that here, as in biology, one could not have the plant without the flower, and one could not have the flower without the plant. The two are linked together from the beginning. We need the physical universe in order to permit the formation of galaxies and stars and planets and evolution of life and the development of mind.

But this universe that we take so for granted just wouldn't be here (on this view) if it weren't for consciousness. It is as if we had pursued the electrical circuit of thought only as far as from the physical universe to man, and we hadn't seen how man himself is wired back into the very existence of the universe; we haven't understood the first thing about it. This is the idea I am trying to state.

CALLAHAN: I hear your words, but I don't know what it means to say that the universe would in some sense not be there.

GILKEY: Put it this way. One can understand you as saying

that the *telos* of the whole thing, that is the goal of everything, as the flower might be said to be the *telos* of the plant, is consciousness as it appears to man.

I think the alternative here is that there wouldn't be any order, such as the plant itself in its own unfolding as a plant, unless there were consciousness continually making order out of the chaos. That is a much more fundamental conception of which time is an example. There must be a metaphysical factor giving order to creativity, flux, in order for there to be any universe.

That is, there is a universe because there is order, not that there is a universe and then there is order. This is the basic point and, therefore, there must in some way be a principle of order that involves something like analogous language.

But you could have consciousness as the possibility of the universe, which is what Mr. Callahan is asking, rather than merely the *telos*, the goal—are you implying both? Fundamental symbols are not the result of thought. They are the invitation to think.

WHEELER: I like so much what you say. One is reminded of Einstein's words, "Actually time and space are modes by which we think and not conditions in which we live." Certainly a central feature of our thinking about the relation between man and the universe is the category of time. We are accustomed to say, the universe began first, and then, much later, man came into being. From this point of view it looks as if the universe could have existed without mind ever coming into being. However, this level of discussion uses the category of time, which is really only an approximate category. Nowhere is the category of time more questionable than in discussions of the beginning and the end of the universe. This limited applicability of the concept of time is an astonishing consequence of the modern quantum analysis of the dynamics of spacetime geometry. Quantum mechanics tells us that the idea of time is neither a primordial nor even an exact concept. "Time" is only an approximate concept. It is a derived concept. It is forbidden even in principle to use the term "time" when one talks about events that occur at very small distances, or very near the beginning of the universe or near its end. Then the idea of time loses all meaning. The terms "before" and "after" no longer make sense.

We have to deal in new and unfamiliar terms when we concern ourselves with the "beginning of time" and "the end of time." Does that mean that "mind" reaches backward into the beginnings of things and forward to the end of it all, and through this route decides the structure of the universe? The universe gives birth to mind; but mind gives meaning to the universe. In this coupling of mind and universe do we have what our electrical engineering friends would call a self-excited circuit?

DILLENBERGER: Yesterday you said something about the collapse of the universe, meaning the end of the laws of physics, or at least their applicability. Does that imply also the collapse of logic and mathematics? Are logic and mathematics more fundamental than the laws of physics?

WHEELER: That is a very deep question. The great Göttingen mathemetician David Hilbert enunciated a famous program to reduce all of mathematics to logic. Then came terrifying discoveries in the realm of logic. Any system of axioms rich enough to give arithmetic inevitably leads to certain propositions which, while making sense, are undecidable, unless one arbitrarily decrees the truth or falsity of each such proposition as a new axiom—in which case one is confronted with new undecidable propositions. Hilbert's program collapsed. Mathematics goes on, of course; but the mathematical community is disoriented as to the foundations for this mathematics.

Less important perhaps for us or physics or the universe than any details of today's mathematics is the central idea in the revolutionary new developments in logic: "self-reference." Self-reference enters logic in a technical way. For example, the statement is interesting that "Proposition No. 67 is undecidable." This statement becomes immensely more interesting, and an example of self-reference, when it itself turns out to be Proposi-

tion No. 67. Among all instances of self-reference is there any
with more implications for our place in the scheme of things
than this: "The universe gives birth to consciousness; and consciousness gives meaning to the universe"? Is this the great ice-
berg of reality of which today in the quantum principle we see
only the tiny tip? How can we best test such views? Can we
hope to derive quantum mechanics from the principle of self-
reference as one long ago derived the energy of a moving electron from the principle of relativity?

THOMPSON: Thank you very much. It is a few minutes after
ten, and we're sorry you have to leave.

DISCUSSION: SCIENCE AND THE FOUNDATION
OF ETHICS

THOMPSON: Our first paper this morning will be given by Dr.
Callahan, Director of the Institute of Society, Ethics and the
Life Sciences.

CALLAHAN: Paper is the wrong word. I want to respond to
Professor Wheeler's address yesterday, and I would like to throw
out some thoughts—some are random, but I'd like to present one
as a thesis.

In a curious sense, and particularly after hearing Professor
Wheeler this morning, I came to feel that, despite all the talk
of the Copernican Revolution—which supposedly displaced man
from the center of the universe—he was really coming back to
say that indeed man is still the center of the universe. If this
doesn't lead to a sense of new arrogance, at least it is possible to
say that we needn't feel so humble, as people are supposed to
have felt in the aftermath of Copernicus.

Let us start with the premise that man is indeed the center

of the universe, if only because we don't know that there are any other beings anywhere. If we then ask what is the nature of this man, I would come back to the very traditional notion that man is essentially a rational animal; the greatest thing about man is his rationality.

I would then argue that if the characteristic mark of man is his reason, the whole problem of relationships between the observer and the observed, man's participation in what he knows, ought, I think, to be seen as man's increasing need to genuinely understand the world "out there." In a very strong sense the world is more than the human community, and, most importantly, in some sense or other there is genuinely something out there apart from man.

I think one of the bad historical effects of the Copernican Revolution in displacing man from the center of the universe was to make him a much more parochial kind of being, one who began increasingly to see the purpose of science as primarily knowledge for the sake of power and control. It is almost to say that since we are not at the center of the universe, the goal now is no longer to understand the heavens, but to understand how we can manipulate this small part of the universe we occupy. The rise of the social sciences then made man more parochial still, particularly when they tried to understand human social reactions in terms of control of human beings. It is almost as if the only important reality to man is the human community.

I would hope that religion in the future would move beyond that. The key task for man is not only to understand his place in the universe, but even more to understand what indeed the universe is. One disastrous thing that has happened, I believe, in theology in general, and in the churches specifically, is that the churches have been turned into a kind of therapeutic community, where the main drive has been to enable people to grapple with themselves, their own consciences, their social, political, and psychological problems. The older notion of theology is really to understand the first and last things, to get some hold of final truth, of which the problems of the human community are only a part.

I feel in this sense very reactionary. I think that, as important as it is to understand the relationship between the observer and the observed, or between the human community and physical nature out there, the really important point still is that there is something quite separate from us. I am struck by the fact that there is, indeed, a very sharp division between subject and object. There is no possibility of manipulating the whole physical universe. The most we can do is to try to grasp what this universe means, quite apart from whether we can control it or not.

Now right away that is a kind of shock to those human beings who have become used to thinking of knowledge as power and manipulation. But I also suspect that the breaking down of the subject/object relationship has made it very difficult for us finally to get a good grasp on what man is.

This brings me to the major point I want to make: If there is one crucial problem in the relationship between science and theology in the future, it is going to be the problem of ethics. What is the good life for man?

Professor Wheeler raised the question of how one seeks the foundation of ethics. I think the basic choice facing us is whether ethics is a kind of an *ex nihilo* creation by human beings simply for the sake of survival, where human beings have to develop moral codes and guidelines to enable them to survive together; or whether, indeed, if there is a nature out there, that nature finally provides some guides for the creation of guidelines, and whether there is something in which to root an ethic other than survival needs.

I suspect that we may be forced to a much older classical medieval notion that there is, indeed, a nature, man is a part of that nature, and that nature is teleological in some very fundamental sense. The problem then for man is to come to understand his own nature, and that problem becomes one of trying to get an insight into nature as it really is. I think a worthwhile line of pursuit might be to see if, indeed, in that nature out there we can get some guidelines on the creation of ethical structures.

DAVIS: Could you expand on this notion of possibly finding sources of ethics out there? Where out there?

CALLAHAN: By an attempt to understand philosophically, first of all, the nature of matter. Secondly, more narrowly and parochially, understanding the philosophical implications of evolution; to what extent has evolution driven man in some direction that then begins to entail certain kinds of ethical values?

The "out there" is—put in broadest terms—that there exists more than consciousness, and that our consciousness works on something other than consciousness. When Professor Wheeler tells us about the black hole, this tells us something about the nature of the universe. The important thing is somehow to see, even in the development of an ethical system, how one can take that kind of thing into account. All I am doing is laying out a research program; I am not answering your question.

DAVIS: You did answer it in part by suggesting that a deeper knowledge of evolution would help guide us in developing a system. Certainly I agree with you. But I am not sure that that would be so distinct from an alternative, which would say that our ethical system is there because it has survival value.

CALLAHAN: I would want to build an understanding of evolution into a larger understanding of the nature of the universe; if you just start from evolution, you may not have been starting back quite far enough. A question is, why is there evolution rather than non-evolution? A common fault is starting too late in the game conceptually.

BALTAZAR: Dr. Davis, might not your view of instinct as the basis of morality within an evolutionary framework be the same as what Dr. Callahan is talking about philosophically in terms of the concept of nature?

DAVIS: It would if the aspects of nature that you were now taking into account were biological aspects, specifically evolution. But I find it a little hard to see how the nature of the physical universe can really serve as a source of help in developing a better system of ethics beyond creating, shall we say, the humility of realizing how small a role we play.

Isn't this one of the dangers in the world today, that the smaller man is made to feel, the harder it is to have him take seriously the kind of ideals that underlie ethical systems?

CALLAHAN: In Professor Wheeler's presentation I heard hints that in some sense or other, the universe is not crazy. There is a drive and a direction, a kind of order in the universe. If this is the case, our problem is to see if we can really discern this order; in some sense then we are part of this order, and that begins to give some positive foundation for an ethic.

It is better to survive than not to survive; everybody would agree on that. But if you say, survival for what? that gets to be a hard question, and at that point you ask questions that go beyond evolution.

DAVIS: Darwin talks about survival of the species, and that's very different from survival of the individual. Survival of a social species involves patterns of activity very different from those that are concerned with just the survival of the individual. Today we are starting to think of protection of the environment as a moral obligation. This is intimately linked with feelings about wanting our children and our children's children to have a world to live in that is worthwhile. There are deep biological roots for that kind of thing.

CALLAHAN: What I find interesting about the whole environmental discussion, at least in terms of ethics, is that so much of Western ethics has been highly individualistic; even when one talks about social good, this really translates into the good of individuals. But when you begin talking about species survival and if you begin thinking of future generations, you have to bring in some notions of human good that are in some way transcendant.

DAVIS: Modern molecular biology has provided extraordinarily direct evidence for evolution, in the form of increasing differences in DNA sequences paralleling increasing differences in bodily form and function. This development has removed any vestige of doubt that Darwin's theory was essentially correct. Given that, how do you react to the proposition that perhaps the most valuable approach for students of ethics today would be to explore as deeply as possible the implications of evolution? Many of the earlier interpretations have been very shallow.

CALLAHAN: I would quite agree. But I guess you still have the problem of exploring what is the context of evolution or for what end. I find that it is very difficult to get clear answers as to why the species ought to go on.

Who could care if ten thousand years from now there were no human species left? Would it make any basic differences? I think there is no decent answer to that question unless one has a larger theory of the nature of reality. Why is it good that there is a flower?

POLLARD: There is a sense in the total picture from the emergence of a universe out of a black hole and the emergence of laws. First it was just hydrogen, and that wasn't a very exciting universe. So you have to cycle it through stars, supernovas, to get carbon and nitrogen and oxygen, in order to achieve.

Now man is certainly an achievement. It is just fantastically extraordinary, when you start out with an information code and in four billion years it elaborates itself to man. Man is a product of this tremendous history, and that itself constitutes a kind of obligation to be man.

HAMILTON: As I listened to John Wheeler this morning, he suggested the direction of the universe is to another black hole. What ethical significance that has for man I am not sure, but man seems to be faced with his ultimate destruction.

Let us consider Christianity for a moment. Does that give you any guidance as to what your responsibility to your grandchildren is?

CALLAHAN: A minimal obligation might be that there should be grandchildren.

HAMILTON: Why?

CALLAHAN: Simply because, following Dr. Pollard's thought, it is somehow part of the overall achievement of the universe that there be man, and one has an obligation to keep this achievement going. I think the only way of finally answering this kind of question is to see man as an intrinsic good in the universe. In a curious sense, one could say that without man the universe would be worse off, though you can't say that human beings would, by virtue of not existing, be worse off.

DAVIS: I would like to come back to the question of why man should try to survive and offer a simpler explanation.

Yesterday, Professor Heisenberg said that, with respect to morality and social behavior, there are certain things we should simply accept as givens and not question because he could not conceive of a society functioning without certain moral rules. I suspect that his having lived through the degradation of Nazism in Germany must have given strong feelings about what happens when you start questioning moral values. Similarly, I would take our instinct for survival as a given. We would not be here unless evolution had provided us with genes that ensured that instinct, and it would seem unnatural to deny or try to suppress it—no matter how depressed we may be about the present state of our cultural evolution, and regardless of whether or not we can prove that our existence is "good." We may, indeed, be programming ourselves for extinction, inadvertently; but, it would be unnatural to do so deliberately.

If we have this deep-seated instinct to survive it seems to me that the problem of morality is trying to persuade people that unless you behave in such and such a way toward your neighbors or toward the environment, your grandchildren and great-grandchildren will have a lot tougher time surviving. That could be one foundation stone for an ethic.

CALLAHAN: It seems to me that one very unique trait of human beings is that they frequently put other values higher than the value of survival. The *quality* of the survival can become as important as the sheer fact of survival. I think that you cannot build an ethic just on survival, but you must have some decent sense of what other values are to be integrated with survival.

DAVIS: I agree completely. Evolution is far from a complete basis for an ethical system, but to the extent that it is relevant, isn't there room for a great deal of exploration of what we mean by survival? When the heroes of Masada were on a plateau attacked by the Romans, the last ones finally committed suicide and left a moving message that they would rather die than be slaves under the Romans. They didn't survive, but their message encouraged the survival of other members. It is the survival of the species, ultimately, and not the survival of individuals, that counts in evolution.

POLLARD: Evolution teaches ambiguous lessons. The vast majority of the species haven't survived. A few have survived a phenomenally long time, like the cockroach—three hundred million years.

THOMPSON: If survival is a value, how does one choose among various strategies of survival, between the natural survival of the fittest and the cultural fitting of the survivors in terms of genetic engineering?

DAVIS: It is very complicated. We are now in a position in which no other species ever was before: We are beginning to understand our own mechanism of inheritance, and to be aware of the biological as well as the environmental factors that influence our survival. We are the first species that has the possibility of deliberately influencing selection within it. It's a very dangerous power, and it scares us all. But sooner or later our species is going to start to look at the question of some kind of deliberate selection; and I suspect the need to limit the total population may bring the problem to a head sooner than we think.

In fact, we are in a rather curious situation. Throughout evolution species have evolved most rapidly under conditions where their environments have changed most severely. In the last few

Apollo 17 view of the earth, December 1972. (Courtesy of the National Aeronautics and Space Administration.)

centuries, man has developed the power to change his environment more rapidly than the environment of any species has changed before. But at the same time, modern developments in medicine, and our ethical patterns, lead us to promote the survival of all individuals, as far as possible. I am not questioning this; I am just describing the very curious paradox that we face.

CALLAHAN: Then we have to project a really gloomy view for this evolutionary process. What you are saying is that we have the possibility of intervening in the process, but with no clear notion of what is indeed good.

You have to have a large theory of the universe to say survival is good. We could make a radical case to say that everybody alive today will simply have no more children, we will just live a riotous life of enjoyment, use up everything and then all die, but we would all somehow feel that is wrong in some deep sense.

DAVIS: That's because we have an instinct for survival of the species, so it does feel wrong. But in addition, having the technical possibility of intervening consciously in our evolution does not mean that we have to do so. As long as we have no clear consensus on what is good it would seem prudent not to intervene. And, in fact, eugenics is hardly a live issue today, except for the prevention of well-defined hereditary diseases.

DISCUSSION: ARE THERE ETHICAL LIMITS
TO SCIENTIFIC DISCOVERY?

THOMPSON: Shall we proceed to Professor Shinn's talk, and then bring all these questions into focus again?

SHINN: I am primarily a student of social ethics, but one greatly interested in scientific discovery, and so I think it is faithful to the topic—the nature of scientific discovery and the future of interaction between science and religion—to look at a rather specific question with a long history: Are there ethical limits that can be imposed on scientific research and discovery?

To ask the question may open a lot of old wounds, because there is a horrendous history in which religious groups have inhibited or even suppressed scientific discovery in ways we here would all disown. We tend to identify ourselves as the lovers of truth, opposed to the censors and tyrants. All of us here have some kind of deep commitment to freedom of inquiry and the encouragement of search for truths. As I look at this, I see that this commitment in itself has an ethical and a religious quality.

I recall a conversation a few years back on this subject with

Isidor Rabi, a famous Nobel laureate in physics. He put it this way, "Science simply operates on the faith that knowledge is good and that ignorance is something to be overcome." He went on to say, "You can't really vindicate this faith empirically. It is a faith." And Rabi related it to the first chapter of the Bible, "Let there be light." Sharing that faith, he advocated an unquenchable desire for knowledge and the search for it.

To a very great extent, I want to identify myself with that. What has caused me to rethink it lately is a number of questions that have come out of the scientific community. I'll start with Marshall Nirenberg, another Nobel laureate, distinguished for his achievements in deciphering the genetic code. He made the statement, "When man becomes capable of constructing his own cells, he must refrain from doing so until he has sufficient wisdom to use this knowledge for the benefit of mankind." * Here he is suggesting a restraint, not on inquiry, but on the uses of what comes out of inquiry, on action. I suppose everyone would endorse some restraints along that line.

But more remarkable, a quite different proposal comes from, of all people, Herman Kahn, who has said, "We may well need an index of forbidden knowledge." The phraseology comes from the old Vatican practice, an Index of Forbidden Books. These amazing words are in an interview in the *New York Times Magazine* of 20 June 1971. As an example, Kahn gives knowledge of genetic engineering. He further points out that we have established, in effect, a kind of index of forbidden knowledge in the nuclear area, where we know certain things that we don't share even with our allies.

Kahn's argument is this: "Whatever the intellectual dangers of an index, they pale in comparison with the danger of not having one. A society that hasn't the moral capacity to absorb new knowledge without putting itself in moral peril has to have

---

* Marshall Nirenberg, "Will Society Be Prepared?" *Science*, vol. 157 (1969), p. 633.

restraints imposed upon it." This came from a source that I hadn't quite expected, but it gives me the courage to conjecture about it, not to endorse it.

Another thing that has happened recently, among a very small part of the scientific community, has been the deliberate withdrawal from research, because some scientists have thought that society was not capable of handling dangerous new knowledge. Catherine Roberts, the microbiologist, did that. She came to the point where she said: "This kind of inquiry not only unlooses dangerous knowledge, but it is in its very nature so dehumanizing that I want no more to do with it." *

Bernard Davis and I were present when James Shapiro declared his intention of renouncing his scientific career, at the AAAS in Boston in 1969. He had been part of the Harvard team that isolated a gene, and he chucked it all, at least temporarily. His particular language, which I wrote down on the occasion, is that so long as men like Nixon and Agnew prescribe what is going to be done with scientific knowledge, we scientists should quit giving them the materials for their bad decisions.

I do not at all agree with Shapiro's reasoning on the occasion. I have checked this idea out with a number of scientists, including some survivors of the Nazi holocaust; and some of them said that, while they do not agree with Shapiro, they would agree that there are social situations in which they would refuse to do research, because they would feel the society would so inevitably misuse it. This is not saying the knowledge is evil. It is saying that it is better not to have the knowledge than to be almost certain it will be used for evil purposes.

It is against that background that I want to advance five possible answers to the question; "Are there ethical limits on scientific research and discovery?"

The first possible answer is no, there are not. There should be no restraints; censorship and suppression are inherently harmful,

---

* Catherine Roberts, *The Scientific Conscience* (New York, 1967).

and freedom of inquiry, the active promotion of inquiry should go right ahead. I'd say that 95 percent of me vibrates with that. The pernicious examples on the other side make me quite sympathetic to unfettered inquiry. But then I have some questions.

Answer two, which I hear from a number of friends in the scientific community, is to distinguish between basic research and applied research, to put no moral limits on basic research, but some definite moral limits on applied research. The most convincing example that they usually come up with is germ warfare. They would, for moral reasons, not participate in applied research aimed at bacteriological warfare. But this would mean no inhibition of basic research.

The problem I find with this is that I, as a nonscientist, don't know how to draw the line between basic research and applied research, and when I press scientists on it, I find that often they don't either. You can easily figure some examples that are one category or the other, but what about the borderline?

When I pushed one scientist, who has done a lot of Government work, on this issue, he estimated that three-quarters of what went on at Fort Detrick, for the sake of research on bacteriological warfare, was indistinguishable from what is going on at the National Institutes of Health. There was one-quarter of it that you can say clearly belonged to the Fort, rather than the Institutes of Health, but three-quarters was quite comparable. Right now there are widespread accusations of cheating on bacteriological warfare, because, in the name of learning how to prevent disease, you can be concocting schemes for spreading disease.

Granted all these difficulties, I would still be inclined to think, as an ethicist, that there is some value in the distinction between basic and applied research, with some moral restraints on the latter. At the very least the distinction may heighten consciousness of the moral issues, and it may have some practical consequences.

The third possible answer: Even on basic research there

should be some limitations. We can pretty quickly think of some we would probably all agree on. Some kinds of basic research that would produce intellectually interesting and maybe practically desirable knowledge are too cruel to people or infringe on human dignity. Anybody can think up experiments to find out what the human tolerance of pain is, but there would be some moral inhibition in practicing this on a group of specimens.

Or take some ethical restraints that don't involve physical pain. Everybody I know who is interested in the relation between biology and culture has a kind of lurking desire to run off a big-scale controlled experiment on identical twins: to separate some identical twins at birth, to keep others together, and to compare the results. Now we piece together little fragments of information from accidental sources, but you could run a controlled experiment here and learn a lot from it. So far we have never done this kind of infringement on human dignity: to separate identical twins for purposes of gaining knowledge that would be scientifically useful but morally intolerable. Thus human dignity imposes some limitations on research. This is not because the knowledge desired would be bad or evil. It is simply not worth the human and ethical cost of the experiment.

Now a fourth question, certainly related to the third, concerns risks. That is, some kinds of research require risks that might not hurt human dignity or inflict cruelty, but they might. We are always weighing this matter of risk, because there is no such thing as a risk-free life, but there are some risks that are immoral to apply to people. The atmospheric testing of nuclear weapons has not been renounced by all nations, but a good many have renounced it because of risks connected with it.

The area of risk can become a moral consideration. In the area of medical experimentation, this gets quite complicated at some points. The general rule, of course, on medical experimentation is to require informed consent of the subject. But then there are all these cases where the subject can't give informed consent—infants, for example. At the moment I am in discussion with the National Institutes of Health, trying to shape up a policy here. Take one case we are working on now, fetal experimentation. Suppose, following an abortion, the fetus is alive. The doctors make their best estimate that this life cannot be preserved. Maybe it will persist for ten or fifteen minutes. Can we run an interesting experiment here? Any inflicting of pain is probably quite minimal. Is this an infringement on human dignity?

Or let's take another one. NIH has to set up guidelines on this, so it is not theoretical. A woman has determined to have an abortion. It is to her benefit to postpone it through the risky period. How about running an experiment on this fetus in her body to see how it will react to certain medical treatments? How about something with a risk potential of thalidomide, in which you are not going to hurt a human being who is going to have a life ahead of him, because this abortion is going to take place anyhow?

There is, among medical people, lawyers, and ethicists who have discussed this case, a great reluctance to do the experiment. Is this connected with instinct? Is it connected with an irrational squeamishness? Is it connected with some sense that life is sacred and that even if the fetus is not going to come to full humanity, you don't play around with it just for the sake of knowledge? I have great difficulty, not in knowing what my initial feelings are, but in trying to ground those in a judgment that I would want to stand by and defend. Yet I am sure that concern for human dignity puts some limits on research.

Now my fifth and final point is a simpler one. It has to do with priorities. Most of the time we are not deciding, shall we do this research or shall we outlaw it? We are deciding, shall we finance it? Here there are great issues of priorities. These are partly chosen just for scientific reasons; some things are worth financing because they have much more possibility of discovery than others. But some choices have to do with a whole set of cultural values and attitudes. Shall we put a lot of money into research on heart transplants, which probably by their nature can help only a small number of people? Or shall we put the money into other things?

The issue is not prohibitions, but public policy. It simply reminds us that every decision to undertake research, at least expensive research, involves assignment of some of the resources of the society, and our valuations enter in. These valuations are not solely scientific, but are cultural, and they may come out of sheer prejudices or may come out of deep convictions.

I have suggested five possible answers, and I have suggested my attitudes on each. I cannot come to any one conclusion, except that the old question that I had once put on the shelf as solved—there should be no ethical, moral, or religious restraints on research—needs some re-thinking. The old clichés, here as elsewhere, have to be reviewed once again.

APOSHIAN: You haven't brought in the risk-benefit calculation. Today everyone is concerned about experimentation on the human fetus. If this were ten, fifteen years ago, when, if it were certain that there were a good chance that polio could be treated and cured by means of experimenting with fetuses, it would have been an entirely different ball game.

SHINN: If I'm betting on a state lottery, the risk-benefit calculus is a very clear one. I know just what it is. The minute you intrude a consideration like human dignity or the sanctity of life, you greatly change your risk-benefit calculus.

A scientist might subject fruit flies to radiation to increase the mutation rate, partly just for what he'd learn, or partly on the assumption that 999 mutations will be worse, and one will be a gain. You don't feel immoral about this. But you are not going to do this with human beings. You have a very different risk-benefit calculus, in particular when it's done without consent. Everybody who thinks about this issue now has that horrendous Tuskegee syphilis experiment somewhere in the back of his mind. In the social context, the risk-benefit calculus is always applied to those who aren't in the position to protect themselves. This business of human dignity or the sacredness of life just upsets what would apply in other cases.

APOSHIAN: But where the dignity comes in is very difficult to see when it's applied to a fetus that is going to live 15 minutes.

CALLAHAN: Part of the controversy is that they can be kept alive longer for the sake of the experiment, too. That complicates it.

DAVIS: I am worried that the strength of our reactions to sins of commission, in the testing of drugs and other new therapeutic devices, may bring to a standstill further advances in therapy. No new drug has ever been produced that hasn't involved risk for the first people, and a great many drugs have been dropped after a great many people were harmed. But we wouldn't have medicine without somebody assuming those risks. As horrified as I am by the Tuskegee incident and as unhappy as I am about orphans in asylums, I am also worried about a society that might stop medical progress.

HAMILTON: Suppose you calculate a serious risk and decide that you would not wish to undertake those kind of experiments. But let's say that down in South America they were not prohibited and then suppose that the information became of some use to the understanding of the gestation process—would you advocate our using this information gained elsewhere?

SHINN: When the information is found and is useful, you use it.

HAMILTON: Is it as clear as that? I have a hunch the more you get into this question, the more you'd find it to be a difficult one.

APOSHIAN: This has always bothered me. For many years, and even today, most of the drugs that are developed by American pharmaceutical companies are first tested out on humans outside of this country, in the underdeveloped countries.

I know of a case about four years ago where the birth-control pill by one of the most eminent pharmaceutical companies in our country was almost ready to go on the market, and was finally taken off the market because the Mexican women who had been given the birth-control pills had ended up with bladder cancers.

SHINN: I was not saying I would endorse doing this in South

America in order to get the knowledge. But it is like admiring an ancient Greek object of art that was produced by slaves; we don't approve of slavery and you wouldn't repeat this thing, but it is beautiful.

HAMILTON: Let me come back at you, because the problem is that once you begin the practice of accepting research information from unethical situations, you encourage more. It is not like museum art, which is only in the past. You are relating to the present and to the future, and there is a feedback both ways. If a man's research, which you wouldn't approve of on ethical grounds because of the nature of the experiments, produces a great burst of new knowledge and is accepted, he becomes a scientific hero. There can be thus a relation in terms of providing future motivations if you accept existing or past information.

CALLAHAN: I want to take the other side of this argument on the use of knowledge, because I think this whole question has to be pushed back to much more fundamental questions: Ought we to continue to think that even knowledge *per se* is a good thing? Roger, you made your question easy by really dealing with applied research. But is it really the case that all knowledge, even fundamental knowledge with no evil consequences, is, *per se* a good thing?

The whole Western tradition is that the more you know, the better off you are, even if there's nothing you can do with the particular knowledge. Now it seems to me that we are reaching a point, particularly with the medical progress, where we might say that there is a diminishing human gain to the additional scientific knowledge. The greatest evil in the world may not be disease, if you will—the greatest evil may be a certain kind of moral defect. Is basic knowledge necessarily always good for human beings? The bias has always been that it is ipso facto good. I prefer to say that it's neutral.

APOSHIAN: Are you proposing, therefore that a moratorium should be declared on all new knowledge? You don't know whether that knowledge is going to be good or bad until after you have the knowledge.

CALLAHAN: I think it is an important question. Looking back historically, it has been pretty clear that all knowledge is a kind of double-edged sword.

DAVIS: A double-edged sword in a much deeper sense. I would suggest that the greatest ethical problem for the world raised by medical discoveries has not been the experiments done on patients without informed consent—though that is a very real moral problem. The real problem has been the role of modern advance in medicine as the main source of world overpopulation.

THOMPSON: It seems there is always a culturally implicit system of values that is unconsciously taken for granted; even survival can be difficult if in fact seeking knowledge for medical progress has interfered with the survival systems of the human species because that progress has prolonged life, raised populations, and eliminated the automatic self-destruction. We always go back into the kind of loop where we cannot really base everything on survival.

DAVIS: Unless we analyze it more deeply. If we are smart enough to agree that survival of the species depends on limiting its density then Zero Population Growth is one answer. Fewer people to survive will survive better.

THOMPSON: But the choices we are going to have to make to enable us to choose between fitting to survival and the survival of the fittest is not itself a philosophical system of survival. It is a system of choice, discrimination, and reason, and a system of values that is independent. They are on philosophically different levels.

DAVIS: Yes, I understand that. Dr. Shinn raised some rather broader questions about knowledge as a whole. In a sense the limitations we have just been talking about, with respect to medical ethics, are not really a novel idea; in practice there have always been some restraints—for example, with respect to cruelty. Herman Kahn's idea, however, is a more novel one: that there are areas in which we could get scientific information but we had better not, because we are better off without it.

CALLAHAN: I'm trying to think of an instance in terms of survival. Suppose we could have the basic knowledge of the human aging process such that people could live an extra 30, 50, 70 years. Once you have that kind of knowledge, it is very difficult not to take advantage of it and use it. You might make the case to support and undertake that line of research on survival alone, but what would it do to the social structure and the population and everything else?

APOSHIAN: Suppose we could design a machine that would enable you to tell what another person is thinking, literally. As you meet your best friend, or your worst enemy, on the street, you would immediately know what he was thinking about you or anything else.

What would this do to our society? I would think that we would almost have a mass war on a one-to-one basis. But it would be very interesting scientifically if we could design that machine.

HAMILTON: What is likely to happen is that such a machine will be developed, be highly expensive, and then the countertechnology will also be developed. Then you will have another apparatus which will interfere with your thoughts being received, and you will get into the whole logistics of missiles and anti-missiles. That seems to me to be a far more likely future.

MCCLURKEN: Even more than that is the question of who is able to afford those machines.

APOSHIAN: Or more importantly, who has it first.

DAVIS: I don't think that kind of machine is going to be developed, but I think that it would be very interesting if we could have a modern, sophisticated electronic version of the lie detector and have that visible on TV, in any public controversy or political presentation!

INDEX

Italicized pages indicate illustrations